P9-DUV-352

Hanging Out, Messing Around, and Geeking Out

The John D. and Catherine T. MacArthur Foundation Series on Digital Media and Learning

Engineering Play: A Cultural History of Children's Software by Mizuko Ito

Hanging Out, Messing Around, and Geeking Out: Kids Living and Learning with New Media by Mizuko Ito, Sonja Baumer, Matteo Bittanti, danah boyd, Rachel Cody, Becky Herr-Stephenson, Heather A. Horst, Patricia G. Lange, Dilan Mahendran, Katynka Z. Martínez, C. J. Pascoe, Dan Perkel, Laura Robinson, Christo Sims, Lisa Tripp, with contributions by Judd Antin, Megan Finn, Arthur Law, Annie Manion, Sarai Mitnick, David Schlossberg, and Sarita Yardi

Inaugural Series Volumes

These edited volumes were created through an interactive community review process and published online and in print in December 2007. They are the precursors to the peer-reviewed monographs in the series.

Civic Life Online: Learning How Digital Media Can Engage Youth, edited by W. Lance Bennett

Digital Media, Youth, and Credibility, edited by Miriam J. Metzger and Andrew J. Flanagin

Digital Youth, Innovation, and the Unexpected, edited by Tara McPherson

The Ecology of Games: Connecting Youth, Games, and Learning, edited by Katie Salen

Learning Race and Ethnicity: Youth and Digital Media, edited by Anna Everett

Youth, Identity, and Digital Media, edited by David Buckingham

HANGING OUT, MESSING AROUND, AND GEEKING OUT

Kids Living and Learning with New Media

Mizuko Ito
Sonja Baumer
Matteo Bittanti
danah boyd
Rachel Cody
Becky Herr-Stephenson
Heather A. Horst
Patricia G. Lange
Dilan Mahendran
Katynka Z. Martínez
C. J. Pascoe
Dan Perkel
Laura Robinson
Christo Sims
Lisa Tripp

with contributions by
Judd Antin, Megan Finn, Arthur Law, Annie Manion,
Sarai Mitnick, David Schlossberg, and Sarita Yardi

The MIT Press
Cambridge, Massachusetts
London, England

First MIT Press paperback edition, 2013

For information about special quantity discounts, please e-mail special_sales@mitpress.mit.edu.

This book was set in Stone Sans and Stone Serif by SNP Best-set Typesetter Ltd., Hong Kong.

Printed and bound in the United States of America.
Library of Congress Cataloging-in-Publication Data

Ito, Mizuko.
Hanging out, messing around, and geeking out : kids living and learning with new media / Mizuko Ito.
    p. cm. — The John D. and Catherine T. MacArthur Foundation Series in Digital Media and Learning
Includes bibliographical references and index.
ISBN 978-0-262-01336-9 (hc. : alk. paper) — 978-0-262-51854-3 (pb. : alk. paper)
1. Mass media and youth—United States. 2. Digital media—Social aspects–United States. 3. Technology and youth—United States. 4. Learning—Social aspects.
I. Title.
HQ799.2.M352I87   2010
302.23'108350973—dc22
2009009932

10 9 8 7 6 5

To the memory and ongoing legacy of Peter Lyman.
His vision, passion, and leadership have guided this project
and animated its spirit of interdisciplinary, collaborative work.

Contents

Series Foreword

In recent years, digital media and networks have become embedded in our everyday lives and are part of broad-based changes to how we engage in knowledge production, communication, and creative expression. Unlike the early years in the development of computers and computer-based media, digital media are now *commonplace* and *pervasive*, having been taken up by a wide range of individuals and institutions in all walks of life. Digital media have escaped the boundaries of professional and formal practice, and the academic, governmental, and industry homes that initially fostered their development. Now they have been taken up by diverse populations and noninstitutionalized practices, including the peer activities of youth. Although specific forms of technology uptake are highly diverse, a generation is growing up in an era where digital media are part of the taken-for-granted social and cultural fabric of learning, play, and social communication.

This book series is founded upon the working hypothesis that those immersed in new digital tools and networks are engaged in an unprecedented exploration of language, games, social interaction, problem solving, and self-directed activity that leads to diverse forms of learning. These diverse forms of learning are reflected in expressions of identity, how individuals express independence and creativity, and in their ability to learn, exercise judgment, and think systematically.

The defining frame for this series is not a particular theoretical or disciplinary approach, nor is it a fixed set of topics. Rather, the series revolves around a constellation of topics investigated from multiple disciplinary and practical frames. The series as a whole looks at the relation between youth, learning, and digital media, but each might deal with only a subset of this constellation. Erecting strict topical boundaries can exclude some of the most important work in the field. For example, restricting the content of the series only to people of a certain age means artificially reifying an age boundary when the phenomenon demands otherwise. This becomes particularly problematic with new forms of online participation where one important outcome is the mixing of participants of different ages. The same goes for digital media, which are increasingly inseparable from analog and earlier media forms.

The series responds to certain changes in our media ecology that have important implications for learning. Specifically, these are new forms of media *literacy* and changes in the modes of media *participation*. Digital media are part of a convergence between interactive media (most notably gaming), online networks, and existing media forms. Navigating this media ecology involves a palette of literacies that are being defined through practice but require more scholarly scrutiny before they can be fully incorporated pervasively into educational initiatives. Media literacy involves not only ways of understanding, interpreting, and critiquing media, but also the means for creative and social expression, online search and navigation, and a host of new technical skills. The potential gap in literacies and participation skills creates new challenges for educators who struggle to bridge media engagement inside and outside the classroom.

The John D. and Catherine T. MacArthur Foundation Series on Digital Media and Learning, published by the MIT Press, aims to close these gaps and provide innovative ways of thinking about and using new forms of knowledge production, communication, and creative expression.

Acknowledgments

The research for and writing of this book was a collective effort that involved a wide network of individuals and institutions beyond those named as authors and contributors. The late Peter Lyman was a principal investigator on the project on which this book reports, and he defined the vision and direction for this project as well as forming the team that started it off. Michael Carter provided leadership as a principal investigator and as the heart and soul of the project, and he held the team together through many challenges. We are also grateful to Barrie Thorne, who stepped in as principal investigator, offering guidance and support including crucial input on our writing and analysis.

The project was funded by the John D. and Catherine T. MacArthur Foundation as part of the digital media and learning initiative. We would particularly like to thank our program officer and director of education at the foundation, Constance M. Yowell, and vice president of human and community development, Julia M. Stasch.

This was a multi-institutional project that was guided by the administrative and research staff at multiple research centers. At the University of California, Berkeley, the project was housed at the Institute for the Study of Social Change and benefited from the technical support of UC Berkeley's School of Information. At Berkeley, we would like to thank Shalia McDonald, Janice Tanigawa, Diane Harley, Kathleen Kuhlmann, and Evelyn Wong for their help in administering and managing the project. At the University of Southern California, the project was housed at the Annenberg Center for Communication and the Institute for Multimedia Literacy at the School of Cinematic Arts. We are grateful to Mariko Oda, Josie Acosta, Steve Adcook, Chris Badua, Willy Paredes, and Chris Wittenberg for guidance with and support for the project at USC.

In addition to the authors and contributors to this report, we had many research assistants and collaborators who enriched this project along the way. Max Besbris, Brendan Callum, Allison Dusine, Sam Jackson, Lou-Anthony Limon, Renee Saito, Judy Suwatanapongched, and Tammy Zhu were research assistants as well as vital informants and experts in all things digital and youth. We also benefited from working with our collaborators on this project, Natalie Boero, Carrie Burgener, Juan Devis, Scott Carter, Paul Poling, Nick Reid, Rachel Strickland, and Jennifer Urban. The Berkman Center for Internet and Society at Harvard University, the Pew Internet & American Life Project, the LAUSD Arts Education Branch, and the Wallis Annenberg Initiative also were institutional collaborators in this research. Karen Bleske, in addition to careful copyediting of the entire book, provided invaluable help in integrating the many different voices and styles of the contributors. Eric Olive was our web guru who helped get our work out to the online universe. At the MIT Press, the book was in the capable editorial hands of Doug Sery, Katie Helke, and Mel Goldsipe.

Our work has benefited from the wise counsel of many colleagues, more numerous than we can fully name here. We would like to acknowledge those who participated in the occasions that we organized to get formal feedback on our work in progress. An early draft of this book was reviewed by John Seely Brown, Paul Duguid, Jabari Mahiri, Daniel Miller, Katie Salen, Ellen Seiter, and Barry Wellman. Their comments resulted in considerable changes to this document that have both sharpened the arguments and made it more intelligible to diverse audiences. As we were conducting our research, we arranged for periodic meetings and conversations so we could be in dialogue with scholars we knew would inform our work. In addition to those who reviewed this book, we would like to thank those who participated and generously shared their insights and perspectives: Sasha Barab, Brigid Barron, Suzy Beemer, Linda Burch, Lynn Schofield Clark, Michael Cole, Brinda Dalal, Dale Dougherty, Penelope Eckert, Nicole Ellison, James Paul Gee, David Goldberg, Shelley Goldman, Joyce Hakansson, Eszter Hargittai, Glynda Hull, Lynn Jamieson, Henry Jenkins, Joseph Kahne, Amanda Lenhart, Jane McGonigal, Ellen Middaugh, Kenny Miller, Alesia Montgomery, Kimiko Nishimura, John Palfrey, Nichole Pinkard, Alice Robison, Ryan Shaw, Lissa Soep, Reed Stevens, Deborah Stipek, Benjamin Stokes, Pierre Tchetgen, Doug Thomas, Avril Thorne, and Margaret Weigel.

Finally, we would like to thank the many individuals, families, organizations, and online communities that welcomed us into their midst and educated us about their lives with new media. Although we cannot name all the individuals who participated in our study, we would like to express our gratitude to those whom we can name who facilitated our access to various sites and who acted as key "local" experts: Vicki O'Day for introducing Heather to Silicon Valley families; Tim Park, Carlo Pichay, and zalas for being Mizuko's *senpai* in the anime fandom; Enki, Wurlpin, and all of KirinTheDestroyers for taking Rachel under their wing; Tom Anderson, who helped danah get access to MySpace; the people of YouTubia who spoke with Patricia and shared their videos; and all the youth media, middle-school, and high-school educators who opened their doors to us.

Notes on the Text

This book is a synthesis of three years of collaborative, ethnographic work conducted through a project funded by the John D. and Catherine T. MacArthur Foundation: Kids' Informal Learning with Digital Media.

Early in the planning of this book, we made a decision not to structure it as a traditional edited volume, nor as a book singly written by a principal project investigator. Instead, this book was written in a highly distributed collaborative process that aimed to integrate both the ethnographic material and the analytic insights of all the project's researchers involved at the time of its writing. We thought this approach was most in line with the spirit of collaborative, interdisciplinary inquiry that has guided our project from its inception. Each chapter has one or more lead authors who took responsibility for the writing, but every chapter incorporates material and input from a wide range of coauthors and the case studies that they represent. In line with this stance, we use a collective voice to describe this work, even in chapters with only one lead author. We did not always reach complete consensus on all aspects of this book, but there was agreement among the coauthors that we would take collective ownership.

Although Mizuko Ito took the lead in the writing of this book, the three other principal investigators, Peter Lyman, Michael Carter, and Barrie Thorne, provided indispensable leadership and support for this project. In addition, we have integrated ethnographic material from former project members, who are named as contributors to this book. The full range of people who have contributed to this three-year project and this book are mentioned in the acknowledgments.

The case studies and approaches that the coauthors brought to the writing have been diverse, but we have agreed on certain representational conventions to provide some consistency in our writing:

• Unlike in more traditional forms of ethnography, the descriptions in this book draw from a wide range of case studies conducted by a large team of ethnographers. When a research participant is quoted or identified, we indicate which case study the material comes from and the name of the fieldworker who conducted the interview or the observation. We use short identifiers (e.g., Horst, Silicon Valley Families) for the studies to avoid cluttering the text. A table of short titles, full study titles, and study researchers is included in Appendix III.

• Full descriptions of the framework for the projects are described in the appendices. More detailed descriptions of the twenty-three individual research studies conducted by members of the Digital Youth Project between the years of 2005 and 2008 are provided online at http:// digitalyouth.ischool.berkeley.edu/projects.

• The various case studies were conducted using different data-collection methodologies, and we have varying degrees of access to contextual information about our participants. In every case, if we know the information, then we have indicated age, gender, and what each participant self-identified as his or her racial or ethnic identity. If this information is not indicated, then it means that we did not know the information for this participant due to the constraints of the particular case study. For example, in many of the studies that focus on online interest groups, interviews were conducted over the phone or through online chat. In most cases, we derived this information from self-reports in background questionnaires we administered in advance of most of our formal interviews. Although we do not see race as a key analytic category in our work, there are times when we think it is relevant to our description, and we thought that if racial or ethnic identity were to be mentioned for some number of participants, then we needed to be symmetrical in our treatment and indicate racial identity for all respondents for whom we did have this information.

• We have used pseudonyms in most cases when referring to our research participants. In many, but not all, cases our participants chose these pseudonyms. In the case of some media producers, these names correspond with their creator identities or screen names in their respective interest groups, an approach that we think honors the reputations and investments of time that many of our participants work very hard to develop. When participants specifically requested it, we have used their screen names or their real-life names. When real names or screen names are used, we indicate this by a footnote in the text.

INTRODUCTION

Digital media and online communication have become a pervasive part of the everyday lives of youth in the United States. Social network sites, online games, video-sharing sites, and gadgets such as iPods and mobile phones are now well-established fixtures of youth culture; it can be hard to believe that just a decade ago these technologies were barely present in the lives of U.S. children and teens. Today's youth may be engaging in negotiations over developing knowledge and identity, coming of age, and struggling for autonomy as did their predecessors, but they are doing this while the contexts for communication, friendship, play, and self-expression are being reconfigured through their engagement with new media. We are wary of the claims that there is a digital generation that overthrows culture and knowledge as we know it and that its members' practices are radically different from older generations' new media engagements. At the same time, we also believe that current youth adoption of digital media production and "social media"[1] is happening in a unique historical moment, tied to longer-term and systemic changes in sociability and culture. While the pace of technological change may seem dizzying, the underlying practices of sociability, learning, play, and self-expression are undergoing a slower evolution, growing out of resilient social structural conditions and cultural categories that youth inhabit in diverse ways in their everyday lives. The goal of this book is to document a point in this evolutionary process by looking carefully at how both the commonalities and diversity in youth new media practice are part of a broader social and cultural ecology.

We write this book in a moment when our values and norms surrounding education, literacy, and public participation are being challenged by a shifting landscape of media and communications where youth are central actors. Although today's questions about "kids these days" have a familiar ring to them, the contemporary version is somewhat unusual in how strongly it equates generational identity with technology identity.[2] There is a growing public discourse (both hopeful and fearful) declaring that young people's use of digital media and communication technologies defines a generational identity distinct from that of their elders. In addition to this generational divide, these new technology practices are tied to what David Buckingham (2007, 96) has described as a "'digital divide' between in-school and out-of-school use." He sees this as "symptomatic of a much broader phenomenon—a widening gap between children's everyday 'life worlds' outside of school and the emphases of many educational systems." Both the generational divide and the divide between in-school and out-of-school learning are part of a resilient set of questions about adult authority in the education and socialization of youth. The discourse of digital generations and digital youth posits that new media empower youth to challenge the social norms and educational agendas of their elders in unique ways. This book questions and investigates these claims. How are new media being taken up by youth practices and agendas? And how do these practices change the dynamics of youth-adult negotiations over literacy, learning, and authoritative knowledge?

Despite the widespread assumption that new media are tied to fundamental changes in how young people are engaging with culture and knowledge, there is still relatively little research that investigates how these dynamics operate on the ground. This book reports on a three-year ethnographic investigation of youth new media practice that aims to develop a grounded, qualitative evidence base to inform current debates over the future of learning and education in the digital age. Funded by the John D. and Catherine T. MacArthur Foundation as part of a broader initiative on digital media and learning, the study represents a $3.3 million investment to contribute to basic knowledge in this emerging area of research. The project began in early 2005 and was completed in the summer of 2008, with the bulk of fieldwork taking place in 2006 and 2007. This effort is unique among qualitative studies in the field in the breadth of the research and the number of case studies that it encompasses. Spanning twenty-three different case studies conducted by twenty-eight researchers and collaborators, this study sampled from a wide range of youth practices, populations, and online sites, centered on the United States. This book has a broad descriptive goal of documenting youth practices of engagement with new media, and a more targeted goal of analyzing how these practices are part of negotiations between adults and youth over learning and literacy.

This introduction sets the stage for the body of the book, which is organized by domains of youth practices that cut across our various case studies. We begin with a discussion of existing research on youth new media practice and describe the contribution that our project makes to this body of work. We then introduce the conceptual frameworks and categories that structure our collective analysis and description.

Research Approach

Although a growing volume of research is examining youth new media practice, we are still at the early stages of piecing together a more holistic picture of the role of new media in young people's everyday lives. In the United States, a number of survey-based studies have been documenting patterns of technology uptake and the spread of certain forms of new media practice (Griffith and Fox 2007; Lenhart et al. 2007; Rainie 2008; Roberts, Foehr, and Rideout 2005), and they provide a reference point for understanding broad trends in media engagement. We understand from this work that youth tend to be earlier adopters than adults of digital communications and authoring capabilities, and that their exposure to new media is growing in volume, complexity, and interactivity (Lenhart et al. 2007; Lenhart et al. 2008; Roberts and Foehr 2008; Roberts, Foehr, and Rideout 2005). Research across different postindustrial contexts also suggests that these patterns are tied to broader trends in the changing structures of sociability, where we are seeing a move toward more individualized and flexible forms of engagement with media environments. Researchers have described this as a turn toward "networked society" (Castells 1996), "networked individualism" (Wellman and Hogan 2004), "selective sociality" (Matsuda 2005), the "long tail" of niche media (Anderson 2006), or a more tailored set of media choices (Livingstone 2002). Youth practices have been an important part of the drive toward these more networked, individualized, and diversified forms of media engagement.

In addition to these quantitative indicators, there is a growing body of ethnographic case studies of youth engagement with specific kinds of new media practices and sites (examples include Baron 2008; Buckingham

2008; Ito, Okabe, and Matsuda 2005; Ling 2004; Livingstone 2008; Mazzarella 2005). Although the United Kingdom has funded some large-scale qualitative studies on youth new media engagements (Livingstone 2002; Holloway and Valentine 2003), the United States has not had comparable qualitative studies that look across a range of different populations and new media practices. What is generally lacking in the literature overall, and in the United States in particular, is an understanding of how new media practices are embedded in a broader social and cultural ecology. While we have a picture of technology trends on one hand, and spotlights on specific youth populations and practices on the other, we need more work that brings these two pieces of the puzzle together. How are specific new media practices embedded in existing (and evolving) social structures and cultural categories?

In this section of the introduction, we describe how our work addresses this gap, outlining our methodological commitments and descriptive focus that have defined the scope of this book. The first goal of this book is to document youth new media practice in rich, qualitative detail to provide a picture of how young people are mobilizing these media and technologies in their everyday lives. The descriptive frame of our study is defined by our ethnographic approach, the study of youth culture and practice, and the study of new media.

Ethnography

Using an ethnographic approach means that we work to understand how media and technology are meaningful to people in the context of their everyday lives. We do not see media or technology as determining or impacting society, culture, or individuals as an external force with its own internal logic, but rather as embodiments of social and cultural relationships that in turn shape and structure our possibilities for social action and cultural expression (see Bijker, Hughes, and Pinch 1987; Edwards 1995; Hine 2000). It follows that we do not see the content of the media or the media platform (TV, books, games, etc.) as the most important variables for determining social or cognitive outcomes. For example, we look at how video-game play is part of youth social lives, where it is situated in the home, how parents regulate play with the games, and how youth identify with the content and characters. We see outcomes not only in whether a child has identified with or learned media content but also in such things as how they are able to negotiate social status among peers, gain autonomy from parents, or acquire expertise in related domains such as knowledge seeking on the Internet. The strength of this approach is that it enables us to surface, from the empirical material, what the important categories and structures are that determine new media practices and learning outcomes. This approach does not lend itself to testing existing analytic categories or targeted hypotheses but rather to asking more fundamental questions about what the relevant factors and categories of analysis are. We believe that an initial broad-based ethnographic understanding, grounded in the actual contexts of behavior and local cultural understandings, is crucial to grasping the contours of a new set of cultural categories and practices.

We describe media and technology as part of a broader set of social structures and cultural patterns. We have organized our description based on practices and contexts that structure youth engagement with new media—friendship, intimacy, family, gaming, creative production, and work. A focus on these foundational social practices enables us to describe changes in youth social lives and culture while being attentive to the continuities with prior practice and structure. In the service of this broad descriptive goal we describe the continued relevance of gender and class in determining new media practice. Our focus, however, is on the issue of age and generational identity as structuring new media engagements. We look both internally at youth culture and the divisions among different youth as well as at the negotiations between youth and adults. How does new media engagement relate to different categories of youth culture and identity? To what extent are new media part of the definition—or, conversely, a disruption—of a generational identity? How are new media practices mobilized in the negotiations between adults and youth, particularly over learning and socialization? Any generation gap we might find in new media literacy and practices needs to be understood in its cultural diversity and specifics.

Our case studies have included diverse studies of youth in particular local communities, studies of after-school youth media programs, as well as studies of youth practices centered on online sites or interest groups. These include fans of Harry Potter and Japanese animation; video-game players; hip-hop creators; video bloggers; and participants on YouTube, MySpace, and Facebook. By looking at a range of populations and youth practices, we were able to combine in-depth textured description of specific group dynamics with collaborative analysis of how these different groups define themselves in relation to or in opposition to one another. We describe these studies and the specifics on data collection and joint analysis in chapter 1, "Media Ecologies." Our material covers both "mainstream" practices of new media use that are widely distributed among U.S. teens as well as more subcultural and exceptional practices that are not as common but represent emerging and experimental modes of technical and media literacy. In this, our work resembles other ethnographic studies that look at the relationships between different kinds of childhood and youth subcultures and identity categories (Eckert 1989; Milner 2004; Thorne 1993), but we focus on the role of new media in these negotiations. To the extent possible, we have also situated our ethnographic cases and findings in relation to the quantitative work in the field. Through this approach, we have worked to mediate the gap between the textured, qualitative descriptions of new media practices and analysis of broader patterns in social, technical, and cultural change.

Youth

Foundational to our descriptive approach is a particular point of view and methodological approach in relation to youth as a social and cultural category. In our research and writing we take a sociology-of-youth-and-childhood approach, which means that we take youth seriously as actors in their own social worlds and look at childhood as a socially constructed, historically variable, and contested category (Corsaro 1997; Fine 2004; James and Prout 1997; Wyness 2006). Adults often view children in a forward-looking way, in terms of developmental "ages and stages" of what they will become rather than as complete beings "with ongoing lives, needs and desires" (Corsaro 1997, 8). By contrast, the "new paradigm" in the sociology of childhood (James and Prout 1997) sees that children are active, creative social agents who produce their own unique children's cultures while simultaneously contributing to the production of adult societies and that "childhood—that socially constructed period in which children live their lives—is a structural form" (Corsaro 1997, 4). This structural form has varied historically and is interrelated with other structural categories such as social class, gender, and race (Corsaro 1997; James and Prout 1997). In keeping with this sociology-of-youth-and-children approach, we move beyond a simple socialization model in which children are passive recipients of dominant and "adult" ideologies and norms, and instead we deploy what Corsaro calls an Interpretive Reproduction model. In this model children collectively participate in society, in which children "negotiate, share, and create culture with adults and each other" (Corsaro 1997, 18). In doing so we seek to give voice to children and youth, who, while they have not been absent in social-science research, have often not been heard (James and Prout 1997).

Our work has focused mostly on youth in their middle-school and high-school years, between the ages of twelve and eighteen. As we have indicated, we have made our best effort at examining the diversity among youth, rather than suggesting that youth share a monolithic identity. As described in chapter 1, we have also engaged, to a lesser extent, with parents, educators, and young adults who participate or are involved in structuring youth new media practices. The category of youth and youth culture is coconstructed by adults and young people (Alanen and Mayall 2001). We capture what is unique about the contexts that youth inhabit while also remaining attentive to the ways in which new media practices span different age cohorts. In addition to their role in provisioning and regulating youth new media ecologies, adults are important coparticipants in youth new media practices. In fact, one of the important outcomes of youth participation in many online practices is that they have an opportunity to interact with adults who are outside of their usual circle of family and school-based adult relationships. The age populations that we look at are keyed to the specifics of the particular case study. In studies that focus on mixed-age interest groups, we have a significant proportion of young adults, while studies that focus on family life or school-based cohorts focus more exclusively on teens and their relationships to parents and teachers. An ethnography of youth insists on attention to both the focal object of youth culture and to the adult cultures that have a formative and pervasive influence.

Readers will see the subjects of this research referred to by a variety of age-related names—children, kids, youth, teens, adolescents, young people, and young adults. In keeping with an ethnographic approach we try to use terms that our respondents use themselves, but given that youth do not commonly refer to themselves in age-graded categories (Thorne 1993), we frequently must impose categories. To that end, for respondents age thirteen and under, the general cutoff age for the term "children" (Wyness

2006), we usually use the word "kids" and, perhaps less often, "children." While "kids" might seem a pejorative term, researchers have documented that this is the term they often use to refer to themselves; as Barrie Thorne noted in her research on schoolchildren, one of her respondents "insisted that 'children' was more of a put-down than 'kids'" (Thorne 1993, 9). For participants between the ages of thirteen and eighteen we usually use the category of "teen" or "teenager" and, less frequently, the more biologically oriented "adolescent." We do this to note that teenagers are, now, a slightly different social category. Teens have more agency than children, develop more elaborate peer cultures, self-consciously construct public and private selves, and challenge conventions of adult life (Fine 2004). We refer to those between the ages of nineteen and thirty as "young adults," and we use the term "young people" to refer broadly to both young adults and teens. "Youth" is the category we reserve for when we are referring to the general cultural category of youth, which is not clearly age demarcated but which centers on the late teenage years.

While age-based categories have defined our object of study, we are interested in documenting how these categories are historically and culturally specific, and how they are under negotiation. Age gradations in Euro-American and other postindustrial countries are perhaps more salient and structuring than they have been at any point in history, as age gradation now has emerged as a way to define entire populations of people (Chudacoff 1989). Youth culture—since its midcentury inception by Talcott Parsons (Eckert 1989; Gilbert 1986)—has been characterized by being set apart from adulthood, defined by the process of "becoming" and "leisure" (Chudacoff 1989). Removing youth from the workforce and home left them with large amounts of leisure time with their own "peers," or age cohorts. More recently, researchers have documented how youth have been limited in their access not only to the workplace but also to other forms of public participation, including mobility in public places (Buckingham 2000; Lewis 1990; Livingstone 2002). Youth occupy more age-segregated institutions than they have in recent history (Chudacoff 1989) and have more cultural products that are targeted to them as specific age demographics (Cross 1997; Frank 1997; Kline 1993; Livingstone 2002; Seiter 1993). The ghettoization of youth culture also leads to its construction as social problem, a generational space in which society channels fears and anxieties (Cohen 1972; Corsaro 1997; Gilbert 1986; Lesko 2001). The current debates over the digital generation are the latest instantiation of these public hopes and fears surrounding youth; as they have in recent history, media continue to play a central role in the contestations over the boundaries and definitions of youth culture and sociability. While we have not conducted a historical or longitudinal study, we see our current snapshot of youth new media engagement as part of this longer trajectory in the definition of youth as a historically specific social and cultural category.

New Media

Popular culture and online communication provide a window onto examining youth practice in contexts where young people feel ownership over the social and cultural agenda. The commitment to taking youth social and cultural worlds seriously has been applied to media studies by a growing number of researchers who have looked at how children engage with media in ways responsive to the specific conditions of childhood. In contrast to much of media-effects research, these qualitative studies see children and youth as actively constructing their social and cultural worlds, not as innocent victims or passive recipients of media messages (Buckingham 1993; Jenkins 1998; Kinder 1999; Seiter 1993). By taking children and youth popular culture seriously, this body of work argues against the trivialization of children's media culture and sees it as a site of child- and youth-driven creativity and social action. While we recognize the ways in which popular culture has provided a site for kids to exercise agency and authority, we think it is important to keep in view the central role of commercial entities in shaping children and youth culture. Media industries have been increasingly successful in constructing childhood culture in ways that kids uniquely identify with (Banet-Weiser 2007; Seiter 1993, 2005). In her analysis of Nickelodeon, Sarah Banet-Weiser describes how the channel constructs a form of "consumer citizenship." She writes, "This recent attention to children as consumers has as much to do with recognizing a particular political economic agency of children as it does to the unprecedented ways in which children are constituted as a commercial market" (Banet-Weiser 2007, 8). The development of children's agency in the local life worlds of home and peer culture is inextricably linked to their participation as consumer citizens.

Within their local life worlds, popular culture can provide kids with a space to negotiate issues of identity and belonging within peer cultures

(Chin 2001; Dyson 1997; Ito 2006; Seiter 1999a). In the case of interactive media and communications technology, the constitutive role of youth voice and sociability is further accentuated in what Henry Jenkins (1992; 2006) has described as a "participatory media culture" and Mizuko Ito (2008b) has described in terms of "hypersociality" surrounding media engagement. In looking at Pokémon, for example, David Buckingham and Julian Sefton-Green (2004) have argued that although all media audiences are in some ways "active," interactive and sociable media such as Pokémon "positively require activity." With teens, this participatory approach toward new media has been channeled into networked gaming and social media sites such as MySpace, Facebook, or YouTube, which have captured the public limelight and added fuel to the discourse of a digital generation. The active and sociable nature of youth new media engagement argues for an ethnographic approach that looks at not only the content of media but also the social practices and contexts in which media engagement is embedded. While we are cautious about assuming a natural affinity between youth and participatory forms of media engagement, it is clear that youth participation in these media forms is high, and that interactive and networked media require particular methodological commitments.

We use the term "new media" to describe a media ecology where more traditional media such as books, television, and radio are intersecting with digital media, specifically interactive media and media for social communication (Jenkins 2006). As described in chapter 1, we are interested in the convergent media ecology that youth are inhabiting today rather than in isolating the specific affordances of digital-production tools or online networks. We have used the term "new media" rather than terms such as "digital media" or "interactive media" because the moniker of "the new" seemed appropriately situational, relational, and protean, and not tied to a specific media platform. Just as in the case of youth, who are always on the verge of growing older, media are constantly undergoing a process of aging and identity reformulation in which there is a generation of the new ready to replace the old. Our focus is on media that are new at this particular historical moment. Our difficulty in naming a trait that defines the media we are scrutinizing (interactive, digital, virtual, online, social, networked, convergent, etc.) stems from the fact that we are examining a constellation of media changes, in a move toward more digital, networked, and interactive forms, which together define the horizon of "the new."

Our work has focused on those practices that are "new" at this moment and that are most clearly associated with youth culture and voice, such as engagement with social network sites, media fandom, and gaming. In contrast to sites such as Linked In and match.com or much of the blogging world, sites such as MySpace, Facebook, YouTube, and LiveJournal and online gaming have a high degree of youth participation, and youth have defined certain genres of participation within these sites that are keyed to a generational identity. We can also see this cultural distinction at play in the difference between email and instant messaging as preferred communication tools, where the older generation is more tightly identified with the former. The ways in which age identity works in these sites is somewhat different from how more traditional media have segmented youth as a distinct market with particular cultural styles and products associated with it. Instead, the youth focus stems from patterns of adoption, the fit with the particular social and communicative needs of youth, and how they take up these tools to produce their own "content" as well as traffic in commercial popular culture. In these sites, it is not only youth consumption that is driving the success of new Internet ventures but also their participation (or "traffic") and production of "user-generated content." In describing these as youth-centric sites and communication tools, we mean that they are culturally identified with youth, but they can be engaged with by people of all ages. We are examining the cultural valences of certain new media tools and practices in how they align with age-based identities, but this does not mean that we believe that youth have a monopoly on innovative new media uses or that youth-centric sites do not have a large number of adult participants.

New media researchers differ in the degree to which they see contemporary new media practices as attached to a particular life stage or more closely tied to a generational cohort identity. For example, in looking at mobile phone use, Rich Ling and Brigitte Yttri (2006) have argued that communicative patterns are tied to the particular developmental needs of adolescents who are engaged in negotiations over social identity and belonging. Naomi Baron (2008) also examines the relation between online communication and changes to reading and writing conventions. She sees youth uptake of more informal forms of online writing as part of a broader set of social and cultural shifts in the status of printed and written communication. Ultimately, the ways in which current communication practice will lead to resilient cultural change is an empirical question that can be answered only with the passage of time, as we observe the aging of the current youth cohort. If history is any guide, however, we should expect at least some imprint of a generation-specific media identity to persist. The aim of our study is to describe media engagements that are specific to the life circumstances of current youth, at a moment when we are seeing a transition to what we describe in this book as widespread participation in digital media production and networked publics. At the same time, we analyze how these same youth are taking the lead in developing social norms and literacies that are likely to persist as structures of media participation and practice that transcend age boundaries. For example, we have seen text messaging expand from a youth demographic to encompass a broader age range, and the demographics of media such as gaming and animation gradually shift upstream.

Finally, the new media practices we examine are almost all situated in the social and recreational activities of youth rather than in contexts of explicit instruction. In this, our approach is in line with a growing body of work in sociocultural learning theory that looks to out-of-school settings for models of learning and engagement that differ from what is found in the classroom (Cole 1997; Goldman 2005; Hull and Schultz 2002b; Lave 1988; Lave and Wenger 1991; Mahiri 2004; Nocon and Cole 2005; Nunes, Schliemann, and Carraher 1993; Rogoff 2003; Singleton 1998; Varenne and McDermott 1998). Our approach also reflects an emerging consensus that the most engaged and active forms of learning with digital media happen in youth-driven settings that are focused on social communication and recreation. As Julian Sefton-Green (2004, 3) has argued in his literature review *Informal Learning with Technology Outside School*, educators must recognize that much of young people's learning with information and communication technologies happens outside of school. "This recognition requires us to acknowledge a wider 'ecology' of education where schools, homes, playtime, and library and the museum all play their part." By focusing on recreational and social media engagement in the everyday contexts of family and peer interaction, we fill out the picture of the range of environments in which youth learn with new media and prioritize those social contexts that youth find most meaningful and motivational. In this, we see our work as addressing an empirical gap in the literature as well as addressing the need to develop conceptual frameworks that are keyed to the changing landscape of new media engagement.

Our primary descriptive task for this book is to capture youth new media practice in a way that is contextualized by the social and cultural contexts that are consequential and meaningful to young people themselves, and to situate these practices within the broader structural conditions of childhood that frame youth action and voice. In this, we draw from an ethnographic approach toward youth studies and new media studies. This commitment to socially and culturally contextualized analysis is evident also in the thematic and conceptual frameworks that guide our analysis of participation, learning, and literacy.

Conceptual Frameworks

Through our collaborative analysis, we have developed a series of shared conceptual frameworks that function as threads of continuity throughout this book's chapters. Our work is guided by four key analytic foci that we apply to our ethnographic material: participation, publics, literacy, and learning. Our primary descriptive research question is this: How are new media being taken up by youth practices and agendas? Our analytic question follows: How do these practices change the dynamics of youth-adult negotiations over literacy, learning, and authoritative knowledge?

In keeping with our focus on social and cultural context, we consider learning and literacy as part of a broader set of issues having to do with youth participation in public culture (Appadurai and Breckenridge 1988; 1995). We draw from existing theories that are part of the "social turn" in literacy studies, new media studies, learning theory, and childhood studies. The 1980s and 1990s saw the solidification of a new set of paradigms for understanding learning and literacy that emphasized the importance of social participation and cultural identity, and that moved away from the previously dominant focus on individual cognition and knowledge acquisition. This social turn has been described in terms of new paradigms of situated cognition (Brown, Collins, and Duguid 1989; Greeno 1997; Lave 1988), situated learning (Lave and Wenger 1991), distributed cognition (Hutchins 1995), and New Literacy Studies (Gee 1990; Street 1993). We see a counterpart in the new paradigm of childhood studies and the recognition among media scholars of the active agency of media audiences, as we describe in the previous section. We tailor these approaches to our specific interdisciplinary endeavor and our objects of inquiry that are at the intersection of these different fields.

While the social turn in learning and literacy studies is now well established, there is relatively little work that applies these frameworks to learning in the context of networked communication and media engagement. Further, though situated approaches to learning and literacy engage deeply with issues of cultural diversity and equity, they tend not to see generational and age-based power differentials as a central analytic problematic in the same way that the new paradigm in childhood studies does. We see the topic of youth-centered new media practice as a site that can bring these conversations together into productive tension. New media are a site where youth exhibit agency and an expertise that often exceeds that of their elders, resulting in intergenerational struggle over authority and control over learning and literacy. Technology, media, and public culture are shaping and being shaped by these struggles, as youth practice defines new terms of participation in a digital and networked media ecology. We have developed an interdisciplinary analytic tool kit to investigate this complex set of relations among changing technology, kid-adult relations, and definitions of learning and literacy. Our key terms are "genres of participation," "networked publics," "peer-based learning," and "new media literacy."

Genres of Participation

One of the key innovations of situated learning theory was to posit that learning was an act of social participation in communities of practice (Lave and Wenger 1991). By shifting the focus away from the individual and to the broader network of social relationships, situated learning theory suggests that the relationships of knowledge sharing, mentoring, and monitoring within social groups become key sites of analytic interest. In this formulation, people learn in all contexts of activity, not because they are internalizing knowledge, culture, and expertise as isolated individuals, but because they are part of shared cultural systems and are engaged in collective social action. This perspective has a counterpart within work in media studies that looks at media engagement as a social and active process. A notion of "participation," as an alternative to internalization or consumption, has the advantage in not assuming that kids are passive, mere audiences to media or educational content. It forces attention to the more ethnographic and practice-based dimensions of media engagement as well as querying the broader social and cultural contexts in which these activities are conducted.

Henry Jenkins has put forth the idea of "participatory media cultures," which he originally used to describe fan communities in the 1970s and 1980s, and which he has recently revisited in relation to current trends in convergence culture (1992; 2006). Jenkins traces how fan practices established in the TV-dominated era have become increasingly mainstream because of the convergence of traditional and digital media. Fans not only consume professionally produced media but they also produce their own meanings and media products, continuing to disrupt the culturally dominant distinctions between production and consumption. More recently, Jenkins has taken this framework and applied it to issues of learning and literacy, describing a set of twenty-first-century skills and dispositions that are based on different modes of participation in media cultures (Jenkins 2006). In a complementary vein, Joe Karaganis (2007) has proposed a concept of "structures of participation" to analyze different modes of relating to digital and interactive technologies. In our descriptions of youth practice, we rely on a related notion of "genres of participation" to suggest different modes or conventions for engaging with new media (Ito 2003; 2008b). A notion of participation genre addresses similar problematics as concepts such as habitus (Bourdieu 1972) or structuration (Giddens 1986), linking activity to social and cultural structure. More closely allied with humanistic analysis, a notion of "genre," however, foregrounds the interpretive dimensions of human orderliness. How we identify with, orient to, and engage with media is better described as a process of interpretive recognition than a process of habituation or structuring. We recognize certain patterns of representation (textual genres) and in turn engage with them in social, routinized ways (participation genres).

In this book, we identify genres of participation with new media as a way of describing everyday learning and media engagement. The primary distinction we make is between friendship-driven and interest-driven genres of participation, which correspond to different genres of youth culture, social network structure, and modes of learning. By "friendship-driven genres of participation," we refer to the dominant and mainstream practices of youth as they go about their day-to-day negotiations with friends and peers. These friendship-driven practices center on peers youth encounter in the age-segregated contexts of school but might also include friends and peers they meet through religious groups, school sports, and other local activity groups. For most youth, these local friendship-driven networks are their primary source of affiliation, friendship, and romantic partners, and their lives online mirror this local network. MySpace and Facebook are the emblematic online sites for these sets of practices. We use the term "peer" to refer to the people whom youth see as part of their lateral network of relations, whom they look to for affiliation, competition, as well as disaffiliation and distancing. Peers are the group of people to whom youth look to develop their sense of self, reputation, and status. We reserve the term "friend" to refer to those relations that youth self-identify as such, a subset of the peer group that individual youths have close affiliations with. By "friendship-driven," we refer even more narrowly to those shared practices that grow out of friendships in given local social worlds. The chapters on friendship and intimacy focus on describing these friendship-driven forms of learning and participation.

In contrast to friendship-driven practices, with interest-driven practices, specialized activities, interests, or niche and marginalized identities come first. Interest-driven practices are what youth describe as the domain of the geeks, freaks, musicians, artists, and dorks—the kids who are identified as smart, different, or creative, who generally exist at the margins of teen social worlds. Kids find a different network of peers and develop deep friendships through these interest-driven engagements, but in these cases the interests come first, and they structure the peer network and friendships, rather than vice versa. These are contexts where kids find relationships that center on their interests, hobbies, and career aspirations. It is not about the given social relations that structure kids' school lives but about focusing and expanding an individual's social circle based on interests. Although some interest-based activities such as sports and music have been supported through schools and overlap with young people's friendship-driven networks, other kinds of interests require more far-flung networks of affiliation and expertise. As we discuss in the chapters on gaming, creative production, and work, online sites provide opportunities for youth to connect with interest-based groups that might not be represented in their local communities. Interest-driven and friendship-driven participation are high-level genre categories that orient our description as a whole. Individual chapters go into more depth on the specific genre conventions of their domain.

Certain forms of participation also act to bridge the divide between friendship-driven and interest-driven modes. In chapter 5, we describe how more friendship-driven modes of "hanging out" with friends while gaming can transition to more interest-driven genres of what we call recreational gaming. Similarly, in chapter 6, we describe how the more friendship-driven practices of creating profiles on social network sites or taking photos with friends can lead to "messing around" in the more interest-driven modes of digital media production. In chapter 1, we identify a genre of participation of "messing around" with new media that in some cases can mediate between genres of "geeking out" and "hanging out." Conversely, we have seen how interest-driven engagements can lead to deep and abiding friendships that might eventually transcend the particular focus of interest and provide a social group for socializing and friendship for youth who may not have been deeply embedded in the more popularity- and friendship-driven networks in their local school or community. Transitioning between hanging out, messing around, and geeking out represents certain trajectories of participation that young people can navigate, where their modes of learning and their social networks and focus begin to shift. Examining learning as changes in genres of participation is an alternative to the notion of "transfer," where the mechanism is located in a process of individual internalization of content or skills. In a participatory frame, it is not that kids transfer new media skills or social skills to different domains, but rather they begin to identify with and participate in different social networks and sets of cultural referents through certain transitional social and cultural mechanisms. It is not sufficient to internalize or identify with certain modes of participation; there also needs to be a supporting social and cultural world.

Rather than relying on distinctions based on given categories such as gender, class, or ethnic identity, we have identified genres based on what we saw in our ethnographic material as the distinctions that emerge from youth practice and culture, and that help us interpret how media intersect with learning and participation. By describing these forms of participation as genres, we hope to avoid the assumption that these genres attach categorically to individuals. Rather, just as an individual may engage with multiple media genres, we find that youth will often engage in multiple genres of participation in ways that are situationally specific. We have also avoided categorizing practice based on technology- or media-centric parameters, such as media type or measures of frequency or media saturation. Genres of participation provide ways of identifying the sources of diversity in how youth engage with new media in a way that does not rely on a simple notion of "divides" or a ranking of more- or less-sophisticated media expertise. Instead, these genres represent different investments that youth make in particular forms of sociability and differing forms of identification with media genres.

Networked Publics

When we consider learning as an act of social participation, our analytic focus shifts from the individual to the broader social and cultural ecology that a person inhabits. Although we all experience private moments of learning and reflection, a large part of what defines us as social beings and learners happens in contexts of group social interaction and engagement with shared cultural forms. Engagement with media (itself a form of mediated sociability) is a constitutive part of how we learn to participate as culturally competent, social, and knowledgeable beings. Although studies of learning in out-of-school settings have examined a wide range of learning environments, these approaches have been relatively silent as to how learning operates in relation to mass and networked media. With some exceptions (Mahiri 2004; Renninger and Shumar 2002; Weiss et al. 2006), contexts of social interaction and public behavior tend to be imagined as local, copresent encounters such as in the case of apprenticeship or learning in the home or street; work in media studies has largely been in a parallel (though often complementary) set of conversations. The focus on situated learning in contexts of embodied presence has been an important antidote to more traditional educational approaches that have focused on kids' relationships to abstract academic content, often through the abstraction of educational media, but it has stood in the way of an articulation of situated-learning theory in relation to mediated practices. Our work here, however, is to take more steps in applying situated approaches to learning to an understanding of mediated sociability, though not of the school-centered variety. This requires integrating approaches in public-culture studies with theories of learning and participation.

Arjun Appadurai and Carol Breckenridge suggest the term "public culture" as an alternative to terms such as "popular culture" or "mass culture" to link popular-culture engagement to practices of participation in the public sphere. They see public culture studies as a way of understanding "the space between domestic life and the nation-state—where different social groups (classes, ethnic groups, genders) constitute their identities by their experience of mass-culture mediated forms in relation to the practices of everyday life" (Appadurai and Breckenridge 1995, 4–5). We draw from this framing and situate it within this current historical moment, where we are seeing public culture, as it is experienced by a growing number of U.S. teens, migrating to digitally networked forms. In this context, youth are participating in publics constituted in part by the nation-state, and also by commercial media environments that are along the lines of the "consumer citizenship" that Banet-Weiser (2007) has theorized. We use the term "networked publics" to reference the forms of participation in public culture that is the focus of our work. The growing availability of digital media-production tools, combined with online networks that traffic in rich media, is creating convergence between mass media and online communication (Benkler 2006; Ito 2008a; Jenkins 2006; Shirky 2008; Varnelis 2008). Rather than conceptualize everyday media engagement as "consumption" by "audiences," the term "networked publics" foregrounds the active participation of a distributed social network in the production and circulation of culture and knowledge. The growing salience of networked publics in young people's daily lives is part of important changes in what constitutes the relevant social groups and publics that structure young people's learning and identity.

This book delves into the details of everyday youth participation in networked publics and into the ways in which parents and educators work to shape these engagements. As danah boyd discusses in her analysis of participation on MySpace, networked publics differ from traditional teen publics (such as the mall or the school) in some important ways. Unlike unmediated publics, networked publics are characterized by their persistence, searchability, replicability, and invisible audiences (boyd 2007). With friendship-driven practices, youth online activity largely replicates their existing practices of hanging out and communicating with friends, but these characteristics of networked publics do create new kinds of opportunities for youth to develop their public identities, connect, and communicate. The chapters on friendship and intimacy describe these dynamics by examining how practices such as Friending, public social drama, flirting, and dating are both reproduced and reshaped by online communication through social network sites, online chat, and mobile communication. These technologies facilitate new forms of private, intimate, and always-on communication as well as new forms of publicity where personal networks and social connections are displayed to broader publics than have traditionally been available locally to teens.

In addition to reshaping how youth participate in their given social networks of peers in school and their local communities, networked publics open new avenues for youth participation through interest-driven networks. In contrast to friendship-driven networked publics, the interest-driven varieties generally do not adhere to existing formal institutions such as school or church, nor are they locally bound. Through sites such as YouTube, fan forums, networked gaming sites, LiveJournal communities, deviantART, or youth media centers, youth can access publics that are engaged in their particular hobby or area of interest. These more specialized and niche publics are settings where youth can connect with other creators or players who have greater expertise than they do, and conversely, where they can mentor and develop leadership in relation to less experienced participants. They are also networks for distributing, publicizing, and sometimes even getting famous or paid for the work that they create. These dynamics of interest-driven networked publics, and the new kinds of peer relations that youth find there, are the focus of our chapters on gaming, creative production, and work.

The relation between friendship-driven and interest-driven networked publics is complex and grows out of the existing status distinctions of youth culture. Although kids with more geeky and creative interests continue to be marginal to the more mainstream popularity and dating negotiations in school, our work does indicate some shifts in the balance of how kids engage with these different networks. Unlike the older generation, today's kids have the opportunity to engage in multiple publics—they can retain an identity as a "popular" kid in their local school networks and on MySpace while also pursuing interest-driven activities with another set of peers online. Although the majority of kids we spoke to participate primarily in friendship-driven publics, we also saw many examples of kids who maintain a dual identity structure. They might have multiple online profiles for different sets of friends, or they might have a group of online gaming friends who do not overlap with the friends they hang out with in school. Although our study does not enable us to identify whether the balance is shifting in terms of how kids participate in different publics, we have identified that there is an expanded palette of opportunity for kids to participate in different kinds of publics because of the growth of the networked variety.

Peer-Based Learning

Sociocultural approaches to learning have recognized that kids gain most of their knowledge and competencies in contexts that do not involve formal instruction. A growing body of ethnographic work documents how learning happens in informal settings, as a side effect of everyday life and social activity, rather than in an explicit instructional agenda. For example, in describing learning in relation to simulation games, James Paul Gee (2008, 19) suggests that kids pick up academic content and skills as part of their play. "These things, which are in the foreground at school, come for free, that is, develop naturally as the learner solves problems and achieves goals." In *School's Out!*, an edited collection of essays documenting learning in home, after-school, and community settings, Glynda Hull and Katherine Schultz (2002a, 2) ask, "Why, we have wanted to know, does literacy so often flourish out of school?" They describe the accumulating evidence documenting how people pick up literacy in the contexts of informal, everyday contexts, and it is often difficult to reproduce those same literacies in the more formalized contexts of schooling and testing. We see our focus on youth learning in contexts of peer sociability and recreational learning as part of this research tradition. Our interest, more specifically, is in documenting instances of learning that are centered on youth peer-based interaction, in which the agenda is not defined by parents and teachers.

Our focus on youth perspectives, as well as the high level of youth engagement in social and recreational activities online, determined our focus on the more informal and loosely organized contexts of peer-based learning. We discuss the implications for learning institutions in the conclusion of this book, but the body of the book describes learning outside of school, primarily in settings of peer-based interaction. As ethnographies of children and youth have documented, kids learn from their peers. While adults often view the influence of peers negatively, as characterized by the term "peer pressure," we approach these informal spaces for peer interactions as a space of opportunity for learning. Our cases demonstrate that some of the drivers of self-motivated learning come not from the institutionalized authorities in kids' lives setting standards and providing instruction, but from the kids observing and communicating with people engaged in the same interests and in the same struggles for status and recognition that they are.

Both interest-driven and friendship-driven participation rely on peer-based learning dynamics, which have a different structure from formal instruction or parental guidance. Our description of friendship-driven learning describes a familiar genre of peer-based learning, in which online networks are supporting those sometimes painful but important lessons in growing up, giving kids an environment to explore romance, friendship, and status just as their predecessors did. In an environment where there are fewer and fewer spaces for kids to hang out informally in public space, these online friendship-driven networks are critical contexts for these forms of learning and sociability. Rather than construe these dynamics negatively or fearfully, we can consider them also as an integral part of developing a sense of personal identity as a social being. Peer-based learning relies on a context of reciprocity, in which kids feel they have a stake in self-expression as well as a stake in evaluating and giving feedback to one another. Unlike in more hierarchical and authoritative relations, both parties are constantly contributing and evaluating one another. Youth both affiliate and compete with their peers.

Like friendship-driven networks, interest-driven networks are also sites of peer-based learning, but they represent a different genre of participation, in which specialized interests are what bring a social group together. In both cases, however, the peer group becomes a powerful driver for learning. The peers whom youth are learning from in interest-driven practices are not defined by their given institution of school but rather through more intentional and chosen affiliations. When kids reach out to a set of relations based on their interests, what constitutes a peer starts to change because of the change in a young person's social network. In the case of kids who have become immersed in interest-driven publics, the context of who their peers are changes, as does the context for how reputation works, and they get recognition for different forms of skill and learning.

Youth are increasingly turning to networked publics as sites for peer-based learning and interaction that are not reliant on adult oversight and guidance. Among the reasons that youth participation in these networked publics is so high is that they are an alternative to publics that the adult authorities in their lives have control over, and they provide opportunities for private conversation with peers. Commercial media industries have a complicated role in these dynamics. Ever since the growth of a youth-oriented commercial and media culture in the past century, children and youth have been marketed to as a unique demographic, with cultural products and identity categories that are distinct from those of their elders (Cross 1997; Frank 1997; Kline 1993; Livingstone 2002; Seiter 1993). The growing influence of peers from a similar age cohort in determining social values and cultural style (Milner 2004; Willis 1990) has grown in tandem with these broader cultural shifts in defining a distinct youth culture (Frank 1997), or "kid power" (Banet-Weiser 2007; Seiter 1993). Although the contemporary media ecology is characterized by the growing centrality of user-generated content, commercial media are still central to youth culture, and Internet companies are becoming a formidable force in structuring the conditions under which youth connect with their peers. This takes the form of technology design decisions, marketing decisions, and policy constraints that are placed on the industry. Although we do not focus on the role of commercial industry in structuring youth peer interactions, we understand that commercial culture and commercial online spaces and services are lending support to youth-centered peer cultures and communication, often at the expense of institutions such as school and family.

New Media Literacy

The negotiations among kids, parents, educators, and technologists over the shape of youth online participation is also a site of struggle over what counts as legitimate forms of learning and literacy. Any discussion of learning and literacy is unavoidably normative. What counts as learning and literacy is a question of collective values, values that are constantly being contested and negotiated among different social groups. Periods of cultural and technological flux open up new areas of debate about what should count as part of our common culture and literacy and what are appropriate ways for young people to participate in these new cultural forms. Education designed by adults for children also has an unavoidably coercive dimension that is situated in a systemic power differential between adults and children. The moral panic over youth new media uptake is also part of this power differential, as adults mobilize public support to direct children away from social forms and literacies that they find threatening and dangerous. Changes in social, cultural, economic, and technological landscapes are often accompanied by anxieties and questions as to what skills need to be learned and taught for subsequent generations to be able to participate in public life, as students, citizens, consumers, and workers.

In our work, we are examining the current practices of youth and querying what kinds of literacies and social competencies they are defining as a particular generational cohort experimenting with a new set of media technologies. We have attempted to momentarily suspend our own value judgments about youth engagement with new media in order to better understand and appreciate what youth themselves see as important forms of culture, learning, and literacy. Those studying literacies within the New Literacy Studies framework have used ethnography as a way of understanding the socially constructed dimensions of literacy, whether studying in school or out-of-school contexts (Collins 1995; Gee 1990; Hull and Schultz 2002b; Street 1993, 1995). This work, in both its anthropological roots in the work of Brian Street and its sociolinguistic roots in the work of James Paul Gee, sees any discussion of literacy as an inherently ideological one. Definitions of literacy are embedded in institutions, broader cultural dimensions, and power. The emphasis has continually been on the local practices associated with the uses of reading and writing and how these are not determined by text, technology, or media, nor are they determined in a top-down manner. Those who may seem in weaker positions often appropriate and transform the agendas of those who may seem in more dominant positions of power. While we are aware that there may be "limits to the local" in the understanding of literacies as practices (Brandt and Clinton 2002), we believe that it is crucial to examine literacy as a set of standards that are under continuous development and negotiation through social activity. In this, our work is in line with that of other scholars (e.g., Chávez and Soep 2005; Hull 2003; Mahiri 2004) who explore literacies in relation to ideology, power, and social practice in other settings where youth are pushing back against dominant definitions of literacy that structure their everyday life worlds.

We see a moving horizon of what counts as new media as the horizon of what those who study technologicial systems have described as a window of "interpretive flexibility." Theorists who have described the social construction of technological systems have posited that when new technologies enter the social stage, there is a period of flexibility in which different social actors mobilize to construct the new meaning of a technological artifact (Bijker, Hughes, and Pinch 1987). Through time, and through contestations among different actors, the meaning and shape of an artifact is gradually stabilized and black boxed. Though the meaning of a technological artifact can later be reopened with the introduction of new facts or new social actors, generally there is a period in the historical evolution of new technologies in which there is heightened public debate and social negotiations about a technology's shape and meaning. The new media that we are examining in this book, and the related generational struggles over the shape of culture, norms, and literacy, are emblematic of this moment of interpretive flexibility. While what is being defined as "new media literacy" is certainly not the exclusive province of youth, unlike in the case of "old" literacies, youth are playing a more central role in the definition of these newer forms. In fact, the current anxiety over how new media erode literacy and writing standards could be read as an indicator of the marginalization of adult institutions that have traditionally defined literacy norms (whether that is the school or the family).

Researchers have posited a variety of ways to understand and define new media literacy. For example, David Buckingham comes from a tradition of media education and considers new media literacy as a twist in the debates over media literacy that have been, until recently, focused on television (Buckingham 2003; Buckingham et al. 2005). Kathleen Tyner (1998) considers media literacy as well as technical literacy in her discussion of literacy in a digital world. James Paul Gee (2003) sees gaming as representing new modes of learning of certain semiotic domains, and in his recent work on twenty-first-century skills Henry Jenkins (2006) applies his insights about active media participation to an analysis of new media literacy. One of the more general statements of literacy that is pertinent to considering new media literacy is The New London Group's (1996, 63) work on multiliteracies. It sees a growing palette of literacy forms in relation to an "emerging cultural, institutional, and global order: the multiplicity of communication channels and media, and the increasing saliency of cultural and linguistic diversity."

Our work is in line with this general impetus toward acknowledging a broader set of cultural and social competencies that could be defined as examples of literacy. However, our work does not seek to define the components of new media literacy or to participate directly in the normalization of particular forms of literacy standards or practice. Rather, we see our contribution as describing the forms of competencies, skills, and literacy practices that youth are developing through media production and online communication to inform these broader debates. More specifically, we have identified certain literacy practices that youth have been central participants in defining: deliberately casual forms of online speech, nuanced social norms for how to engage in social network activities, and new genres of media representation, such as machinima, mashups, remix, video blogs, web comics, and fansubs. Often these cultural forms are tied to certain linguistic styles identified with particular youth culture and subcultures (Eckert 1996). The goal of our work is to situate these literacy practices within specific and diverse conditions of youth culture and identity as well as within an intergenerational struggle of literacy norms. Although the tradition of New Literacy Studies has described literacy in a more multicultural and multimodal frame, it is often silent as to the generational differences in how literacies are valued. In our work, we suggest that not only are new media practices defining forms of literacy that rely on interactive and multimedia forms but they also are defining literacies that are specific to a particular media moment, and possibly generational identities. Although some of the literacy practices we describe may be keyed to a particular life stage, new media literacies are not necessarily going to "grow up" to conform to the standards of their elders but are likely to be tied to foundational changes in forms of cultural expression.

Overview of Chapters

The chapters that follow are organized based on what emerged from our material as the core practices that structure youth engagement with new media. Unlike the specific case studies that individual researchers will address in independent publications, these chapters are efforts to synthesize across different cases and youth populations. Throughout the book, we include a series of illustrative numbered sections that provide more detailed descriptions of specific youth and cases. With this format, we have tried to provide general summative findings that do justice to the breadth of our research while also providing some of the detailed description that is the hallmark of ethnographic writing.

Chapter 1, "Media Ecologies," frames the technological and social context in which young people are consuming, sharing, and producing new media. The chapter introduces the various locations in which we conducted our research and our methods of data collection and collaborative analysis. The second half of the chapter introduces three genres of participation with new media that are an alternative to common ways of categorizing forms of media access: hanging out, messing around, and geeking out.

The following two chapters focus on mainstream friendship-driven practices and networks. Chapter 2, "Friendship," examines how teens use instant messaging, social network sites, and mobile phones to negotiate their friendships in peer groups that center on school and local activity groups. These are the dominant forms of sociality in teen communication. Familiar practices of making friends—gossiping, bullying, and jockeying for status—are reproduced online, but they are also reshaped in significant ways because of the new forms of publicity and always-on communication.

The discussion of friend-centered practices is followed by the chapter on intimacy, which also examines practices that are a long-standing and pervasive part of everyday youth sociality. The chapter discusses how teens use online communication to augment their practices of flirting, dating, and breaking up. The dominant social norm is that the online space is used to extend and maintain relationships, but that first contact should be initiated offline. While these norms largely mirror the existing practices of teen romance, the growth of mediated communication raises new issues surrounding privacy and vulnerability in intimate relationships.

Chapter 4, "Families," also takes up a key given set of local social relationships by looking across the diverse families we have encountered in our research. The chapter describes how parents and children negotiate media access and participation through their use of physical space in the home, routines, rules, and shared production and play. The chapter also examines how the boundaries of home and family are extended through the use of new media.

The final three chapters of the book focus primarily on interest-driven genres of participation, though they also describe the interface with more friendship-driven genres. Chapter 5, "Gaming," examines different genres of gaming practice: killing time, hanging out, recreational gaming, mobilizing and organizing, and augmented game play. The goal of the chapter is to examine gaming in a social context as a diverse set of practices with a range of different learning outcomes.

Chapter 6 examines creative production, looking across a range of different case studies of youth production, including podcasting, video blogging, video remix, hip-hop production, fan fiction, and fansubbing. The chapter follows a trajectory of deepening engagement with creative production, beginning with casual personal media production and then discussing how youth get started with more serious commitments to creative work and how they improve their craft, specialize, collaborate, and gain an audience.

The final chapter, "Work" examines how youth are engaged in economic activity and other forms of labor using new media. The chapter suggests that new media are providing avenues to make the productive work of youth more visible and consequential. We showcase some of the innovative ways that kids are mobilizing their new media skills and talents, including online publishing, freelancing, enterprises, and various forms of nonmarket work.

The conclusion, in addition to highlighting the key findings of this book, discusses the implications of this research for parents, educators, and policy makers.

Notes

1. We use the term "social media" to refer to the set of new media that enable social interaction between participants, often through the sharing of media. Although all media are in some ways social, the term "social media" came into common usage in 2005 as a term referencing a central component of what is frequently called "Web 2.0" (O'Reilly 2005 at http://www.oreillynet.com/pub/a/oreilly/tim/news/2005/09/30/what-is-web-20.html) or the "social web." All these terms refer to the layering of social interaction and online content. Popular genres of social media include instant messaging, blogs, social network sites, and video- and photo-sharing sites.

2. A wide variety of terms have been coined to link generational identity to digital and information technologies. Some examples include Don Tapscott's (1998) "net generation," the Kaiser Family Foundation's report on "Generation M" (for media) (Roberts, Foehr, and Rideout 2005), Mark Prensky's (2006) work on "digital natives," and John Beck and Mitchell Wade's (2004) "gamer generation." See Buckingham (2006) for a critique of the discourse of "digital generations."

1 MEDIA ECOLOGIES

Lead Authors: Heather A. Horst, Becky Herr-Stephenson, and
Laura Robinson

I get up in the morning and I just take a shower and eat breakfast and then I go to school. No technology there. And then when I come home—I invited a friend over today and we decided to go through my clothes. My dad saw the huge mess in my room. I had to clean that up, but then we went on the computer. We went on Millsberry [Farms]. And she has her own account too. So she played on her account and I played on mine, and then we got bored with that 'cause we were trying to play that game where we had to fill in the letters and make words out of the word. That was so hard. And we kept on trying to do it and we'd only get to level two and there's so many levels, so we gave up. And we went in the garage and we played some GameCube. And that was it, and then her mom came and picked her up. I came back in, played a little more computer (tried to get that word game and tried to get more points), and, but I got bored with that and so I went in my room and I listened to a tape. And then I ate dinner and you came . . .

—Geo Gem, age 12 (Horst, Silicon Valley Families)

In the spring of 2006, Heather Horst interviewed Geo Gem, a twelve-year-old girl who attends a public middle school in Silicon Valley, California. The youngest of two children in a biracial family (white and Asian-American), Geo Gem twirled her long dark hair while she talked about all the things she was "into": playing piano, singing, volleyball, the rain forest, and playing games on the computer or the GameCube in the family's media room, a space in the converted garage. Although Geo Gem's family lives in a wealthy area of the San Francisco Bay Area, the media and technology she uses every day do not necessarily reflect the family's economic status. The "kids' computer" is a secondhand desktop computer that sits in the living room and the GameCube is dated. Moreover, Geo Gem's parents decided not to buy cable in an effort to shelter their kids from what they thought was the brash commercialization and high costs of cable television. While Geo Gem has accepted the fact that she can watch only the occasional movie on the family DVD player, she notes that this often presents problems when her friends come over, "since they usually watch cable." Instead of watching television, Geo Gem plays games such as basketball, online games, and the GameCube. For Geo Gem, her media ecology, and the learning that takes place within her home environment, seems unremarkable; she moves fluidly between sitting in her bedroom with her friend going through the clothes in her closet and hanging out playing GameCube after school or sitting down for an hour to try to get to the next level on Millsberry Farms. Although it is unlikely that Geo Gem would describe her after-school activities with media as "learning" in the same way that she might describe school-work or piano lessons (see Seiter 2007), Geo Gem's home environment, the institution of the family, rules, and a variety of other factors constitute her everyday media ecology and her social and cultural context for learning.

Young people in the United States today are growing up in a media ecology where digital and networked media are playing an increasingly central role. Even youth who do not possess computers and Internet access in the home are participants in a shared culture where new social media, digital media distribution, and digital media production are commonplace among their peers and in their everyday school contexts. As we outline in the introduction, we see technical change as intertwined with other forms of historically specific social and cultural change as well as resilient structural conditions, such as those defined by age, gender, and socioeconomic status. We emphasize that there is a diversity of ways in which U.S. youth inhabit a changing and variegated set of media ecologies. We also recognize that the ways in which U.S. youth participate in media ecologies are specific to contextual conditions and a particular historical moment. In line with our sociocultural perspective on learning and literacy, we see young people's learning and participation with new media as situationally contingent, located in specific and varied media ecologies. Before we begin our description of youth practice, we need to map what those ecologies of media and participation look like. That is the goal of this chapter.

We use the metaphor of ecology to emphasize the characteristics of an overall technical, social, cultural, and place-based system, in which the components are not decomposable or separable. The everyday practices of youth, existing structural conditions, infrastructures of place, and technologies are all dynamically interrelated; the meanings, uses, functions, flows, and interconnections in young people's daily lives located in particular settings are also situated within young people's wider media ecologies. We also take an ecological approach in understanding youth culture and practice. As we suggest in the case of interest-driven and friendship-driven participation, these are not unique social and cultural worlds operating with their own internal logic, but rather these forms of participation are defined in relation and in opposition to one another. In this way, we extend the understanding of media ecologies used in communication studies (e.g., McLuhan 1964/1994; Meyrowitz 1986; Postman, 1993), which has focused primarily on "media effects," to studies of the structure and context of media use. Similarly, we see adults' and kids' cultural worlds as dynamically co-constituted, as are different locations that youth navigate such as school, after-school, home, and online places. The three genres of participation that we introduce in this chapter—"hanging out," "messing around," and "geeking out"—are also genres that are defined relationally. The notion of "participation genre" enables us to emphasize the relational dimensions of how subcultures and mainstream cultures are defined; it also allows us to use an emergent, flexible, and interpretive rubric for framing certain forms of practice.

In this chapter, we frame the media ecologies that contextualize the youth practices we describe in later chapters. By drawing from case studies that are delimited by locality, institutions, networked sites, and interest groups (see appendices), we have been able to map the contours of the varied social, technical, and cultural contexts that structure youth media engagement. This chapter introduces three genres of participation with new media that have emerged as overarching descriptive frameworks for understanding how youth new media practices are defined in relation and in opposition to one another. The genres of participation—hanging out, messing around, and geeking out—reflect and are intertwined with young people's practices, learning, and identity formation within these varied and dynamic media ecologies.

Box 1.1   Media Ecologies: Quantitative Perspectives
Christo Sims

Here I contextualize our ethnographic data by connecting our work to quantitative measures collected in several recent large-scale surveys of American youth media practices. Such surveys strikingly demonstrate the pervasive, and seemingly increasing, prevalence of media in the daily lives of American youth. In 2005, the Kaiser Family Foundation published data from a nationally representative survey of eight- to eighteen-year-olds showing that most American youth lived in households where media technologies were varied and numerous. On average, the youth in its sample lived in households with 3.5 televisions, 2.9 VCRs or DVD players, 2.1 video-game consoles, and 1.5 computers (Rideout, Roberts, and Foehr 2005). Additionally, the Kaiser Family Foundation survey found that more than 80 percent had access to cable or satellite television. More recently, the Pew Internet & American Life Project conducted a survey that showed 94 percent of all American teenagers—which it defines as twelve- to seventeen-year-olds—now use the Internet, 89 percent have Internet access in the home, and 66 percent have broadband Internet access in the home (Lenhart et al. 2008). In 2008, the USC Digital Future Project reported that broadband was now used in 75 percent of American households (USC Center for the Digital Future 2008). Additionally, Pew reported that in the fall of 2007, 71 percent of American teenagers owned a mobile phone and 58 percent had a social network site profile (Lenhart et al. 2008). In a 2006 survey, Pew found that 51 percent of teens owned an iPod or MP3 player (Macgill 2007). In addition to access, these studies tend to emphasize the frequency with which American youth engage media, many of which have become part of daily life. The Kaiser Family Foundation study found that young Americans spend on average 6.5 hours with media per day: almost 4 hours a day with TV programming or recorded videos, approximately 1.75 hours per day listening to music or the radio, roughly one hour a day using the computer for nonschool purposes, and about 50 minutes a day playing video games (Rideout, Roberts, and Foehr 2005). Pew's 2007 survey found that daily 63 percent of teens go online, 36 percent send text messages, 35 percent talk on a mobile phone, 29 percent send IMs, and 23 percent send messages through social network sites.

The Pew, Kaiser, and USC studies each report on the increasing prevalence of new media—notably the Internet and the mobile phone. Pew reports a steady increase in teen Internet use, from 73 percent in 2000, to 87 percent in 2004, to 95 percent in 2007, and a rapid increase in mobile phone ownership, going from 45 percent in 2004 to 71 percent in 2007 (Lenhart, Rainie, and Lewis 2001; Lenhart, Madden, and Hitlin 2005; Lenhart et al. 2008). Yet while new media have increased in popularity, they have not, according to the Kaiser report, displaced other types of media, nor have they led to an increase in the overall amount of time teens spend with media.[1] The authors of the Kaiser report suggest that this is because youth engage with more than one type of media at the same time, reading a magazine while watching TV, for example. Furthermore, the Kaiser report found that media engagement does not crowd out time spent with parents, pursuing hobbies, or doing physical activity. Rather, those who engaged in high amounts of media reported spending more time on average with family, hobbies, and physical activity (Rideout, Roberts, and Foehr 2005).

When compared to participants in these surveys, our survey participants[2] appear, on average, to be more engaged with new media than national averages. While Pew's 2007 survey found that 63 percent of American teens go online daily, 75 percent of our surveyed participants reported going online daily and 85 percent reported going online at least a few times a week. Additionally, only 1 percent of our survey participants had never been online, whereas Pew's 2007 survey found a nonuse rate of 6 percent.[3] In terms of daily communications, our survey participants again outpace those found by Pew in the fall of 2007: IM (Digital Youth Project (DY) 50 percent, Pew 29 percent), text messaging (DY 43 percent, Pew 36 percent), talking on a mobile phone (DY 56 percent, Pew 35 percent), and using a social network site (DY 46 percent, Pew 23 percent).[4] If our survey participants tend to be more engaged with media than the national average, it would not be surprising because our sites and participants were often chosen based on having already demonstrated some affiliation with new media. This was particularly true of the online and/or interest-driven sites.

While the national surveys by Pew, Kaiser, and USC tend to illustrate widely pervasive engagement with media, they also highlight ways in which media access and use vary according to demographic distinctions in age, gender, socioeconomic status, and ethnicity. In terms of variations that correspond to age divisions, Pew's fall 2007 survey found that a significantly higher proportion of older teens (defined as fifteen- to seventeen-year-olds) go online daily, own mobile phones, and communicate daily via mobile phone calls, text messages, IMs, and messages through social network sites (Lenhart et al. 2008). With respect to gender distinctions, the same Pew survey found that a significantly greater proportion of teenage girls than boys owned mobile phones and communicated daily via text messaging, talking on mobile phones, talking on landlines, sending IMs, and messaging through a social network site (Lenhart et al. 2008). The Kaiser survey found that girls spent significantly more time than boys listening to music and significantly less time than boys playing video games (Rideout, Roberts, and Foehr 2005).

In terms of variation in measures of access and use that corresponded to distinctions in socioeconomic status—often measured as based on household income and/or the level of parental education obtained—Pew's 2007 survey and Kaiser's survey both found that youth living in the most economically disadvantaged households had significantly lower rates of Internet access in the home and tended to rely on nonhome locations, such as schools and libraries, to access the Internet. In the case of the Pew survey, 70 percent of teens living in households with an income of less than $30,000 per year had Internet access in the home whereas 99 percent of teens living in households with earnings of $75,000 per year or more had such access (Lenhart et al. 2008). Both Pew and Kaiser found that youth from higher-income households go online more frequently than youth from lower-income households—39 percent of teens living in households earning less than $30,000 per year go online daily whereas 75 percent of teens from households earning more than $75,000 per year go online daily (Lenhart et al. 2008; Rideout, Roberts, and Foehr 2005). In 2007 Pew also found that teens from more well-off households are significantly more likely to own mobile phones. Finally, in terms of variations that correspond to distinctions in ethnic identifiers, Pew's 2007 survey and Kaiser's survey both found that minorities (blacks and Hispanics) were significantly more likely to rely on nonhome locations to access the Internet (Lenhart et al. 2008; Rideout, Roberts, and Foehr 2005). Additionally, Pew found that a significantly greater share of white teens went online daily than black teens, reporting 67 percent and 53 percent, respectively. Last, Pew found a significant difference in the proportion of white teens who had broadband access in the home when compared to broadband access in black and Hispanic households—70 percent, 56 percent, and 60 percent, respectively.

Some aspects of these national surveys shed light on some of the themes noted in this book: namely the friendship-driven and interest-driven practices. In terms of friendship-driven practices, the most illustrative survey data are those that indicate patterns of ownership, access, and use of communication technologies such as mobile phones, IM, and social network sites. While the current indicators used by Pew and others do not differentiate when teenagers use these technologies to communicate with friends versus communicate with family members and other members of the youth's social world, a few trends are worth noting.[5] For one, Pew's 2007 survey finds that both gender and age distinctions map to significant differences in several factors related to communications. Girls and older teens are more likely to own a mobile phone than boys and younger teens; additionally, both girls and older teens are significantly more likely to make a mobile phone call, send a text message, send an IM, or send a message through a social network site (Lenhart et al. 2008). Another noteworthy trend indicated by the Pew data is what Lenhart and her colleagues (2007) refer to as "super communicators." The term is meant to refer to the finding that those who communicate using multiple technologies and channels—phone calls, text messages, IMs, social network sites—not only communicate more in aggregate than teens who use fewer channels but they also tend to communicate more frequently within each channel.

Regrettably, there are fewer survey data for making comparisons to what we have characterized as interest-driven practices. USC's 2008 Digital Future Report surveyed some activities that could, but do not necessarily, indicate interest-driven practices. In its survey it asks about participation in, and attitudes about, online communities, which it defines as "a group that shares thoughts or ideas, or works on common projects through electronic communication only" (USC Digital Future Report Highlights 2008, 8). While the overall percentage of respondents who reported participating in an online community was relatively small—15 percent of all respondents—the authors note that this rate has more than doubled in three years. Of those who participate, more than half reported that the community related to a hobby. Many of the interest-driven practices we account for in this report could be seen as reasonably fitting this definition, but a few problems limit a more direct mapping. For one, we show examples of interest-driven participation that does not take place solely, or at all, through electronic communications. Additionally, the USC Digital Future report surveys adults and youth. While participation in online communities is on the rise, a majority of adults with children reported being uncomfortable having their children participate in online communities—65 percent reported feeling uncomfortable whereas only 15 percent felt comfortable. This last indicator suggests that spreading youth participation in online venues for interest-driven participation will likely require a change of attitude among adult populations.

Genres of Participation: Hanging Out, Messing Around, and Geeking Out

How does young people's social and cultural participation shape new media engagement, interest, and expertise? Throughout this project, our challenge has been to develop frameworks that help us understand youth participation in different social groups and cultural affiliations, a framing that is in line with approaches that see knowledge and expertise as embedded in social groups with particular media identities. For example, James Paul Gee (2003) has suggested that gaming is part of the construction of "affinity groups," where insiders and outsiders are defined by their participation in a particular semiotic domain. Similarly, a communities-of-practice approach to learning posits that the development of knowledge and expertise is deeply integrated with being part of social groups engaged in joint activity (Wenger 1998). In order to understand these forms of group practice and identity, studies need to take into account an individual's media engagement as well as the properties of social groups and cultural identity. While quantitative studies (see box 1.1) can help us situate an individual's media engagement with specific media and technologies, we provide an ethnographic accounting of shared practices and cultural categories that structure youth new media participation.

"Hanging out," "messing around," and "geeking out" describe differing levels of investments in new media activities in a way that integrates an understanding of technical, social, and cultural patterns. It is clear that different youth at different times possess varying levels of technology- and media-related expertise, interest, and motivation. The genres of participation that emerged from our research can be viewed as an alternative to existing taxonomies of media engagement that generally are structured by the type of media platform, frequency of media use, or structural categories such as gender, age, or socioeconomic status. Quantitative studies customarily categorize people according to high and low media use, which is then analyzed in relation to different social categories or outcomes of interest. For example, the Kaiser Foundation report on "Generation M" (Rideout, Roberts, and Foehr 2005) looks at how differing amounts of media exposure time relate to individual measures such as age, educational status, race and ethnicity, school grades, or personal contentedness. Our approach is closer to those of qualitative researchers who take a more holistic approach to media engagement by focusing on how social and cultural categories are cut from the same cloth as media engagement, rather than looking at them as separate variables. For example, Holloway and Valentine (2003) suggest the categories of "techno boys," "lads," "luddettes," and "computer competent girls" to understand how gender intersects with computer-based activity and competence. Sonia Livingstone (2002) suggests the categories of "traditionalists," "low media users," "screen entertainment fans," and "specialists" to relate frequency of engagement with specific media types to certain forms of social and cultural investments. However, all these taxonomies are based on categorizing individuals in relation to certain practices. By contrast, our genre-based approach emphasizes *modes of participation* with media, not categories of individuals.

The distinction between a genre-based approach centered on participation and a categorical approach based on individual characteristics is significant for a number of reasons. First, it enables us to move away from the assumption that individuals have stable media identities that are independent of contexts and situations. In our work, we have observed how many youth craft multiple media identities that they mobilize selectively depending on context; they may be active on Facebook and part of the party scene at school, but they may also have a set of friends online focused on more specific interests related to gaming or creative production. Second, the notion of genre moves away from a focus on media platform (TV, computers, music, etc.) and shifts our attention to the crosscutting patterns that are evident in media content, technology design, as well as in the cultural referents that youth mobilize in their everyday communication. Finally, genre analysis relies on what we believe is an appropriately interpretive model of analyzing social and cultural patterns. Rather than suggesting that we can clearly define a boundary between practices in a categorical way, genres rely on an interpretation of an overall "package" of style and form. Genres of participation take shape as an overall constellation of characteristics, and are constantly under negotiation and flux as people experiment with new modes of communication and culture. In this way, it is a construct amenable to our particular methods and approach to looking at a dynamic and interrelated media ecology. Our approach is ecological rather than categorical. In the remainder of this chapter, we turn our attention to the three genres of participation, hanging out, messing around, and geeking out, in an effort to define and describe how these genres emerge through youth practice.

Hanging Out

The interdisciplinary literature on childhood and youth culture has established that coming of age in American culture is marked by a general shift from given childhood social relationships, such as families and local communities, to peer- and friendship-centered social groups. Although the particular nuances of these relationships vary in relation to ethnicity, class, and particular family dynamics (Austin and Willard 1998; Bettie 2003; Eckert 1989; Epstein 1998; Pascoe 2007a; Perry 2002; Snow 1987; Thorne 1993), the vast majority of the middle-school and high-school students we interviewed expressed a desire to "hang around, meet friends, just be"

(Bloustein 2003, 166), as much and as often as possible, as part of their burgeoning sense of independence. Given the institutional restrictions and regulations placed on young people by schools, teachers, parents, and neighborhood infrastructures, kids and teenagers throughout all our studies invested a great deal of time and energy talking about and coordinating opportunities to "hang out." In the first part of this section, we examine how youth mobilize new media communication to construct spaces for copresence where they can engage in ongoing, lightweight social contact that moves fluidly between online and offline contact. We continue by discussing the ways in which new media content, such as music and online video, becomes a part of young people's social communication. Finally, we consider how youth use new media to be present in multiple social spaces, hanging out with friends in online space while pursuing other activities concurrently offline.

Getting Together and Being Together As we describe in this book's introduction, contemporary teens generally see their peers at school as their primary reference point for socializing and identity construction. At the same time, they remain largely dependent on adults for providing space and new media and they possess limited opportunities to socialize with peers and romantic partners without the supervision of adults. Young people move between the context of the school, where they are physically copresent but are limited in the kinds of social activities they can engage in, and the context of the home, where they have more freedom to set their social agendas but are not usually copresent with their peers. Parental and official school rules, availability of unrestricted computer and Internet access, competing responsibilities such as household chores, and transportation frequently complicate efforts toward hanging out. Young people who have ready access to mobile phones or the Internet, view online communication as a persistent space of peer sociability where they exercise autonomy for conversation that is private or primarily defined by friends and peers. Although in most cases they would prefer to hang out with their friends offline, the limits placed on their mobility and use of space means that this is not always possible.

Chapters 2 and 3 describe the many mechanisms that youth mobilize to keep in ongoing contact with their peers through social media. By moving between the browsing of social network profiles, instant messaging

(IM), and phone conversations, youth experience a sense of hanging out with their peers that is unique to online interaction, but that also has many parallels to how kids hang out offline. The more passive and indirect mode of checking people's status updates on Facebook or MySpace, or exchanging lightweight text messages indicating general status ("I'm so tired," "just finished homework"), are examples of "ambient virtual co-presence" that in many ways approximates the sharing of physical space (Ito and Okabe 2005b). Through these modalities, youth keep tabs on one another. At other times, youth engage in more sustained and direct conversation, such as when they start an IM chat or initiate a telephone call. C. J. Pascoe's box 1.3, "You Have Another World to Create," for example, discusses the ways in which a participant in her "Living Digital" study, Clarissa, coordinates hanging out with friends and her girlfriend through MySpace and LiveJournal and how she negotiates hanging out with an expanded friend base within an online role-playing game. By flexibly mobilizing different networked communications capabilities, young people circumvent some of the limits that prevent them from hanging out with their friends.

When young people want to get together and hang out (for both online and offline meetings), they typically go online first, since that is where they are most likely to be able to connect. For example, Java, a white twelve-year-old living in the suburbs, describes how she will first get permission from her mom, and then use email or IM to find a friend and ask her over. "Well, if I just want a friend over I'll ask my mom and she'll say yes or no. And if she says yes, then I'll call them or ask them online or email them or something." After that, she and her friends must coordinate with a parent to drive them to each other's homes (Sims, Rural and Urban Youth). Even when kids are independently mobile (e.g., if they can drive, or if they live in a more urban context where public transportation is available), online media still remain the place where they find and connect with their friends. For example, Champ, a nineteen-year-old Latino who lives in Brooklyn, New York, with his mom and two sisters, discussed with Christo Sims how hanging out has changed since the incorporation of MySpace within his peer group:

Champ: I guess before, before it was MySpace is, like, you just go outside, whoever you bump into, you bump into 'em. Whatever, you gotta do what you gotta do. And, now, computer, like, you go talk to the people and like,

"Oh, what you doing?" "You wanna do this?" "All right. So, I'll be over there in ten minutes, five minutes."

Christo: And that's mostly on MySpace? You can see if they're online now or something like that?

Champ: Yeah, like I was saying, online under their names. And, it has like a little computer there. Click on their page and then like, "Yo, I was about to come outside." And, if [I] tell you "coming out, wanna meet up?"

Java and Champ use new media to help orchestrate face-to-face hanging out, but their examples also reveal how proximity, or neighborhood, affects their ability to get together. In rural and suburban California, young people must mobilize parents and their vehicles for hanging out with friends who are separated by greater distances, at least until teens are old enough to drive or have friends who drive. By contrast, urban youth such as Champ live close to friends and rely less on their parents for transportation because they can take advantage of a more durable transportation system such as that in New York. Champ and other urban youth more readily move between online and offline sociality. In most of the cases we have seen, youth rely to some extent on networked communication to facilitate arranging offline meetings, these networked sites and communication devices becoming an alternative hanging out site in its own right.

Sharing, Posting, Linking, and Forwarding   When teens are together online and offline, they integrate new media within the informal hanging out practices that have characterized peer social life ever since the postwar era and the emergence of teens as a distinct leisure class (Snow 1987). As we describe in the introduction, this era saw a growth in the number of teens who attended high school and the emergence of a distinctive youth culture that was tightly integrated with commercial popular cultural products targeted to teens. The growth of an age-specific identity of "teenagers" or "youth" was inextricably linked with the rise of commercial popular culture as young people consumed popular music, fashion, film, and television as part of their participation in peer culture (Cohen 1972; Frank 1997; Gilbert 1986; Hine 1999). While the content and form of much of popular culture has changed in the intervening decades, the core practices of how youth engage with media as part of their hanging out with peers remains resilient. In relation to gaming, Ito (2008b) has described how children and youth traffic in popular media referents as part of their everyday sociability. She describes how contemporary media mixes such as Pokémon enable kids to develop identities in peer culture in relation to customizable, interactive media forms. This "hypersocial" social exchange is more generally a process through which people use specific media as tokens of identity, taste, and style to understand and display who they are in relation to their peers. While hanging out with their friends, youth develop and discuss their taste in music, their knowledge of television and movies, and their expertise in gaming, practices that become part and parcel of sociability in youth culture.

One of the most common ways that kids hang out together with media is listening to music, a practice that stands as a source of affinity among friends. In fact, rock and roll was a central piece of the emergence of youth culture (Snow 1987). Technologies for storing, sharing, and listening to music are now ubiquitous among youth. Indeed, only 2 percent of the youth we interviewed reported *not* owning a portable music player. In addition, digital music formats are increasingly dominant. Among our respondents, 88 percent reported downloading music or videos over the Internet and 74 percent reported that they had shared files (music or other) over the Internet. Two practices related to music were particularly prominent among the teens in our study: First, teens frequently displayed their musical tastes and preferences on MySpace profiles and in other online venues by posting information and images related to favorite artists, clips and links to songs and videos, and song lyrics. Second, sharing and listening to music continues to be an important practice and something that teens do together when they are hanging out. For example, sixteen-year-old Sasha, a teenager from Michigan who participated in danah boyd's interviews (Teen Sociality in Networked Publics), outlines how acquiring music is an important part of hanging out in her life because she can get free music from her friends. "I use like the iTunes store, but I don't have any more money, so I just go over to my friends' houses and plug in to their computer and get songs off of there." Sites such as MySpace often extend this kind of music-driven sociability online, where young people can add music to their own profiles and view one another's musical preferences. As Mae Williams, a sixteen-year-old teen in Christo Sims's study of rural California (Rural and Urban Youth), explains, "That's the one thing MySpace is good for, is that you can actually browse through music pretty easily. And so you can select a genre and you can go through other people's [profiles] and sometimes if I see a name that keeps popping up, I'll be like, 'Oh, this guy must be halfway good.'" As with earlier forms of music sharing, the digital music on iPods and MySpace profiles are still about the sharing of media and media tastes with friends and local peers. Digital technologies enhance these practices by making music more readily available to youth for listening and sharing in a wider variety of contexts.

Many teens also view new media as something to do while they are hanging out with their friends. One example of hanging out with media can be found in box 1.2, in which Lisa Tripp describes the media ecology of Michelle, a twelve-year-old girl from Los Angeles who uses television, online media, and books for entertainment when she is hanging out at home with her mother or with friends. Like other youth, Michelle uses MySpace to connect with friends when they cannot hang out in person. As discussed at length in chapter 5, boys often prefer to play games when they are together. A white ten-year-old boy, dragon, who was part of Heather Horst's study of Silicon Valley Families, illustrates that hanging out together in a game is important when friends are in different locations and time zones. At the time of his interview, dragon had recently moved from the U.S. East Coast to the West Coast. While he was making friends at his new school, he regularly went online after school to play RuneScape on the same server as his friends back east. In addition to playing and typing messages together, dragon and his friends also use the phone to call each other using three-way calling,. Dragon then places the phone on speakerphone, filling the house with the sounds of ten-year-old boys arguing and yelling about who killed whom, why one person was slow, and reliving other aspects of the game.

Box 1.2   Michelle
Lisa Tripp

Michelle Vargas lives in the San Fernando Valley region of Los Angeles. She is a twelve-year-old girl, just finishing the seventh grade at Cameron Middle School, where Lisa Tripp and Becky Herr-Stephenson conducted fieldwork (Los Angeles Middle Schools). Michelle is being raised by her mother, Rose, who immigrated to the United States from El Salvador years before Michelle was born. The two share a bedroom in an immaculately clean apartment and rent their second bedroom to a cousin. Rose works as the apartment manager for the complex where they live, and sometimes she cleans houses on the weekends. She describes herself as both a strict and loving mom. Rose explains, *"Me gusta que [Michelle] ande conmigo. Yo soy con ella como su amiga, su hermana, su mamá, todo. Así lo siento yo."* ("I like her [Michelle] to be with me. I am like her girlfriend, her sister, her mom, all of that. That's the way I feel.") When Michelle is not at school, she spends most of her free time at home. Sometimes on weekends she helps her mom at work, or the two do other things together, such as go to a birthday party or stay home and watch a DVD. A recent favorite movie was *Grease,* which she and her mom have watched in both English and Spanish.

Michelle is not allowed to watch TV on school days, with two exceptions. She can watch the news if she wants to and, every night after dinner, she and her mom have a special date to watch *La Tremenda,* a popular Spanish-language soap opera, or *telenovela.* At the end of the school week the TV restrictions are lifted. As Michelle explains, "On Fridays, my mom can't tell me nothing, because I'm watching TV!"

Michelle likes watching mainstream "kid shows" such as *Phil of the Future, That's So Raven, Danny Phantom, The Suite Life of Zach and Cody,* and *Hannah Montana,* as well as "little kid" shows such as *Winnie the Pooh* and *Blue's Clues.* She is also a major fan of *High School Musical* and considers teen idol Zac Efron her absolute favorite. Her friends are also fans of the shows, and sometimes she will call one of her friends and say, "Turn it on, turn it on," so they can watch a TV show at the same time. When Michelle gets the chance to go online for fun, her favorite thing to do is play games based on these shows, especially the maze games on the Disney Channel website.

Michelle listens to music around the house while hanging out in her room or doing chores and when she is in the car riding around with her mom. She has a CD player but longs for an iPod, and she claims to like "any kind of music, except country." She gets most of her music by downloading it from the Internet, either buying it from iTunes or getting it for free from LimeWire (see figure 1.1). She often burns music on CDs to give to her friends—many of whom either do not have a computer or do not know how to burn CDs. She says she sometimes feels "too lazy" to help them, however, so they have to wait.

Michelle is also an avid reader. She keeps a bookshelf in her bedroom stocked with young-adult literature. The books come from her mom's boss, who regularly gives the family hand-me-down books. Michelle tries to read for about an hour before bed every night. This sets her apart from the rest of her friends, who engage in little to no pleasure reading. Michelle has a learning disability and reads at approximately a third-grade level, and she takes her time reading a book. When she comes across a word she does not

Figure 1.1
Michelle looking around online. Photo by Lisa Tripp, 2006.

understand, she writes down the word and asks her teachers at school for help. Some of her recent favorite books include *Thoroughbred: A Horse Called Wonder, Sideways Stories from Wayside School,* and *Harry Potter,* which took her about three months to read.

Rose helps Michelle with reading and doing homework to the extent that she can, but she speaks limited English and studied up to only the eighth grade in her native El Salvador. This makes providing homework help difficult. Rose bought a computer and pays for high-speed Internet, all to help Michelle complete school assignments. At the same time, Rose worries a lot about Michelle visiting websites such as MySpace, where she fears her daughter might get in to trouble, talk to strangers, or be the target of sexual predators. She also worries that Michelle will waste time playing online games instead of doing her homework. As a result, the computer is kept in the living room, where Rose can keep an eye on what Michelle is doing and, if Rose has to leave the house, she often takes the modem with her to keep Michelle from going online unsupervised. Sometimes when Rose is not looking, Michelle sneaks online to one of her favorite sites. When she gets caught, she yells back at her mom, "I'm not doing anything wrong!"

Several of Michelle's friends have MySpace pages, and Michelle has one too. From Michelle's perspective, the site is fun because it allows her another way to talk to her friends. She likes leaving messages for her friends on MySpace, or reading messages they have left for her, and sometimes she likes to type back and forth with them and talk on the phone at the same time. Michelle thinks her mom's fears about the Internet are misplaced and that her mom is just overreacting to scare stories on the news. "I just type to my friends. That's all I do," she explains. "Like, I don't talk to people I don't know."

On other occasions, mother and daughter use the computer for more collaborative endeavors. Rose likes to send email to a friend in El Salvador and to her twenty-six-year-old son, who lives in Texas, but she does not know how to do it without help. According to Rose, she types her own email messages and then asks her daughter, *"Hija ven: ¿cómo le tengo que hacer aquí?"* ("Hey, come here: What do I need to do here?") Michelle then helps her send the email. More recently, Michelle has been giving her mom lessons on how to pay bills online and how to create birthday cards. Rose explains, *"Ella me ha enseñado a usar todo lo de la computadora . . . todo que ha aprendido en la escuela."* ("Michelle has taught me how to do everything on the computer . . . everything she has learned at school.")

For Rose, not knowing as much about the computer as Michelle produces a great deal of anxiety and leads her to closely supervise and often limit her daughter's time online, particularly for "hanging out" and "messing around." Thus while Michelle is able to go online outside school more readily than most of her classmates (because she has home Internet access), her mother's concerns ultimately lead to Michelle having less time online for open-ended exploration and self-directed inquiry than might otherwise be possible.

At school this year Michelle has been part of a special program in which students create media art projects, such as graphic art images and short videos. The program has given Michelle her first chance to use PowerPoint and iMovie, and she already has learned enough to help other students learn the software. The class was Michelle's favorite, and she thinks that creating media projects for a school project "just helps her learn better." At the same time, she still had difficulty with the reading and writing part of the process, such as doing research online and writing a script for her video. "I did not like that part," she explains. "It was so boring." It is likely that Michelle found parts of the media production process in school "boring" because they were teacher-driven exercises, designed to achieve goals mandated by the school curriculum and teacher lesson plans. Unlike how Michelle and her classmates typically engage in "youth-driven" practices with media, at school they have much less input into defining the goals and content of their media production work. Outside school, Michelle loves taking photos of her friends and family on her mom's mobile phone, and some day she would like to make more videos with her friends . . . but just of them hanging out together. She says she will "skip the script writing part."

During the course of our three-year study, many of the American teenagers we interviewed also became regular viewers of short videos and television programs on sites such as YouTube. Although most youth still watch television shows on a television set, there has been a rapid growth of TV-show viewing on YouTube. In her study "Self-Production through YouTube," Sonja Baumer describes how watching television shows on YouTube differs from traditional viewing because of the overlay of social information and networks, enabling viewers to engage in a kind of lightweight hanging out with other viewers, even if they may not be spatially or temporally copresent. YouTube videos are contextualized by YouTube participants who provide a layer of opinion and linking that differs from the ways in which television has traditionally been organized by channels and networks. As KT, an eighteen-year-old male from suburban California, describes: "I go to the most-viewed page. . . . Mostly I want to know what's up, what's cool, like what was funny on the *Colbert Report* yesterday, and it is just there. You can browse and look for stuff. Awesome!" Similarly, "When I start watching YouTube, I cannot stop. Each video takes me to another video. . . . It takes me to the author's profile page. . . . I like to click on related videos that YouTube gives you on the side, you know what I mean. . . . There are always pointers to other videos."

We see this hypersocial mode of video viewing in a more immediate and socially interactive way when youth view videos together offline. Video downloads and sites such as YouTube mean that youth can view media at times and in locations that are convenient and social, provided they have access to high-speed Internet. At the after-school center where Dan Perkel, Christo Sims, and Judd Antin observed students in their study, "The Social Dynamics of Media Production," they began seeing youth gathering in front of a computer during downtime, watching episodes of *Family Guy* on YouTube. For college students in dorm rooms, the computer often became the primary TV-viewing mechanism. High bandwidth connections mean that there is little need for the added expense and clutter of a TV purchase. Ryan, a seventeen-year-old white working-class student in high school in urban California who participated in C. J. Pascoe's "Living Digital" study, describes hanging out with his friend John while they were on a school-sponsored ski trip. He describes how they went online together and "pretty much just grabbed videos, and laughed at a bunch of shock stuff," meaning videos that involved "death, and crazy accidents, and people like, torture

cams and stuff like that, just because I've never been exposed to that."
Ryan was able to share his reactions to these extreme videos with a friend
at an opportune moment when they returned to their rooms for the night
after a school-sanctioned outing. In effect, access to rich, networked media
enables youth to engage in social activity around video in the diverse settings of their everyday lives. This ready availability of multiple forms of
media in diverse contexts of daily life means that media content is increasingly central to everyday communication and identity construction.

Work-Arounds, Back Channels, and Multitasking Unlike other genres
of participation we discuss in which individuals justify that the activities
are "productive" and/or possess the potential for secondary skills, the
practice of hanging out is usually not seen by parents and teachers as
supporting productive learning. Many parents, teachers, and other adults
we interviewed described kids' and teenagers' inclination toward hanging
out as "a waste of time," a stance that seemed to be heightened when
hanging out was supported by new media. Not surprisingly, teenagers
reported considerable restrictions and regulations tied to hanging out in
and through new media. Sites such as MySpace, which are central to
hanging out genres of participation, are often restricted by parents and
blocked in schools. In their examination of schools in Southern California
(Los Angeles Middle Schools), Lisa Tripp and Becky Herr-Stephenson find
that schools generally provide students with the opportunity to log on to
the Internet in a school library before school, during lunch or other free
periods, or after school. While students in schools with media and
technology resources frequently obtain access to the Internet in classrooms
using mobile laptop labs or small centers with three or four desktops in an
area of the classroom, gaining access to the library is a more complex
process of obtaining passes and working in strict silence, and students tend
to use the library infrequently aside from class periods during which the
entire class would visit the library to do research. Moreover, teachers and
schools attempt to determine appropriate use of those resources. The desire
to restrict hanging-out practices at school in favor of keeping students "on
task" while using media and technology for production or research,
combined with concerns about which media and websites are suitable for
citation (e.g., Wikipedia and .edu sites), can prompt teachers and principals
to develop rules about the appropriate use of media structures.

In response to these regulations, teenagers develop work-arounds, ways to subvert institutional barriers to hanging out while in school (see Thorne 1993 on the concept of underground economies in the classroom). C. J. Pascoe (Living Digital) reports that teenagers in her study regularly used proxy servers to get online at school. She also notes that many of the kids she spoke with seemed to know which students were experts at finding available proxy servers. During one of her interviews at California Digital Arts School (CDAS),[6] one teen wanted to show Pascoe his MySpace profile, but he could not because the school's server blocked the site. He spent thirty minutes during the interview tracking down one of the school's experts on proxy servers. Unfortunately, when the proxy expert sat down to log on to the proxy, he discovered that school officials had already blocked the server, forcing him to start a search for a new server. Karl, a fifteen-year-old mixed-race student in San Francisco, attested to the fact that teenagers who want to hang out with their friends will find ways to use MySpace in the school library even though the school bans access to the site. As Dan Perkel (MySpace Profile Production) describes, "while wiggling his fingers in the air in front of an imaginary keyboard, a sly look crosses his face as if to show how sneaky people are and also the big grin on his face as he confirms, 'They can't ban MySpace!'" Karl's general attitude toward bending the rules in the name of maintaining contact with his friends throughout the day is mirrored in Liz's and her boyfriend's use of text messaging. Liz, a sixteen-year-old high-school student who lives in a middle-class suburb in the San Francisco Bay Area, highlights the importance to her friends of back-channel communication:

C.J.:   And so why is texting such a big deal?

Liz: You want to talk in class, but then like you're in different classes and so this is the only way you can talk to them. Or you just aren't allowed to talk in class [and] your friend is sitting next to you, so you text. Or write notes. But nobody writes notes anymore. . . .

Liz's boyfriend: Yeah, it replaced the note.

Liz: Nobody.

C.J.:   There's none of the elaborately folded?

Liz: We sit next to each other, so sometimes we write little notes and then usually the teacher takes it away because we're right in front of them. But we're not even talking about anything. But then if we're across the room then he'll start texting me and I text someone else. And then if you're in other classrooms you definitely need to text. . . . (Pascoe, Living Digital)

Like many of the other participants in our studies, Liz and her boyfriend reveal how hanging out with friends, boyfriends, and girlfriends represents a continuation of practices that have been pervasive among American teenagers in the school setting since the 1950s. Rather than mouthing words behind a teacher's back or secretly passing notes underneath tables and desks at school, texting or sending short messaging services (SMS) on the mobile phone now facilitates communication.

These work-arounds and back channels are ways in which kids hang out together, even in settings that are not officially sanctioned for hanging out. This happens in settings such as the classroom, where talking socially to peers is explicitly frowned upon, as well as at home when young people are separated from their friends and peers. Just as recent studies indicate that "multitasking," or engaging in multiple media activities at the same time, is on the rise among kids (Roberts and Foehr 2008), we note that the teens in our studies are becoming particularly adept at maintaining a continuous presence in multiple social communication contexts. We also see kids hanging out or engaging in multiple social contexts concurrently. Derrick, a sixteen-year-old Dominican American living in Brooklyn, New York, explains to Christo Sims (Rural and Urban Youth) the ways he moves between using new media and hanging out.

Derrick: My homeboy usually be on his Sidekick, like somebody usually be on a Sidekick or somebody has a PSP or something like always are texting or something on AIM. A lot of people that I be with usually on AIM on their cell phones on their Nextels, on their Boost, on AIM or usually on their phone like he kept getting called, always getting called.
Christo: So even when you're just hanging out they're constantly texting and all that?
Derrick: Getting phone calls.
Christo: What . . . to find out what's going on or what do you think they're usually like?
Derrick: Just to meet up with everybody, just to stay in contact.

As Derrick's discussion suggests, even when teenagers and kids are hanging out in a face-to-face group, many feel the need to stay connected to other teens who are not there. The drive to hang out, and the use of new media to coordinate such endeavors, continues even when there may be a copresent, cohesive group. Playing games, making videos, and listening to music may well be the focus when teens are hanging out, yet they may also become part of the background, something to do when teens are waiting for other people to come and other plans to develop. Moreover, there may be multiple activities occurring at the same time while kids and teens are hanging out together. As Christo Sims notes in one of his field notes from "Rural and Urban Youth," "When I was in rural California, I saw a few boys playing a console game, another carrying on an ongoing text-message conversation, and another one making food," all in the same room together. The layering of media and social interaction is part of a changing media ecology that youth inhabit, where they are in persistent touch with friends and intimates through networked communication while accessing popular and commercial media in varied settings. The social desire to share space and experiences with friends is supported now by a networked and digital media ecology that enables these fluid shifts in attention and copresence between online and offline contexts.

Box 1.3   "You Have Another World to Create": Teens and Online Hangouts

C. J. Pascoe

Tall and lithe, white seventeen-year-old Clarissa moves with the grace and the particular upright posture of a ballerina, a lasting effect of her years of participation in dance. Her long blond hair is often braided and woven in a complicated pattern across the nape of her neck. She laughs easily, and she frequently accents her lively eyes by drawing a lacy circular pattern in silver glitter below her left eye. She lives with her parents and two younger siblings in a small unincorporated working-class suburb of San Francisco. Clarissa says that she is not a particularly avid user of technology since she "doesn't even look" at a computer until she gets to school and laments the fact that her mobile phone is so "old school" that she cannot use it to send text messages. Clarissa represents many teens in her casual technology use—using new media as a meeting place, a place to foster romantic relationships, and a place to engage in hobbies. These digital environments have grown increasingly important as pastimes and socializing places for Clarissa because she recently suffered a debilitating leg injury that robbed her of the ability to engage in her first passion, ballet.

Like other teens I have spoken with, Clarissa and her girlfriend, Genevre, play out much of their relationship through digital media. Clarissa and Genevre share online spaces in a variety of ways. They publicly declare their relationship status and affection for one another on their social networking pages, share their passwords, and have created a blog together. Clarissa said that when she first gets home she checks her MySpace page. Her avatar features her girlfriend and her kissing on the bus on the way to their senior picnic. Her list of "Top 8" friends prominently features Genevre in addition to her other close friends. Genevre's presence is threaded throughout the page, from the pictures of Clarissa and her at prom to the notes declaring love and support Genevre leaves for Clarissa.

During our interview, Clarissa expressed surprise when we logged on to her MySpace and saw a new addition to her site, saying to me that her girlfriend must have added it. Clarissa explained that because she shared her password with Genevre, "I have not done my MySpace. It's all my girlfriend, except a very little bit of it. My girlfriend's done all the colors and all that." Recently, Genevre changed Clarissa's website again, altering the background from a ballet dancer's foot en pointe to a background of fanciful colored hearts and transforming the text from a standard font to a whimsical large script. She also changed Clarissa's avatar to a picture of her friends. Flirtatiously, Genevre left a note on the site reading, "So . . . yet again . . . Clarissa was hacked. . . . Her girlfriend was bored and her MySpace was boring, so I spiced it up!"

Beyond the intimacy they created by sharing a password, the couple keeps a blog together on LiveJournal. While the site itself is public, Clarissa says, "I do a lot of private entries that my girlfriend and I can read, because we know each other's passwords." When Genevre took a motorcycle trip for a week, Clarissa said good-bye and wished her well by posting a picture of an elaborate rose accompanied by a poem. In this way the two could remain digitally linked, a way of being together even when they were not.

In addition to her MySpace and LiveJournal sites, Clarissa spends much of her online time on Faraway Lands,[7] her preferred hangout. Clarissa describes Faraway Lands as a "really nice-quality, good, inviting, comfortable, fun place to be." She finds it to be a community of supportive friends who have high writing standards and creativity. Members must write intricate character applications to join the site. These character applications are essentially 25,000-word descriptions of a given character, its race, its history, and its location. For Clarissa, an aspiring writer and filmmaker, this site allows her to use "words like clay to create whatever stories suit your fancy." She finds the community to be a "nurturing" one in which she is "able to fully develop intricate personalities and plots that in computer games, sports, and academics are simply not possible." Faraway Lands is a text-based site where members weave long and detailed tales about their characters' quests and adventures.

In this online hangout Clarissa has made many friends and transcended her local boundaries. While people of all ages are on this site, "most of the people that I've interacted with are in my age group. It's sort of cool 'cause they're far away and sort of fun." On Faraway Lands she is simultaneously in character and out of character as she hangs out and chats on an Internet relay channel. During these chats, she has made friends all over the world, telling me, "I know a guy in Spain now and fun stuff like that." She and her friend from Spain are in the middle of planning a new role play in which his evil character tries to hire one of Clarissa's characters, Saloria, as an apprentice (see figure 1.2).

Clarissa's stories involve themes of fantasy, triumph, and escape. Her character Saloria, for instance, grew up in a poor neighborhood and was raised by a "loving community" rather than a nuclear family. As a teen, Saloria leaves this community to seek her fortune in the wider world. However, she soon realizes that, as a single woman, the world is a dangerous place. Saloriathen decides to live her life as a man "because men have it better. So she spends her days as a man." During the day, as a man, Saloria performs "roadwork around the city. She's a happy-go-lucky charming young fellow." At night "she's a crazy lady who has fun." Clarissa drew on her real-life experience to create Saloria. She recalled fondly stories of adventurous women.

Figure 1.2
Saloria. Photo titled "Little Red Bird" by Cathy Hookey, 2006–2008, http://little-red-pumpkin.deviantart.com.

She "loved those women who would go on these voyages acting like they were boys for months, and months, and months. It was daring and crazy. And I was like, 'I want to do that. That would be fun.'" While this sort of adventuring is not feasible for Clarissa, her characters can live out these fantasies. She sums up Saloria's story by saying, "It just started with that, the freedom of being a boy." Through this particular role play, Clarissa grapples with intense issues of adolescent identity work and imagines her way out of some of the gendered expectations faced by teenage girls.

Faraway Lands also provides a forum in which Clarissa can be creative and hone her writing skills. She and her role-playing friends critique one another's writing and stories. She and a fellow role player from Oregon "had this sort of thing where we were reviewing each other's work all the time 'cause he just wanted all the input he could get." The creative aspect of this site is part of what drew Clarissa to Faraway Lands. "It's something I can do in my spare time, be creative and write and not have to be graded. . . . You know how in school you're creative, but you're doing it for a grade so it doesn't really count?" Unlike in school, where teens live in a world of hierarchical relations—where they are graded, run the risk of getting in trouble, and must obey all sorts of status- and age-oriented rules—in Faraway Lands Clarissa is evaluated on her creativity and artistic ability.

Clarissa struggles with some normal teenage challenges—finding time for her girlfriend, power-struggling with her father, lacking money, and figuring out a path to college—and some unusual challenges—having a disabled brother, being involved in a same-sex relationship, and suffering a severe leg injury. While she might be particular in her use of the Internet as a space to role-play, her story is a compelling one with which to think through possibilities of the Internet as a semipublic, third space for teens to hang out in. These digital spaces are particularly interesting because of the variety of hangout options they afford. As Clarissa illustrates, teens can do public-identity work by setting up sites defining "who they are"; they can maintain and deepen romantic relationships; and they can make new friends, play, be creative, and be treated as competent artistic producers.

Messing Around

The second genre of participation prevalent among American teenagers is what we have termed "messing around." Whereas hanging out is a genre of participation that corresponds largely with friendship-driven practices in which engagement with new media is motivated by the desire to maintain connections with friends, messing around as a genre of participation represents the beginning of a more intense engagement with new media. In the first section on "Looking Around," we focus on the ways in which kids use search engines and other online information sources to find information, a practice we call "fortuitous searching." The second section attends to the importance of "Experimentation and Play" in facilitating learning about the way a particular medium works, particularly through the processes of trial and error. The final section, "Finding the Time, Finding the Place," outlines many of the conditions or environments that are conducive to young people's efforts to engage with new media through illustrations of young people seeking out and taking advantage of the resources available to them at home, at friends' homes, and at after-school programs and in other institutional contexts.

Looking Around   One of the first points of entry for messing around with new media is the practice of looking around for information online. As Eagleton and Dobler (2007), Hargittai (2004; 2007), Robinson (2007), and others have noted, the growing availability of information in online spaces has started to transform young people's attitudes toward the availability and accessibility of information (Hargittai and Hinnant 2006; USC Center for the Digital Future 2004). Among our study participants who completed the Digital Kids Questionnaire, 87 percent reported using a search engine at least once per week, varying from Google to Yahoo! and Wikipedia as well as other more specialized sites for information.[8] The vast majority of the young people we interviewed engaged in "fortuitous searching," a term that distinguishes itself as more open ended as opposed to being goal directed. Rather than finding discrete forms of information, such as the exchange rate between the United States and Great Britain, the color of a particular flower, or the name of the twentieth U.S. president, fortuitous searching involves moving from link to link, looking around for what many teenagers describe as "random" information. As seventeen-year-old Carlos, a Latino from the San Francisco bay area described the process to Dan Perkel (MySpace Profile Production), "I was just going through Google . . . it just gives a lot of websites. So I just started finding these . . . I put Google . . . then it took me to a website and it had a lot of different stuff. . . ."

Despite the seemingly roundabout method of following links described by Carlos, teens' online research can be quite focused. Many searches involve finding information to facilitate the completion of homework and school projects, looking for a "cheat" for a particular game (see chapter 5), or looking for a way to complete a particular task. However, the nature of search engines and the organization of information on search results pages enables teenagers who are interested in a topic to find out more by clicking from one link to another.

Fortuitous searching represents a strategy for finding information and reading online that is different from the way kids are taught to research and review information in texts at school. Students are taught to use tools such as identifying a purpose for reading, activating prior knowledge, predicting the content of the text before and during reading, and summarizing or discussing the text after reading in order to improve their skills in finding and comprehending information in both traditional and online resources (Eagleton and Dobler 2007; Graves, Juel, and Graves 2001). By contrast, fortuitous searching relies upon the intuition of the search engine and the predictive abilities of the reader. Eagleton and Dobler write:

Readers of web texts rely on a similar process of making, confirming, and adjusting predictions. However, not only do web readers make predictions about what is to come in the text (and within other multimedia elements), they also make predictions about how to move through the text in order to find information. When a reader who wants to know more about how to do an olley on a skateboard and clicks on the hyperlink "olley," she is mentally making a prediction that this link will lead her to learn more about this skateboarding trick. (37)

Indeed, participants' skills in navigating large numbers of pages and using appropriate search terms indicate proficiency at predicting the information available to them online.

Kids often will look around online to find material for creative production. For example, we have seen kids use fortuitous searching to find materials for customization, appropriation, and alteration of their MySpace pages. As Perkel (2008) notes, copying and pasting has become a prevalent practice among American teenagers who want to update and alter their MySpace pages (see also chapter 6). Many of the tips or guides for changing a MySpace page (such as embedding images and videos and uploading pictures) are online—on other people's profiles, in online guides, and on the MySpace site itself. Many kids use a variety of search sites' strategies to obtain information about their interests (Robinson, Wikipedia and Information Evaluation). Nineteen-year-old Torus, an Indian Italian who lives in the Los Angeles area, described to Patricia Lange (YouTube and Video Bloggers) how he looks on Wikipedia for information about games he is interested in. "I actually went on recently to learn about one aspect of [a particular type of mod]. There's some card game inside the game and I didn't understand it so I went on Wikipedia and Wikipedia told me, as usual." Similarly, Christo Sims interviewed eighth grader MaxPower, a white fourteen-year-old living in a middle-class area of rural California (Rural and Urban Youth), who expressed a strong interest in music. MaxPower learned about music in some of the traditional ways, such as watching music videos on television. However, after a song or a band piqued his interest, he turned to online sites, searching for a particular band on iTunes, doing a Google search to learn more about the band, or identifying Google images to download a picture for his binder. When he liked what he saw, he sometimes bought music, and if he really liked it, he would burn a copy for his friends.

The youth we spoke to who were deeply invested in specific media practices often described a period in which they discovered their own pathways to relevant information by looking around. Unlike MySpace profiles, where many kids can find local experts, kids with more specialized interests often need to rely on online resources for an initial introduction to a particular area. While the lack of local resources can make some kids feel isolated or in the dark, the increasing availability of search engines and networked publics where they can "lurk" (such as in web forums, chat channels, etc.) effectively lowers the barriers to entry and thus makes it easier to look around and, in some cases, dabble or mess around anonymously. Without having to risk displaying their ignorance, they find that opportunities for legitimate peripheral participation (Lave and Wenger 1991) abound online. For example, SnafuDave,[9] a web comics creator described in box 7.1, explains how he learned many of his initial graphics skills from online tutorials and web forums before becoming an active participant in a web comics community. Similarly, Derrick, a sixteen-year-old teenager born in the Dominican Republic who lives in Brooklyn, New York, looked to online resources for initial information about how to take apart a computer. He explains to Christo Sims (Rural and Urban Youth) how he first looked around online for this topic:

I just searched on Google and I just went to . . . because I bought myself a video card. I had no idea what a video card looked like. I typed in video card image. Before I went to searching for it, image. I wanted to know what it looked like first. I seen different pictures. So Google sometimes gives you different pictures. If you type something in, it gives you . . . So I'm confused. I'm like, "I thought it looks like this but it looks like" . . . so I typed something in and I seen on Google what it looks like. So I looked at mine and I seen exactly where's it at. If you smart you don't got to search out, "How do I put in and put out." It's simple. It's just take the piece out. Have your computer off. Take it out. When you get your new one if it has a fan you can't have your sound card too close to it. So you've got to put your sound card in another slot and I bought myself a sound card too. I had no idea what none of those looked like. I thought a sound card was called a sound disk. I learned a lot on my own that's for computers. . . . Just from searching up on Google and stuff. . . . That's why I like Google.

As Derrick makes clear, looking around online and searching is an important first step to gathering information about a new and unfamiliar area. Although many of these forays do not necessarily result in long-term engagement, youth do use this initial base of knowledge as a stepping-stone to deeper social and practical engagement with a new area of interest. Online sites, forums, and search engines augment existing information resources by lowering the barriers to looking around in ways that do not require specialized knowledge to begin. Looking around online and fortuitous searching can be a self-directed activity that provides young people with a sense of agency, often exhibited in a discourse that they are "self-taught" as a result of engaging in these strategies (see chapter 6). The autonomy to pursue topics of personal interest through random searching and messing around generally assists and encourages young people to take greater ownership of their learning processes.

Experimenting and Play As with looking around, experimentation and play are central practices for young people messing around with new media. As a genre of participation, one of the important aspects of messing around is the media awareness that comes from the information derived from searching and, as we discuss in this section, the desire and (eventually) the ability to play around with media. Often experimentation starts small, such as using digital photo tools to crop, edit, and manipulate images. As Gee (2003) has argued for games and other interactive technologies that have low stakes attached to making mistakes or trying multiple scenarios to solve a problem, messing around also involves a great deal of trial and error. In chapter 5 we argue that the sociability around gaming combines with the affordances of gaming systems to support an ecology of playful experimentation with technology that can often lead to technical and media expertise. This kind of social play and experimentation can happen in the home, as an extension of hanging out with family and friends, as well as online in networked gaming contexts where players join in collaboration and competition through game play, practices that are buttressed by ongoing exchange and collegiality. In fact, much of contemporary gaming is built on the premise that players will engage in a great deal of experimentation on their own in a context of social support. Many key dimensions of game play in complex games are not explicitly spelled out by designers, and players learn about them from other players either directly or through online resources such as fan sites, game guides, and walk-throughs.

Because of the ease of copying, pasting, and undoing changes, digital media-production tools also facilitate this kind of experimentation. The availability of these tools, combined with the online information resources just described, means that youth with an interest and access to new media now possess a rich set of tools and resources with which to tinker and experiment. In chapter 6 we describe how youth media creators typically recount a period of time early in their learning about media production when they were tinkering with new media in a self-taught mode. They often describe getting started by messing around with home videos, modifying photos, or using a program such as Photoshop. Eventually, many of these media producers begin to get more serious about their craft and develop a hobbyist network to support their work. Often these activities start as social hanging out modes of media creation, but young people with an interest in media production sometimes go on to play and experiment with different media beyond simple plug and play. Young people who are successful in learning advanced technology skills through messing around sometimes become experts among their families, friends, teachers, and classmates. Megan Finn describes this position as the "techne-mentor" in box 1.4. Techne-mentors, like guides and digital tools, support learning about technology in informal settings.

Box 1.4 The Techne-Mentor

Megan Finn

In conceptualizing the media and information ecologies in the lives of University of California at Berkeley freshmen, classical adoption and diffusion models (e.g., Rogers [1962; 2003]) proved inadequate. Rather than being characterized by a few individuals who diffuse knowledge to others in a somewhat linear fashion, many students' pattern of technology adoption signaled situations in which various people were at times influential in different, ever-evolving social networks. The term "techne-mentor" is used to help to describe this pattern of information and knowledge diffusion. The term "technology" is generally thought to be partially derived from the Greek word *techne*, which means craftsmanship. Mentor is a figure in the *Odyssey* who advised both Odysseus and Telemachus and is the source of the modern use of the word "mentor." Techne-mentor refers to a role that someone plays in aiding an individual or group with adopting or supporting some aspect of technology use in a specific context, but being a techne-mentor is not a permanent role. The idea of the techne-mentor is useful for expanding conversations about adoption patterns to one of informal learning in social networks.

Growing up, Joan learned about technology on her own and acted as a techne-mentor to her family and friends. Joan started as a techne-mentor when her computer got a virus. She then helped her friends get rid of the virus.

We got this one [virus] on AIM [AOL Instant Messenger] actually. It was on your user profile so whenever you clicked info, it would say, "Ha, ha, ha, I found the picture of insert your name here" and you would click on the link and then you would get this spyware. . . . It took me a day to figure it out. . . . Then I got rid of it for all my friends. It's kind of like a little game. . . . It was a challenge, especially the first virus. . . . I just started getting into [computer] stuff.

Many students such as Joan were often driven to learn about technology on their own when they encountered problems with the technology and did not have other support to learn how to fix them. Other students started learning about computers while trying to get rid of viruses on their families' computers. For example, Ben explained, "I did get a virus once and had to learn how to get rid of it. The damn 'I love you' virus. Gosh, that nailed everybody." Once students such as Ben and Joan figured out how to get rid of a virus, they would often help the people in their social networks get rid of the virus, essentially becoming techne-mentors to others.

Joan also explicitly directed her siblings about how to use technology.

I would teach them [my siblings]. Not so much in middle school but in high school, they're usually, "Do you know how to use Photoshop?" I'll say, "Yeah, do this." . . . Or

"Do you know how to get rid of this spyware?" . . . for my brother at least; my [older] sister has her own tech guy.

Once Joan started at Berkeley, she found a job working for a computing help desk. Through her colleagues at work, Joan picked up a lot of information about best computing practices: "When I got my job, there was this girl at work who did a yearbook and knows everything and so whenever we have a shift, she will teach me all this random stuff." In a work context Joan was mentored by her friends and colleagues, but in other social contexts, such as her family, Joan was a techne-mentor to others. It is important to note the nonstatic nature of the techne-mentor; the status of techne-mentor is relative to the knowledge of others within a social context. The significance of the techne-mentor is that he or she provides information to others without implying absolute expertise.

Joan uses information from the work context where she has found a techne-mentor to help her friends.

I see that they are using it [AIM]. . . . [I say,] "Your AIM starts playing a movie trailer with audio every half hour and it's just annoying." [My friends say,] "My god, I want to get rid of that, can you help me?" and so I'll go on like a downloading site and download GAIM or DeadAIM.

We can see here that when Joan acts as a techne-mentor to her friends, she is not teaching in a traditional way. The techne-mentor interactions are very ad hoc and informal. The mentorship can be in the form of exposure to a technology. Joan, the techne-mentor in this case, has preexisting relationships with those whom she mentors that are much more elaborate than just the techne-mentor/student relationship. It allows her to casually mentor her friends when a technology is not working.

Besides Joan, in the Freshquest study we found many cases of technementors. The kind of roles they played varied from case to case and situation to situation. One one hand, the techne-mentor may simply make someone aware of a technology. On the other hand, he or she may play an integral role in demonstrating the technology practice or even installing the technology and ensuring its status as operational. Sometimes students we interviewed had one primary techne-mentor in their lives, but in turn the students would take on the role when they passed this information on to other groups. In fact, it is this constant flow of information about technology among a student's multitude of social networks that accounts for the fluidity of the role of techne-mentor. In all these socially situated contexts, techne-mentors were an integral part of informal learning and teaching about technology and technology practices.

In chapter 7 we describe how young people who started successful online and digital media ventures enjoyed a certain amount of time and autonomy during which they could try out various modes of working that were different from the standard forms of part-time labor available to teenagers. Indeed, messing around requires a good deal of time for self-directed learning. For example, SnafuDave, the successful web comics artist profiled in box 7.1, described how school provided an important venue for developing his new media skills. While he learned few useful new media skills in his college classes, school did provide him with the time and space to learn on his own. Similarly, Zelan, profiled in box 7.2, described how his interest in new media began with gaming while his parents were prospecting for gold. Eventually, Zelan parlayed his interest in gaming into different forms of technical expertise, and he learned how to take apart and fix game consoles and eventually computers. Now he is a local technical expert and gets paid for his services; he sees his future in a new media–related business.

Messing around is easiest when kids have consistent, high-speed Internet access, when they own gadgets such as MP3 players and DVD burners, and when they have a great deal of free time, private space, and autonomy. However, these are not necessary conditions for messing around. Some of the innovative experimentation in youth's messing around was seen in their circumventing limited media access. Consider, for example, James, a fourteen-year-old from Lisa Tripp and Becky Herr-Stephenson's study (Teaching and Learning with Multimedia). James's parents promised him an iPod as a graduation gift if he completed eighth grade with acceptable grades. With graduation still a few weeks off and his grades in question, James figured out a way to substitute the technology he *did* have for the iPod he was anticipating. James borrowed his aunt's digital camera, on which he could record several minutes of video, and recorded music videos off the television in his bedroom. Getting a good recording took time and several tries, but fortunately for James, he had a few hours at home alone after school before his parents arrived home from work, so he could shut his bedroom door and crank the sound on the television to get a good recording without having to worry about his parents' overhearing questionable lyrics or complaining about the volume. Although the camera's memory card held only two or three songs at one time, it had a headphone jack and fit in James's pocket so no one had to know that it was not an

MP3 player. By messing around and being creative with technology, James was able to find an acceptable interim solution until he could get his iPod. Similarly, Melea, a mixed-race high-school student in San Francisco enrolled in an after-school program, used resources at the after-school center to devise a creative way of getting a custom ringtone for her phone. Dan Perkel describes Melea's ringtone practices:

I saw that Melea had come in, sat down at the adjacent computer, and was using the computer. I realized that she was playing music and getting everyone else to be quiet. She was bent way over next to the Mac's external speakers with her cell phone up to the speaker recording the song that she had put on her MySpace profile. JJ at one point started talking and she shh'd him (later she said in a threatening voice, "If your voice is on that . . ."). She said it was going to be her ringtone. Then she went to the Fergie page on MySpace music. She played the Fergie song. I asked her if this were Fergie from the Black Eyed Peas and she said, "Yes." She played the song and asked herself over and over again . . . "Do I want this song? Do I want this song?" Then she said, "Yes!" and right in the middle hit the record button on her phone (or whatever) and started recording from the speakers again. (Antin, Perkel, and Sims, The Social Dynamics of Media Production)

Melea circumvented economic costs associated with buying ringtones, costs that could have prohibited her from possessing her ringtone of choice. Despite the difficulty of getting a high-quality recording in a noisy computer lab, by recording it from the playback of a MySpace page Melea creatively acquired the media she wanted in her desired format.

Whether in media production, game play, or other mediated contexts, opportunities to experiment, play, and fail with minimal consequence can support young people in developing problem-solving skills and learning to use resources wisely and creatively. As with looking around, the social dimensions of experimentation and play are important, as peers are able to scaffold experiences for one another based on experience and the results of previous experimentation.

Finding the Time, Finding the Place   The ability to mess around requires access to media, technology, and social resources that are not always available to youth. Just as in the case of hanging out, messing around is a genre of participation that is driven by young people's own interests and motivations. It is not always fully provided by the adults who have authority over kids. While schools may provide structured media production programs for youth, these programs are task focused and there is little time for unstructured experimentation and play. Most of the messing around activities that we observed occurred at home with kids who had both well-provisioned media households and an environment where they had certain amounts of free time and whose parents gave them a fair degree of autonomy over their media choices. The dynamics of homes and families are described more in chapter 4. We also found that transitioning to college was often a key moment when kids took the time and space to engage in messing around, particularly if they did not grow up in a home where they were given the freedom to engage in these activities before college. The older participants we spoke to who were highly engaged with media production or gaming generally described falling in with a crowd of friends in college who shared some of these interests.

For young people without access to digital media at home, after-school programs can be an important place for experimentation and play, providing technical and social resources and a time and space for messing around with technology that they do not have at home. Jacob, a seventeen-year-old African-American high-school student in Oakland, is enrolled in a program where he can stay after school to work with computers. He described the program where he had the opportunity to mess around to Dan Perkel (Antin, Perkel, and Sims, The Social Dynamics of Media Production):

So it's fun, because they teach you all these different programs that you had no idea what they were until you get into there. And then they have nice software. They have LCD screens. Every seat, every computer they have fast Internet service, processor. They have nice seats. I mean, the seats aren't like these. I mean, they have nice roll-around comfy sit-back seats where you can just sit back and type. It's comfortable. And then they got tables. And then they got a table where you eat. So they bring out food, like sandwiches, chips, apples, fruit. Nutritious stuff. They don't really serve fast [food] . . . they do have chips, like Doritos, but not sloppy things. And so I learned Photoshop, Flash animation, Dreamweaver, a couple of other programs like Word, Excel. They have all the latest programs. Flash. Our school has Flash [inaudible], but Tech Visions have the new ones—Flash 8 and Dreamweaver 9. And I think it's Photoshop CS and Fireworks. They got all the programs. Anything you need to do to build any kind of website, or any kind of project or picture, they have it.

Jacob recounts with delight how the program provides a whole environment that gives him a sense of empowerment and efficacy; not just the technology but the provisioning of good, nutritious food and comfortable work spaces are all part of the package that draws him to this program.

Messing around happens according to a variety of trajectories and in different settings. Although the youth in our study who had in-home, private, and consistent access to new media (particularly computers and Internet connections) tended to have an advantage in relation to those who had more limited resources, for a number of youth, the most important spaces for messing around took place at school or in after-school settings. For Katynka Martínez's study, "High School Computer Club," Martínez observed a Los Angeles high school where the computer-lab instructor allowed kids to hang out and use the lab for their own self-directed activities. The kids in the computer lab set up the computers so they could engage in networked game play, launched a variety of self-directed media-production projects, and started some small business ventures as described in box 7.3. In many ways, the computer lab was a unique context where kids could gather informally during school breaks and after school to mess around with a comfortable mix of social and technical resources.

Some teens were able to construct their own times and places for messing around in the absence of formal programs, even if they did not have a home context that fully supported these activities. For example, Toni, a twenty-five-year-old living in New York City whom Mizuko Ito (Anime Fans) interviewed over an instant-messaging program, reflected on his experiences as a student coming to the United States from the Dominican Republic and the ways in which he was able to create space to mess around at school. He was first exposed to computers soon after he moved to the United States for middle school and took a computer class. He quickly took an interest in computers and then later went back to the Dominican Republic for a year and attended a computer-training institute, all the while not having computer access at home. When he returned to the United States in ninth grade, he became part of an informal computer club.

Toni: i would stay after school and play around/help the teacher who kept the lab open for students to use

Mizuko: sounds like a cool teacher

Toni: he was except when i printed out the student database he wasn't happy then

Mizuko: lol but sounds like he gave you some freedom to mess around

Toni: yeah, the exposure i got both learning how parts of a computer make the whole and also helping other students was pretty good for me and i sort of do the same kind of thing these days

Today, Toni is an active online participant in the anime fandoms that are the subject of Ito's study, and he is a technology expert for his family. He eventually acquired his first computer in eleventh grade and attended school at a technical university. While Toni's experience of messing around informally at school is not necessarily typical, it speaks to the fact that schools and after-school programs continue to play an important role to many youths for learning about technology. In addition, it illustrates the value of informal learning, unscheduled time, and student-driven inquiry, even in a formal educational environment.

As a collection of practices and a stance toward media and technology, messing around highlights the advantages of growing up in an era of media saturation, interactive media, and social software. Although messing around can be seen as a challenge to traditional ways of finding and sharing information, solving problems, or consuming media, it also represents a highly productive space for young people in which they can begin to explore specific interests and to connect with other people outside their local friendship groups. As noted in the beginning of this section, messing around can be understood as a transitional genre of participation that can mediate between hanging out and geeking out. Kids can move from media engagement that centers on peer sociability to forms that are more interest focused via messing around. Conversely, kids who are participating in more geeky interest-driven activities see messing around as a form of social play in which they engage with their friends around interests and learning. Unlike learning in more structured settings, messing around involves a more open-ended genre of participation, which often hinges on certain modes of sociability and play, along with access to resources on a timely and as-needed basis. As we outline, even youth with well-provisioned media environments can lack the time and social resources to successfully mess around with media. Messing around is therefore a powerful modality of learning that requires a whole ecology of resources, including time and space for experimentation.

Geeking Out

The third genre of participation we have identified is "geeking out." This genre primarily refers to an intense commitment or engagement with media or technology, often one particular media property, genre, or a type of technology. This stance is characteristic of the young people we interviewed who were involved in a media fandom, such as the young people in Mizuko Ito's "Anime Fans" study, in Becky Herr-Stephenson's "Harry Potter Fandom" study, or the more committed gamers who participated in Matteo Bittanti's "Game Play" study. The term "geeking out" can be used to describe the everyday practices of some of the gamers and media producers who participated in our project. In addition to intensive and frequent use of new media, high levels of specialized knowledge attached to alternative models of status and credibility and a willingness to bend or break social and technological rules emerged as two additional features of geeking out as a genre of participation.

Before discussing geeking out in more detail, it is important to note that although "geeking out" describes a particular way of interacting with media and technology, this genre of participation is not necessarily driven by technology. The interests that support and encourage geeking out can vary from offline, nonmediated activities, such as sports, to media-driven interests, such as music, which are larger than the technological component of the interest. That is to say, one can geek out on topics that are not culturally marked as "geeky." We also wish to distinguish here between geeking out and other uses of the word "geek," as an identity category. Whereas notions of geek identity have traditionally been associated with white, affluent, suburban boys (Jenkins 2000; Thomas 2002), our understanding of geeking out as a genre of participation—a way of understanding, interacting, and orienting to media and technology—widens the definition to include activities and people outside established understandings of what it means to identify (or be identified) as a geek. This is not to negate the potential implications of participation for the negotiation and articulation of identity. As we discuss elsewhere, participation, learning, and identity development are contingent within communities of practice. Our point here is to call attention to examples of continued, intensive, and sophisticated interaction and use of new media that might otherwise be overlooked because the person doing it does not fit a preconceived notion of the gender, class, or race of a "geek".

Expertise and Geek Cred   For many young people, the ability to engage with media and technology in an intense, autonomous, and interest-driven way is a unique feature of the media environment of our current historical moment. Particularly for kids with newer technology and high-speed Internet access at home, the Internet can provide access to a huge amount of information related to their particular interests. The chapters on gaming, creative production, and work describe some of the cases of kids who geek out on their interests and develop reputation and expertise within specialized knowledge communities. Geek cred involves learning to navigate esoteric domains of knowledge and practice and being able to participate in communities that traffic in these forms of expertise.

Box 1.5 describes zalas, one highly expert participant in online knowledge cultures who has customized his media engagement in a way that focuses on developing deep expertise in a specific area of interest. Although very few of the youths we spoke to exhibited the kind of informational expertise that zalas did, it was not uncommon to find young people who customized their media environments to facilitate access to specialized knowledge. For example, one of Heather Horst's interviewees in her study "Silicon Valley Families" a fifteen-year-old boy who chose the pseudonym 010101, discussed the way he keeps up with information about his interest in technology by creating a customized Google home page with various RSS (Really Simple Syndication) feeds so he can keep tabs on different sites of interest. In addition to Slashdot, one of the most popular technology news blogs featuring "news for nerds," 010101 regularly reads a variety of technology websites specific to his interest, including MacRumors.com and Engadget.com. His sources of information are sites with high status within the tech geek community, where the credibility of technology information is debated among people who identify as tech experts.

Box 1.5   zalas, a Digital-Information Virtuoso

Mizuko Ito

My first encounter with zalas was through email, through an introduction from another anime fan. I was seeking information about my new study on fansubbing practices, and I was told that zalas was the person I should know. Initially, we corresponded over email, where I peppered him with questions about the fansub community. He seemed to have eyes and ears all across the vast web of the online fandom around anime, not just among the fansub communities. Apparently no question was too esoteric; he could come back with information about the latest anime releases in Japan, the activities of even the most minor fansub groups, and the juiciest gossip on the online forums surrounding Japanese popular culture in both Japan and the United States. I had the good fortune of having zalas, a digital-information virtuoso, as a key informant in my study of anime fans.

After immigrating with his family to the United States from mainland China when he was a child, zalas grew up in a technology-rich household, with two parents who worked with computers. "I got introduced to computers early on. And, also, I just tend to be better at science and math than the arts and English and things like that. I was sort of just drawn to [the computer] because it was like this super, über toy, you know." Both his parents were in graduate school at the time, and he had online access to their VAX machine. Ever since, he kept up with the latest online technologies, moving from AOL Instant Messenger, to Internet relay chat (IRC), and eventually to BitTorrent. He discovered the online anime and fansub scene through his contacts in IRC.

He participates in a wide range of fan activities. He has been involved in a variety of fansub groups and activities, including projects for fansub games and electronic visual novels. He also makes anime music videos (AMVs), is an officer at his university's anime club, and is a frequent speaker at his local anime convention. I have seen zalas give talks on topics as varied as Japanese anime and game-remix videos, fansubbing, and visual novel subtitling. He describes himself as something of an elder in the online anime scene, despite the fact that he is still in his early twenties.

In my interview with zalas, he guided me through some of what was behind the curtain of his information magic. He explains that he is constantly on IRC, logged into multiple channels populated by the information elite of the online anime fandom.

I used to have just one copy of mIRC running that simultaneously connected to all these channels, and every once in a while just scroll through to see which ones have new messages, go to them, see if it's important, if it's not, go to the next one and things like that. But right now I actually have a text-only IRC client that's running on my friend's web server, and I'm connected to about twenty channels on that one. It's actually down from what I'm usually connected to. And that one lights up a little number near the bottom of the screen indicating which channels have new activity, and I'll switch to it and see if it's worthwhile or something.

He has four computers at home: a Windows computer, a Linux computer, a Macintosh desktop computer, and a Macintosh laptop.

So, my Windows computer is there so I can play games. It's—most of my desktop processing stuff and all my video editing and things like that are on [my] Windows computer. My Linux computer is there because I need—sometimes I need a Linux compiler, and it's also there as a server. So, it's serving my source code repositories, and it's—it has a IRC file server on there as well and IRC bot on that or something like that, which controls some channel. And my OS10 one is actually my laptop, which I bring with me. It's kind of like my portable computer . . . I bought it because I wanted to be able to work anywhere, and also I bought it so I can sort of connect to IRC at conferences—at conventions.

Although zalas is an avid consumer of music and television, he rarely accesses this content through standard broadcast channels. He frequents the Japanese streaming-video site Nico Video in addition to using BitTorrent to download anime episodes. IRC is zalas's home base for communication. But in addition to IRC zalas frequently visits information websites and online forums devoted to his hobby. He does not keep a personal blog but prefers to post to shared online forums. He will often scour the Japanese anime and game-related sites to get news that English-speaking fans do not have access to. "It's kinda like a race to see who can post the first tidbit about it."

In addition to his prolific activities as an anime fan, zalas is a graduate student in electrical engineering at one of the top universities in the country. He says that he mainly uses IM for people he has met in school and other real-life contexts, and IRC is for people he "met randomly online." Despite the fact that he is in a high-powered graduate program, zalas says that almost all his online activity centers on his anime- and game-related hobbies. He estimates that he spends about eight hours a day online keeping up with his hobby. "I think pretty much all the time that's not school, eating, or sleeping." Building a reputation as one of the most knowledgeable voices in the online anime fandom requires this kind of commitment as well as an advanced media ecology that is finely tailored to his interests.

Another example of how geeking out relates to finding and producing credible information comes from a number of the gamers with whom we spoke during this project. Particularly when it comes to massively multi-player online role-playing games (MMORPGs), the intensive engagement associated with geeking out as a genre of participation extends beyond participation within the boundaries of the game world and to the para-texts[10] that support and extend the game. Paratexts take many forms, varying from gaming magazines and official guides published by game manufacturers, to player-generated guides and tutorials, to materials more recognizable as fan texts such as fan fiction and fan art. For example, Rachel Cody notes that the players in her study "Final Fantasy XI" used guides, typically on websites but sometimes in books, regularly during game play for information about quests, missions, and crafting. The guides assisted players in streamlining some parts of the game that otherwise took a great deal of time or resources. For example, guides that instructed players on strategies for leveling crafting skills could help players save on the in-game expense of materials by providing tips on the best way to craft items. Cody observed that a few members of the linkshell in her study kept Microsoft Excel files with detailed notes on all their crafting in order to postulate theories on the most efficient ways of producing goods. As Wurlpin,[11] a twenty-six-year-old male from California, told Cody, the guides are an essential part of playing the game. He commented, "I couldn't imagine [playing while] not knowing how to do half the things, how to go, who to talk to."

As Wurlpin and many other players with whom we spoke noted, the information sought from guides is often used to save time, resources, or to draw upon advice from players who have successfully completed a task with which the player is struggling. In this context, user-generated guides often have greater credibility with players because they have been created by other players rather than by the producers of the game. Using and creating player-generated guides is an example of geeking out because it reflects an acceptance of the alternative status economy and markers of credibility that exist in many gaming communities. While not endemic to gaming communities, valuing geek cred is a unique feature of geeking out as a genre of participation and is significantly different from the ways in which information is assessed while messing around.

Status and credibility also remain linked in alternative status economies, which represent another area of blending between interest- and friendship-driven groups. For example, in her study of anime fans, Mizuko Ito observes that fans gravitate toward particular fan sites that have credibility within the community rather than relying on industry-produced sites for information about anime. She notes that fans in specialized creative communities often avoid official discussion forums (those provided by the media producers or otherwise sponsored by the industry), instead looking to specialized fan communities where the knowledgeable fans congregate. For example, fansubbers such as zalas generally prefer to participate in closed IRC groups or specialized forums rather than general fan discussion forums, which they see as catering to less knowledgeable fans.

In interest-driven groups built around technology expertise, media fandom, or electronic gaming, status does not have to align with the hierarchies of status at school, at home, or more general social status. Whereas family, peers, classmates, and others might contribute to a young person's feeling of marginalization for having a particular niche interest, within an interest-driven group the niche interest is what brings people together. Therefore knowing a lot about it, sharing unique information with the group, or producing interesting and high-quality productions (fan fiction, art, fansubs, videos, podcasts, etc.) are highly valued practices.

Rewriting the Rules Rewriting the rules is a practice related to both messing around and geeking out. However, there are important differences in the ways in which the rules are rewritten in each of these genres of participation. Like messing around, which involves an inchoate awareness of the need and ability to subvert social rules set by parents and institutions such as school, geeking out frequently requires young people to negotiate restrictions on access to friends, spaces, or information to achieve the frequent and intense interaction with media and technology characteristic of geeking out. Rewriting the rules in the service of geeking out, however, also involves a willingness to challenge technological restrictions—to open the black box of technology, so to speak. This practice is most often done in the service of acquiring media—either media that are unavailable through commercial outlets (such as anime that has not yet been released in the United States) or media that are unavailable because of the cost of buying it. Geeking out often involves an explicit challenge to existing social and legal norms and technical restrictions. It is a subcultural identity that self-consciously plays by a different set of rules than mainstream society.

Many of the geeking out practices we describe in the chapters on gaming, creative production, and work involve youth engaged in passionate interests who are concurrently innovating in ways that rewrite the existing rules of media engagement. For example, fans of various forms of commercial media have engaged in their own alternative readings of media and created secondary productions such as fan fiction, video mashups, and fan art. These activities are proliferating online, and we capture some of this in chapter 6. Similarly, gaming represents a breeding ground for practices of code hacking, creating and exploiting cheats, and making derivative works such as machinima and game modifications. These forms of geeking out are described in chapter 5.

Geeks also have been at the forefront of alternative regimes of media circulation. Fansubbing bridges fan practices of secondary production and peer-to-peer (P2P) circulation, and it is described further in chapter 7. Despite attention in recent years to large numbers of youth downloading music illegally, more sophisticated downloading—particularly download-ing video—continues to be associated with more intense engagement and commitment to media. Whereas figuring out LimeWire to download songs with friends might be more characteristic of hanging out or messing around, geeking out tends to require more systematic, long-term, and purposeful use of less-common technology to acquire media. As Derrick in Brooklyn, New York, explains to Christo Sims (Rural and Urban Youth):

Christo: So when you surf on the Internet what are some of the things that you are looking for?

Derrick: Well, mostly I look for . . . I ain't going to lie . . . illegal things.

Christo: That's fine.

Derrick: I just search. I just try to get . . . if I seen a movie or I like that movie, I go home, I get the movie.

Christo: You mean just find it and download it?

Derrick: Yeah.

Christo: Do you use like LimeWire or what do you . . .

Derrick: Torrent.

Christo: BitTorrent?

Derrick's friend: He's a computer freak.

What is interesting about the conversation between Christo and Derrick is Derrick's friend's comment. His act of calling Derrick "a computer freak" (even if meant as a joke between friends) indicates that he associates a particular and deviant identity with video file sharing, which is con-sidered geekier than music file sharing. Although the publicity and legal campaign against file sharing has had the effect of curtailing some P2P practices, our discussions with youth indicate that P2P sharing (particularly of music) is still widespread. Youth such as Derrick are becoming more savvy about what practices are likely to get them in trouble socially and legally, and more savvy about how to bend rules in ways that present the least amount of risk. The time and skill involved in subverting legal and technological rules is often quite intensive. For example, Federico, a seventeen-year-old Latino who participated in Dan Perkel's study (MySpace

Profile Production), described the process he goes through to download software:

Federico: Like if I don't want to try to pay for a software that costs a hundred dollars and some, I just go to the website and then I download it. Probably like Nero. There's a new version. I'm like . . . I just look for it on Google or something and see the whole name, what's the name. And then just go over there to the other website and . . . then press okay. Then they'll take you to another website and then they'll go like, you got to download part one, part two, part three . . . whatever. Right after that I go over there and then it takes you to another website and you press "free" and then it takes you whatever minutes, depending on your Internet. And then it opens up and it tells you if you have to put a code. Right after the code you got to put a [inaudible]; that's like another code. And you got to find it in another website. And then right after that you've got to find the serial number that I've got to download. And right after the serial code I got the software.

Dan: How much time does that take . . . the whole process?

Federico: Depending. If I'm trying to download a good software, sometimes I've got to download six parts . . . that's like two, three days.

Getting around the copyright rules and software market is, in this case, quite an intensive exercise, but acquiring the software for free is an incentive for this interviewee to put forth the effort. The commitment to geeking out pays off in this ability to navigate and exploit alternative media ecologies that are counter to the given, mainstream consumer logic of new media.

Having What It Takes   The intensive commitment to new media that is characteristic of geeking out clearly requires access to new media. However, in many of our cases, we have found that technological access is just part of what makes participation possible. Returning to the concept of media ecologies, it is important to emphasize the interaction of different resources in determining access. Family, friends, and other peers in on- and offline spaces become particularly important to facilitating access to the technology, knowledge, and social connections required to geek out. Just as in the case of messing around, geeking out requires the time, space, and resources to experiment and follow interests in a self-directed way.

Furthermore, it requires access to a community of expertise. Contrary to popular images of the socially isolated geek, almost all geeking out practices we have observed are highly social and engaged, although these are not necessarily expressed as friendship-driven social practices. We also have found that families provide a cultural and social context conducive to geeking out. For example, Carolina, a white female creator of AMVs in her twenties who was interviewed by Mizuko Ito in her study of anime fans, learned how to access P2P networks within the context of a family of file sharers. In her interview, she described learning about file sharing with her parents and siblings:

I started out by using search engines to look up what I was seeing on TV, or the manga we had at the bookstore, and that inevitably led me to review sites that [led] me to other series and movies. At the same time, our whole household was discovering peer-to-peer file sharing, so I'm sure you can imagine what that led to :$[12]

Carolina notes that different interests motivated each family member's file-sharing practices. Whereas her parents and sister were most interested in downloading music, Carolina and her brother focused on finding video clips, mainly anime fansubs. Carolina and her brother navigated multiple sites for P2P file sharing. She told Mizuko, "I know my brother has gotten things for me off of IRC, but we also used Napster, [LimeWire], Morpheus, more recently any number of [BitTorrent] clients. . . ." In this case, as well as in some of the cases highlighted in chapters 4 and 6, it is evident that family support and/or participation can be an important source of encouragement and access for geeking out.

Friends form an important support structure, not only in terms of gaining access to hardware or Internet connections when one does not have them at home but also in terms of recommending media, technology, or other resources related to a shared interest. In chapter 5 we describe how friendships built through playing together become a source of technical expertise that often extend beyond game-specific interests. In Katynka Martínez's study (Pico Union Families), she interviewed Dark Queen, a seventeen-year-old eleventh grader who told Martínez that she does not talk about her music, television, or reading preferences with friends in her neighborhood or school or with family members. However, Dark Queen likes to read manga and relies on MySpace friends for reading recommendations. She notes:

It's actually really interesting because they [her MySpace friends who are into manga] have read so many books that I haven't and I would be like—if they would give me a brief summary about like the book they have read or a movie they've seen, an anime movie, we would be like, "Okay. I have to read this book, or I have to see this movie." And I would look for it.

Having access to a community with similar interests allowed Dark Queen to pursue her interest in manga privately and to interact with a community of experts through the exchange of recommendations. In this case, exploring her interest in manga was as much about being a part of the community as it was about accessing the media itself.

Similarly, orangefizzy, a thirteen-year-old Asian-American Harry Potter fan from California and participant in Becky Herr-Stephenson's Harry Potter fandom study, described her experiences as an avid fan-fiction reader and writer on two fan-fiction archive sites. As orangefizzy notes, she prefers the smaller of the two sites because it "has more of a 'community we all know each other' feeling to it than [the larger archive], which is huge." In addition, orangefizzy observes that her decision to post her own work on the smaller archive site was very much influenced by the fact that she got to know other people participating on the site through extended conversations in the site forums. The examples of Dark Queen and orangefizzy illustrate how interest-driven and friendship-driven genres of participation often overlap and become intertwined.

Conclusion

"Hanging out," "messing around," and "geeking out" are three genres of participation we found to be widespread among the American kids and teenagers who participated in our studies. As descriptive frames, the three genres of participation are closely related to the genres of interest-driven and friendship-driven participation that we outline in this book's introduction, although here we have focused on issues of expertise and the intensity of media engagement. Hanging out tends to correspond with more friendship-driven practices and geeking out to the more interest-driven ones, although we have seen cases of kids geeking out on more friendship-driven practices, such as in the case of kids who are intensely into Facebook or MySpace, or when kids engage in video or photo production as part of their hanging out with friends. Messing around is a genre of participation in its own right, but it is also a transition zone along a continuum between geeking out and hanging out and between interest-driven and friendship-driven participation. It describes those modes of media engagement in which kids are tinkering, learning, and getting serious about particular modes or practices, which are often supported by the social networks they have developed in their friendship or interest groups. Taken together, these different genres of participation provide a flexible vocabulary for describing the different ways in which kids engage with new media and how their engagement relates to social participation and identity.

While each genre of participation represents a different stance toward engagement in terms of intensity and level of commitment to new media, we want to emphasize that these practices do not correspond with "types" of young people. Derrick, the sixteen-year-old in Christo Sims's project focused on rural and urban youth, is chronicled in all three genres of participation. In the section on hanging out, Derrick describes hanging out with friends in person and trying to coordinate further plans to hang out by using his mobile phone. In the section that focuses on messing around, Derrick participates in fortuitous searching on Google to build a computer. Finally, in our discussion of geeking out, Derrick downloads movies over BitTorrent, a somewhat obscure application that is used to download media and is often associated with geek culture and identity. This is not to suggest that Derrick is somehow schizophrenic or that he plays different roles. Rather, he is a young man born in the Dominican Republic, now living in a relatively low-income neighborhood in Brooklyn, who moves through the different genres of participation depending upon his motivation and within the constraints of his socioeconomic status, age, and location. When he is with his friends in Brooklyn, Derrick participates in his friendship, or peer, group by strategizing ways to hang out with his friends through the use of their mobile phones. When he wants to gain knowledge about computers and how they work, his engagement with new media more closely involves geeking out and messing around.

Throughout this chapter our primary aim is to map the media ecologies that constitute the lives of our research participants. We suggest that learning and participation with new media needs to be contexualized within a broader social-, cultural-, technical-, and place-based ecology. Our work has approached this problem by examining a diverse range of cases that were selected and delimited according to different criteria, some based on location, others based on online and institutional sites, and others based on interest-based groups. We designed our research to understand the environmental, socioeconomic, and infrastructural dimensions of media use. By sampling in these diverse ways, we have been able to grasp at least some of the variegated ecological factors that structure new media participation. We have suggested that the conceptual construct of genres of participation is one way of extrapolating from this material, which reflects the patterns of engagement of the young people we interviewed. These genres of participation, which are not reductive, retain the ecological context and begin to characterize how different forms of engagement and participation are defined in relation and in opposition to one another. Although our discussion does not focus on issues of the digital divide or the participation gap, we have worked to illustrate the kinds of resources that need to be present in youth's environments for them to participate in certain genres of practice.

In the following chapters, we elaborate upon this ecological frame and the genres of participation we introduce here by delving into specific youth practices. Throughout our descriptions, we use the broad genre distinction between interest- and friendship-driven genres of participation and the specific characteristics of hanging out, messing around, and geeking out, as points of orientation to bring the reader back to the ecological frame we outline here. We delve into some of the specific practices that make up the media ecologies of the young people who participated in our study. Although the subsequent chapters look at specific media practices, our investigation situates these practices within the diverse contexts of young people's lives—homes and neighborhoods, learning institutions, networked sites and spaces, and interest-based groups. We also use the broad distinction between interest-driven and friendship-driven genres of participation as well as the specific characteristics of hanging out, messing around, and geeking out as frames for understanding these practices within a larger media ecology. While individual chapters necessarily focus on specific populations and practices, we hope that when taken as a whole they allow us to retain a sense of context and relationality that has characterized the overall collaborative endeavor of analyzing and writing across a range of case studies, using multiple methods and disciplinary approaches.

Notes

1. The Kaiser report finds that youth spend the same number of hours, approximately 6.5 per day, with media in 2004 as they did in a similar survey conducted in 1999.

2. These comparisons are between national surveys and the share of our participants who completed our survey. Since not all the participants at our various ethnographic sites completed surveys, these figures should not be read as descriptions of our participant population as a whole.

3. We did, however, have 11 percent of participants report going online a few times a month or less. Since Pew reports frequency only in terms of the percent of participants who go online daily, we cannot compare these figures directly.

4. Part of the discrepancy in this final figure could be due to posing the question differently. We asked our participants if they "use a social network site daily," whereas the Pew survey asks whether or not they "send a message through a social network site daily." Since teens can use a site without sending a message, part of our figure probably includes those who visit a social network site daily but do not send messages every day.

5. Boase (2008) has analyzed variation in communication practices based on Pew's survey data of adults. To our knowledge, no similar survey analysis has been conducted of variation in communication among youth.

6. A pseudonym.

7. A pseudonym.

8. Although a variety of search engines are available to digital youth, across different case studies there are frequent references to Google. Some youth use various permutations such as "Googling," "Googled," and "Googler" as normative information-seeking language. The ubiquitous nature of Google may indicate that the idea of "Googling" has been normalized into the media ecology of digital youth such that for many, Googling may be considered synonymous with information seeking itself.

9. "SnafuDave" is a screen name.

10. "Paratext" refers to elements that surround a text. In relation to written texts, examples would be tables of contents or indexes. Mia Consalvo has described the products of the gaming industry—including guides—as a paratext for gaming. For a full discussion of paratexts, please see Consalvo (2007) and Lunenfeld (2000).

11. "Wurlpin" is a real character name.

12. ":$" is an emoticon meaning "embarrassed."

2 FRIENDSHIP

Lead Author: danah boyd

Sitting in a coffee shop in suburban Michigan in June 2007, Tara, a Vietnamese sixteen-year-old, was asked about Facebook. She giggled and said that she had "an addiction" to the site. She had heard from adults that Facebook might be bad, but "like everyone says get a Facebook. You need to get one." She made sure to log in often to check for new messages from friends, read updates about her classmates, and comment on friends' photos. For Tara, this type of participation on a social network site is a critical element of staying socially connected. She is not alone. While the specific tools vary by geography, time, and peer group, the teens we interviewed throughout the United States regularly told us that engaging with social media is important for developing and maintaining friendships with peers. While these teens may see one another at school, in formal or unstructured activities, or at one another's houses, they use social media to keep in touch with their friends, classmates, and peers when getting together is not possible. Skyler Sierra, an eighteen-year-old from Colorado, succinctly articulated the importance of these new media to these teens' social lives when she explained to her mother that "if you're not on MySpace, you don't exist."[1] For many contemporary teenagers, losing access to social media is tantamount to losing their social world.

We found that U.S. youth use a variety of social media to develop and maintain broader communities of peers. Teen practices when using social media mirror those that scholars have documented in other places where teens gather with peers (Eckert 1989; Milner 2004; Skelton and Valentine 1998). Just as they have done in parking lots and shopping malls, teens gather in networked public spaces for a variety of purposes, including to negotiate identity, gossip, support one another, jockey for status, collaborate, share information, flirt, joke, and goof off. They go there to hang out.

By providing tools for mediated interactions, social media allow teens to extend their interactions beyond physical boundaries. Conversations and interactions that begin in person do not end when friends are separated. Youth complement private communication through messaging and mobile phones with social media that support broader peer publics.

In the 1980s, the mall served as a key site for teen sociability in the United States (Ortiz 1994) because it was often the only accessible public space where teens could go to hang out (Lewis 1990). Teens are increasingly monitored, though, and many have been pressured out of public spaces such as streets, parks, malls, and libraries (Buckingham 2000). More recently, networked publics have become the contemporary stomping ground for many U.S. teens. Just as teens flocked to the malls because of societal restrictions, many of today's teens are choosing to gather with friends online because of a variety of social and cultural limitations (boyd 2007). While the site teens go to gather at has changed over time, many of the core practices have stayed the same. The changes we are seeing today are a variant of these core practices, inflected in distinctive ways as youth mobilize social media.

During the course of our study, we watched as a new genre of social media—social network sites (SNSs)[2]—gained traction among U.S. teenagers. While teenagers have many choices of media with which to interact with one another, two large social network sites—MySpace and Facebook—captured the imaginations of millions of U.S. teenagers while we were doing fieldwork in the years 2004 through 2007. Not all teens frequent these sites (Lenhart and Madden 2007), but social network sites became central to many teens' practices. This form of networked public allowed broad peer groups to socialize together while other social media such as instant messaging (IM) and mobile phones allowed teens to interact one-to-one or in small groups. All these tools can be used for a wide variety of purposes, but what we witnessed during our study was that the dominant practices for most youth were friendship-driven and exhibited the genre of participation that we have described in chapter 1 as "hanging out."

This chapter documents how social media are incorporated into teen friendship practices in the context of their everyday peer groups. We emphasize the practices that take place on social network sites because they emerged and took hold during our study as a central gathering spot for

U.S. teens. The material used in this chapter primarily comes from studies that emphasized the friendship-driven practices of youth as they interacted with peers in their school-centered social networks. These studies include those conducted by C. J. Pascoe (Living Digital); Christo Sims (Rural and Urban Youth); Dan Perkel (MySpace Profile Production); Heather Horst (Silicon Valley Families); Katynka Martínez (Pico Union Families); Megan Finn, David Schlossberg, Judd Antin, and Paul Poling (Freshquest); and danah boyd (Teen Sociality in Networked Publics). Unless otherwise stated, the quotes come from danah boyd's study.

This chapter and chapter 3, "Intimacy," focus specifically on the dominant and normative practices of high-school teenagers. For most teens, friendship-driven practices, such as those described in this chapter, play a more central role in structuring new media participation than interest-driven practices. The seemingly popular social media highlighted in this chapter, including MySpace and Facebook, are common tools for friendship-driven practices. While teens invested in both friendship-driven and interest-driven activities may use these services, these sites are emblematic of the genre of friendship-driven participation and support the kind of social relations that center on popularity, romantic relationships, and status. Although sites such as LiveJournal or web forums share much of the functionality of MySpace or Facebook, they inhabit a genre in closer alignment to interest-driven practices. While the dominant practice of teens in MySpace and Facebook conform to a hanging out, friendship-driven genre, kids sometimes also use these practices as jumping-off points to messing around and more "geeked out" interests. Chapter 6 examines the kind of technical and media expertise that youth develop as part of their participation on social network sites.

This chapter focuses on the role that technology plays in establishing, reinforcing, complicating, and damaging friendship-driven social bonds. Emphasizing the role of mediating technologies, this chapter contextualizes practices involving social media within a broader discussion of youth's everyday friendship practices. After outlining a historical and conceptual framework for understanding teen peer-based friendship, the chapter examines how social media intersect with four types of everyday peer negotiations: making friends, performing friendships, articulating friendship hierarchies, and navigating issues of status, attention, and drama. In all these cases, we consider how the unique affordances of contemporary networked publics are inflecting existing peer learning, sharing, and sociability in new ways.

Peers and Friendship

Teen friendship practices in contemporary networked publics need to be understood in relation to the broader contexts of teen sociability as it plays out in U.S. high schools. The current debates over teen participation on MySpace and Facebook are part of a longer history of intergenerational struggle over parental authority, youth culture, and the peer relations fostered in high schools. Sociologists of youth culture identify the 1950s as a pivotal period that saw the emergence of many of the dynamics that define contemporary youth peer culture and adult attitudes toward youth. This period saw a broadening of the base of teens who attend high school, a growth in youth popular and commercial cultures, and the emergence of an age-segregated peer culture that dominated youth's everyday negotiations over status and identity (Chudacoff 1989; Frank 1997; Gilbert 1986; Hine 1999). This period also saw the growth of a new set of intergenerational tensions, evident in the emerging discourse of juvenile delinquency and tied to the recognition that "the American family itself now exercised less influence on the cultural formation of youngsters" (Gilbert 1986, 17). Even as youth were developing a sense of autonomous generational identity with the aid of popular media cultures, their period of financial dependency and segregation from adult roles was expanding as more and more youth attended high school and higher education institutions. Stanley Cohen (1972, 151) writes, "The young are consigned to a self-contained world with their own preoccupations, their entrance into adult status is frustrated, and they are rewarded for dependency."

For contemporary youth, the age-segregated institutions of school, after-school activities, and youth-oriented commercial culture continue to be strong structuring influences. Despite the perception that online media are enabling teens to reach out to a new set of social relations online, we have found that for the vast majority of teens, the relations fostered in school are by far the most dominant in how they define their peers and friendships. In the later chapters of this book, we consider how new media networks enable youth to reach out beyond their given social relations and to engage with intergenerational interest groups and forms of creative production and economic activity that give youth a role in adult social worlds. This chapter, however, focuses on the more mainstream practices of teens that are situated within the more conservative structures of youth sociability, as largely segregated from but dependent on adult social worlds. Within these contexts of normative youth sociability, adults (whether in the role of parent, teacher, or media-technology maker) are generally relegated to the role of provisioning or monitoring youth media ecologies rather than as coparticipants.

The peer relations of children and teens are structured by a developmental logic supported by educational institutions organized by rigid age boundaries. We share a cultural consensus that the ability to socialize with peers and make friendships is a key component of growing up as a competent social being, and that young people need to be immersed in peer cultures from an early age (Newcomb and Bagwell 1996; Berndt 1996). Children are brought into preschools, kindergartens, and elementary schools not only to learn what is traditionally taught and measured in the classroom but also to learn how to develop friendships with peers (Corsaro 1985; Howes 1996). The "personal communities" that youth develop help them negotiate identity and intimacy (Pahl 2000). During the period of adolescence, kids' social worlds become dominated by same-age peers, adult oversight recedes, and the status and popularity battles that we typically associate with middle school and high school take hold. This is the same period when kids transition from a largely homosocial context that dominates elementary school to one that is increasingly defined by performances of heterosexuality (Eckert 1996; Pascoe 2007a; Thorne 1993).

Milner suggests that teens' obsession with status exists because "they have so little real economic or political power" (2004, 4). He argues that hanging out, dating, and mobilizing tokens of popular culture all play a central role in the development and maintenance of peer status. Working out markers of cool in the context of friendship and peer worlds is one of the key ways that youth do gender, race, class, and sexuality work (Bettie 2003; Pascoe 2007a; Perry 2002; Thorne 1993) and engage with teen-specific identity categories such as "jocks and burnouts" (Eckert 1989), "nerds and normals" (Kinney 1993), or "freaks, geeks and cool kids" (Milner 2004). Teens have flocked to social media because they represent an arena to play out these means of status negotiations even when they are away from the school yard. Mediated teen social worlds began with the telephone and continue to today's variegated palette of communications technologies and popular media. Teens use all that is available to craft and display their social identities and interact with their peers. Just as we see in the locker rooms and cafeterias in high schools, online spaces introduce opportunities for kids to display fashion and taste, to gossip, form friendships, flirt, and even harass other peers. While not all teens experience bullying, most struggle with fitting in, standing out, and trying to keep up with what is cool. These dynamics are often described in negative terms, as "peer pressure," but we can also consider them a powerful peer-based learning environment where youth are constructing and picking up social norms, tastes, knowledge, and culture from those around them.

For most teens, social media do not constitute an alternative or "virtual" world (Abbott 1998). They are simply another method to connect with their friends and peers in a way that feels seamless with their everyday lives (Osgerby 2004). Popular social media[3] such as instant messaging, mobile phones, and social network sites are used interchangeably by teens for a variety of friendship-driven practices. At an intimate level, teens use social media to maintain "full-time intimate communities" with their closest friends, just as Misa Matsuda (2005) witnessed in Japan with youth usage of mobile phones. Yet, because of the affordances of media such as social network sites, many teens move beyond small-scale intimate friend groups to build "always-on" networked publics inhabited by their peers. Teens will usually have a small circle of intimate friends with whom they communicate in an always-on mode via mobile phones and IM, and a larger peer group that they are connected to via social network sites. Social media support a wide range of interactions, including those between close friends and those that take place among a broader cohort of peers. Social relations—not simply physical space—structure the social worlds of youth.

The relations and social dynamics that play out in school extend into the spaces created through social media. What takes place online is reproduced and discussed offline (Leander and McKim 2003). When teens are involved in friendship-driven practices, online and offline are not separate worlds—they are simply different settings in which to gather with friends and peers. Conversations may begin in one environment, but they move seamlessly across media so long as the people remain the same. Social media mirror, magnify, and extend everyday social worlds. By and large, teens use social media to do what they have been doing—socialize with friends, negotiate peer groups, flirt, share stories, and simply hang out. At the same time, networked publics provide opportunities for always-on access to peer communication, new kinds of authoring of public identities, public display of connectedness, and access to information about others. In the sections to follow, we describe how these dynamics reinforce existing friendship patterns as well as constitute new kinds of social arrangements.

Box 2.1   Sharing Snapshots of Teen Friendship and Love

Katynka Z. Martínez

It is not uncommon for Stephanie to call Sandra so that they can plan their outfits or hairstyles in anticipation of the next day of school. The two sixteen-year-olds are best friends. They live in a low-income urban area of Los Angeles and attend a public school thirty miles away from home. Stephanie, who identifies as Colombian and Irish, shares a bedroom with her mother. Her twenty-six-year-old brother sleeps in the converted den of their condominium apartment. I met Stephanie at the youth group of a local community center. The center is less than a block away from her home. Stephanie volunteered to take part in a general interview regarding how youth use digital media. She also signed up for a more detailed diary study in which she recorded her use of digital media during the course of two days. Stephanie would receive gift certificates for participating in these interviews. She had the choice of receiving a certificate from iTunes, Amazon.com, or any other online vendor. She opted for a gift certificate from Best Buy, the home-electronics store where she would buy her first digital camera.

Photographs are important artifacts used by youth to capture their participation in teen rituals such as a prom or a *quinceañera* and also to document less formal social escapades with friends. Sandra takes her digital camera to school every day. On the days that she and Stephanie plan their outfits or hairstyles, they make it a point to take photos of themselves that they then post on MySpace. These photos, which they post on their individual profiles, receive many comments from friends. Typical comments include "You look so pretty!" and "This was so much fun!"

Before Stephanie had a digital camera, she would rely on Sandra to take pictures. Stephanie explained, "I have the iPod and she has a digital camera. We just work together." Working together meant that the two girls shared passwords to their Photobucket accounts. Photobucket is an image-hosting and photo-sharing website. Individuals create an online album where they upload photos, videos, and any images they may have found online. Users have the option of setting their album to private (accessible only through a password) or public (accessible to anyone online). Stephanie's and Sandra's Photobucket accounts are set to private, but the girls, as mentioned, have shared their passwords with each other. While Sandra uploads photos that the girls took together, Stephanie searches through public Photobucket albums and uploads images that she may want to share with friends via MySpace. Stephanie accesses Sandra's album, finds pictures of herself, and uploads these onto her own MySpace page. She rarely posts pictures of herself on Friends' pages. However, the images that she finds via public Photobucket albums are eventually posted as comments on her Friends' MySpace pages.

While showing off her Photobucket account, Stephanie proudly proclaimed that she had more than four hundred images in her album. As she described her typical session on Photobucket, it became clear that a shared understanding of friendship and romance was being constructed by her and other Photobucket users:

I save a picture, save a picture, save a picture. How do I decide? Well, the first thing like, you know, girls think about . . . I typed in "love." And then things from *The Notebook* came up. Different things. Then so I liked that so I was like, "Oh, I'll type in 'The Notebook.'" And then I typed in "A Walk to Remember" because, you know, it's another love movie.

Stephanie begins describing her Photobucket activities with the assumption that the first thing girls her age think about is love. After conducting a Photobucket search for the word "love" she finds that many users have tagged the film *The Notebook* with this word. It is not surprising that the film would be associated this way. *The Notebook* won the 2005 MTV Movie Award for Best Kiss, an award that is voted on by MTV viewers. Like those viewers, Stephanie was a fan of the film. However, she also typed in the name of "another love movie," *A Walk to Remember,* and continued typing in modified versions of the word "love" to find additional images. She explained, "If you change the word, it's always different. 'Young love' like to see what comes up. And then I typed in . . . and in 'young love' you saw 'high-school sweethearts.' And then I typed in 'high-school sweethearts.' It all connects."

It does, indeed, "all connect." Sometimes these connections are made by Photobucket users who have used the word "love" to tag snapshots of themselves with their boyfriends or girlfriends. Other times the connection is made by users who use the word "love" to tag stock footage of actors or models displaying trite acts of affection (such as kissing on the beach amid shallow waves). Also common on Photobucket are banners or boxes of text with greetings, sayings, and words of encouragement. For example, a "love" banner states the following in glittered letters: "It only takes a second

2 say I luv u, but a lifetime 2 show it!" Stephanie has many similar banners stored in her Photobucket album and plans to eventually post them on Friends' MySpace pages. She hopes that the "Get Out of Jail Free card" will add humor to the MySpace page of a friend who knows someone who is incarcerated. Stephanie is also storing images for future developments in her friends' lives. She displayed a banner with an inspirational quote and explained, "Like if a guy broke up with my good friend or something, then I'll send her this."

Most of the images in Stephanie's Photobucket album allude to the importance of friendship. For example, one proclaims: "Inside jokes, midnight calls, crazy at night, equals best friends." While going through her album, Stephanie explained, "And then I'll type in 'best friends' and then 'friends' and then 'boyfriend' and then 'girlfriend.' You can go on forever." Sitting and watching Stephanie search for additional images and navigate through the four hundred saved in her photo album, it was easy to see that she very well could "go on forever." The search engine served as a type of thesaurus for Photobucket users. Having witnessed how engrossed she was in these searches, one might wonder if this online quest would also manifest itself in her approach to schoolwork that incorporates online research.

Katynka: And then so do you ever do searches like this, for homework?

Stephanie: For homework?

Katynka: Yeah. Like for a research paper or anything like that?

Stephanie: No.

Katynka: No? Do you use the Internet much for homework or not really?

Stephanie: Kind of. But they make it so hard. Like for English, you can't use Wikipedia. I understand that because whoever could, like, write in whatever. But then they say we can't use websites that have ".com" on the end. Only ".edu." I think they said. Or ".org." So it's hard.

Katynka: Uh-huh. So do you explain the difference to you between ".edu" and . . .

Stephanie: Yeah. For that I will just use, like, the Internet at school because they have this special library thing. I forgot what it's called. I'll show you. "So long and good night," I wrote, I posted on the bulletin. I put: "I'm going to bed now." Because that's when I turn off the computer. "I want what I want." "I want to love somebody like you." "I want to be your favorite hello and your hardest goodbye." "Texting is love." "Cell phone love." "My cell phone is love." "Best friends."

Stephanie never did go to the "special library thing" that she briefly mentioned. Instead, she continued clicking through her album and eventually shared her Photobucket password with me. This openness and collaborative spirit is at odds with her school's approach to online sources of information. The fact that her school has restrictions against referencing Wikipedia frustrates Stephanie but she ultimately understands that the school would take this stance because "whoever could, like, write in whatever." Yet it is precisely this collaborative feature that makes Photobucket so appealing—you are able to see the images that other users have associated with terms such as "love" and "best friends." Many times these images simply reproduce conventional gender roles and a culture of consumption. However, youth are able to pick and choose from among the images and, perhaps most important, contribute their own works—some of which will challenge the representations of teen friendship and love that have been created by outside forces without any understanding of how youth actually negotiate relationships. Youth today are taking portraits at social events, snapping pictures in the halls of their schools, and borrowing from the photo albums of people they've never met. The fact that they draw from all these sources suggests that youth's friendship maintenance is in tune with a discourse of love and friendship that is being widely displayed and (re)circulated.

Making Friends

Teens may select their friends, but their "choice" is configured by the social, cultural, and economic conditions around them (Allan 1998). Studies have shown that most friendships American youth develop are between youth of approximately the same age, in part because of age-stratified school systems and other cultural forces that segregate youth by age (Chudacoff 1989; Montemayor and Van Komen 1980). Likewise, these friendship groups tend to be relatively homogenous (Cohen 1977; Cotterell 1996), resulting in what sociologists call "homophily" (McPherson, Smith-Lovin, and Cook 2001). Homophily describes the likelihood that people connect to others who share their interests and identity. Most of the teens we interviewed tended toward building friendships with others of similar age who shared their interests and values. While teens' friendships were not completely segregated by race, ethnicity, religion, and gender, none of these factors was absent either.

Social media theoretically allow teens to move beyond geographic restrictions and connect with new people. Presumably, this means that participants could develop relations with people who are quite different from them. Research that tests this premise is sparse. One survey of Israeli teens suggests that those who develop friendships online tend toward less homogenous connections than teens who do not build such connections (Mesch and Talmud 2007). While this suggests tremendous possibilities, developing friendships online is not a normative practice, at least not for U.S. teens. Surveys of U.S. teens indicate that most teens use social media to socialize with people they already know or are already loosely connected with (Lenhart and Madden 2007; Subrahmanyam and Greenfield 2008).

Even though MySpace is commonly viewed as a site for networking with new people, teens consistently underscored that this is not what they do. For example, Sabrina, a white fourteen-year-old from suburban Texas, explained that while she uses MySpace, she never uses it to meet new people. "I just find my friends and hang out." Teens emphasized that IM and social network sites were primarily valuable as media for socializing with those they knew from school, worship centers, summer camps, and other activities.

This is not to say that teens do not leverage social media to develop friendships. Teens frequently use social media as additional channels of communication to get to know classmates and turn acquaintances into friendships. Melanie, a white fifteen-year-old from Kansas, explained, "Facebook makes it easier to talk to people at school that you may not see a lot or know very well." She found Facebook to be helpful in getting to know some of her classmates. Social network site profiles can also become valuable tools for learning more about acquaintances. Carlos, a Latino seventeen-year-old, told Dan Perkel (MySpace Profile Production) how MySpace allowed him to learn that a boy who lived up the street was really into skydiving. This prompted a conversation between Carlos and the neighborhood boy, who then invited Carlos to go skydiving, but Carlos was not old enough. While both Melanie and Carlos used social network sites to make friends, these other teens were already members of their social circles; they simply did not know them very well. Teens often use social media to make or develop friendships, but they do so almost exclusively with acquaintances or friends of friends (see figure 2.1).

While the dominant and normative social media usage pattern is to connect with friends, family, and acquaintances, there are some teens who use social media to develop connections with strangers. Some teens— especially marginalized and ostracized ones—often relish the opportunity

Figure 2.1
Teens socializing online and off-line. "MySpacing" photo courtesy of Luke Brassard, 2006, http://www.flickr.com/photos/brassard/138829152.

to find connections beyond their schools. Teens who are driven by specific interests that may not be supported by their schools, such as those described in chapters 5 and 6, "Gaming" and "Creative Production," often build relationships with others online through shared practice. Likewise, many lesbian, gay, bisexual, and transgender (LGBT) teens who feel isolated at school often find social media valuable in making social connections with other LGBT youth (Gray 2009). In addition to these interest- and identity-driven motivations for building connections, some teens connect with strangers precisely because they are strangers. One of the boys Christo Sims spoke with in his "Rural and Urban Youth" study valued the opportunity to talk anonymously with other youth without facing social consequences (see box 3.2). Social media allowed him to discuss intimate matters—such as going through puberty—that would be difficult to bring up in the local context for fear of embarrassing himself and damaging his local—and persistent—reputation. He was not interested in meeting his Internet friends or connecting them to his everyday peer group, but he valued the social support he gained through these connections.

While there are plenty of teens who relish the opportunity to make new connections through social media, this practice is heavily stigmatized. Jessica, a college freshman who participated in the "Freshquest" study, told Megan Finn that she had been very shy in middle school so she started meeting people through IM. While she made a close friend that way, she believes that such connections are rare—"I don't know anyone that has any Internet friends." She also highlights that her classmates think she's "weird" and label her a "freak" for meeting people online.

The stigma that Jessica faces is not simply kid-driven. While there is a stigma for not being able to make friends at school, developing friends online is further vilified by cultural fears that meeting people online is dangerous. The same "stranger danger" rhetoric and "terror talk" that limit youth from interacting with strangers in unmediated public spaces (Levine 2002; Valentine 2004) also have taken hold for online spaces. There are school assemblies dedicated to online dangers, primarily the possibility of sexual predators. Mainstream media, law enforcement, teachers, and parents reinforce the message that interacting with strangers online is risky. While the percentage of teens who have experienced unwanted sexual solicitations has declined through the years (Wolak, Mitchell, and Finkelhor 2006), the fear that youth—and especially girls—are at risk has increased (Cassell and Cramer 2007; Marwick 2008). At a deeper level, the public myths about online "predators" do not reflect the actual realities of sexual solicitation and risky online behavior (Wolak et al. 2008). Not only do unfounded fears limit teenagers unnecessarily but they also obscure preventable problematic behavior (Valentine 2004). During the tenure of our project, we watched as this stigma was amplified by a moral panic that formed around MySpace.

While social media have the potential to radically alter friendship-making processes, most teens use these tools to maintain preexisting connections, turn acquaintances into friendships, and develop connections through people they already know. Social media offer a platform for teens to take friendships to a new level. Those teens who seek new friends through networked media are a minority, often because developing online connections is stigmatized and set against a backdrop of adult fears of stranger danger and mainstream youth norms that center on school-centered sociability. Even against this backdrop, some teens value the opportunity to gain social support that they cannot find locally.

Box 2.2   From MySpace to Facebook: Coming of Age in Networked Public Culture

Heather A. Horst

One of the fundamental shifts in American youth culture revolves around kids' engagement in what has been termed "networked public culture," or "those cultural artifacts associated with 'personal' culture (such as home movies, snapshots, diaries, and scrapbooks) that have now entered the arena of 'public' culture (such as newspapers, cinema, and television)" (Russell et al. 2008). For young adults such as eighteen-year-old Ann, a white teenager living on the outskirts of Silicon Valley, the entrée into networked public culture came through MySpace. Throughout her junior and senior years of high school, Ann was an active MySpace user who uploaded pictures and commented on friends' comments on a daily basis. Ann also participated in what she and her friends called "MySpace parties," or sleepovers that involved dressing up and taking photographs to post on their respective MySpace pages. Ann and her friends enjoyed trying on different clothing, such as short skirts, bra tops, fishnet stockings, or other sexy clothes. They also began to make videos of "funny stuff," such as her friends dancing or imitating celebrities.

After accepting an offer to attend a small liberal-arts college in Washington State, Ann received an invitation from her future dorm's resident assistant (RA) to participate in Facebook, a social network site that (at the time) catered to the college community. Ann's RA sent her an invitation to be a member of the "Crystal Mountain" wing, part of a wider network of ninety dorm residents attending her new college. Ann admitted that in the course of two weeks she was spending hours at a time perusing different people's sites, looking for familiar names and faces and checking out friends of friends. As the summer progressed, Ann increasingly felt that she was becoming "addicted" to Facebook, checking it anytime she had a free moment for status updates (e.g., a change to someone's profile), which was an average of four to five times per day, a typical session lasting about ten minutes. Through this brief, repetitive engagement, Ann started to meet the other students slated to live in her dorm, the most important and exciting of these new connections being her future roommate, Sarah. Describing her fascination with her Facebook page, Ann explained:

And you can see everyone else's dorm room and I have groups. Like everyone in my dorm room is in this group. And you can see all the others . . . and so I can see who my RA is going to be and stuff and so it's really cool. And then I have . . . I can show you my roommate. It's really exciting. So I can see her. And so it . . . I don't know, I can just see a picture of her instead of having to wait and stuff.

During the course of the summer, Ann and her future roommate, Sarah, "poked" each other and sent each other short messages and comments.

Some of these messages were pragmatic, such as when they planned to move into their dorm room, what "stuff" they had, or which classes they planned to take. Alongside using Facebook to facilitate communication, Ann delved into the details of Sarah's Facebook page for insight into what she imagined would be shared interests, the most obvious being her taste in music and media.

> But actually her and I like a lot of the same music, I could tell from her Facebook. And so we were talking about concerts that we've been to this summer and stuff. So I'm sure . . . 'cause she's bringing a TV 'cause she lives in a really, really rich area of Washington. And so I think she's bringing a really nice TV, so I'm like I should probably bring something kind of nice. So I think I'll bring this [iPod speakers] and then we can both hook our iPods up whenever we want. . . . I'm supposed to bring a microwave but I don't think I'll bring a microwave.

More than reflecting shared interests or competitive consumption, Ann's decision about what to bring to college was aligned with a desire to construct an aesthetic balance. Buying new, trendy iPod speakers complements the "really nice TV" Sarah will be contributing to their room. Ann also hoped that the speakers might create an acoustic space wherein Ann and Sarah could hang out and listen to music together. Ann and Sarah decided to upload a few pictures of their bedrooms at home onto their Facebook pages to get a sense of each other's style and tastes. Ann was excited when she looked at the photographs and saw Sarah's signature colors. "I'm brown and pink stuff and she's brown and blue stuff!" Ann surmised that this aesthetic harmony would also signify a harmonious relationship (cf. Clarke 2001; Young 2005).

For Ann, and individuals like her, MySpace and Facebook have played an important role in structuring and sustaining her social worlds, including her ability to imagine her future college life in the dorm and to establish relationships with new individuals and communities. They also have provided Ann with opportunities to understand and assert her own sense of who she is and who she will become in the mediated transition from high school to college. Much like homecoming, prom, and graduation, Facebook, MySpace, and other spaces of networked public culture have now become part and parcel of the coming-of-age process for teenagers in the United States.

Performing Friendships

Small children often seek confirmation of friendships through questions such as "We're friends, right?" (Corsaro 1997, 164). Yet, in everyday life, most youth friendships are never formalized or verified except through implicit social rituals. One of the ways in which social media alter friendship practices is through the forced—and often public—articulation of social connections. From instant-messaging "buddy lists" to the public listing of "Friends" on social network sites, teens are regularly forced to list their connections as part of social media participation. The dynamics surrounding this can directly affect friendship practices.

The articulation of connections in social media serves three purposes. First, these lists operate as an address book, allowing participants to keep a record of all the people they know. Second, they allow participants to leverage privacy settings to control who can access their content, who can contact them, and who can see if they are online or not. Finally, the public display of connections that takes place in social network sites can represent an individual's social identity and status (Donath and boyd 2004).

The practice of creating an "address book" is common across many genres of social media. With email address books and mobile phone contacts lists, the collection of relations is simply meant as a reference tool to help the participant remember another person's email address or phone number. Because these are never made visible nor are people required to approve of address book inclusion, address books are little more than a reference tool.

With IM, buddy lists are both references and the initial site of interaction. Buddy lists display a person's contacts as well as a variety of presence information about online and idle status as well as "away messages" that convey additional personal and contextual information (Baron 2008; Grinter, Palen, and Eldridge 2006). Social network sites take this one step further by displaying the list of connections on a person's profile in a way that is visible to anyone who can view that profile. On social network sites, "Friends" end up serving as a part of a person's self-representation on the site as well as the foundation of access control to certain features (e.g., commenting) and content (e.g., blog posts). Teens use Friends to enact their identity (Livingstone 2008) and imagine the social context (boyd 2006).

"Friends" in the context of social media are not necessarily the same as "friends" in the everyday sense (boyd 2006).[4] Social network sites use the term "Friends" to label all articulated relationships, regardless of intensity or connection type (e.g., family or colleagues). Different challenges are involved in choosing whom to select as Friends. Because Friends are displayed on social network sites, there are social tensions concerning whom to include and whom to exclude. Furthermore, as many IM clients and most social network sites require confirmation for people to list one another,[5] choosing to include someone prompts a "Friend request" that requires the recipient to accept or reject the connection. This introduces another layer of social processing. While teens are developing a set of shared social practices for Friending, the norms for these practices are still in a state of flux and interpretive flexibility, as is characteristic of the early years of adoption of a new technology. Further, the technology capabilities also are evolving in tandem with the development of user practices or norms. Teens' ongoing debate and negotiation over what is socially appropriate, combined with Internet companies' efforts to monitor and regulate these practices, is gradually stabilizing a set of practices for how youth publicly articulate their social relations on social network sites.

Teens have different strategies for choosing whom to mark as Friends. By and large, the teens we interviewed include as Friends those they know—friends, family, peers, and so on. Yet, even within the confines of this general rubric, there is immense variation. Teens may choose to accept requests from peers they know but do not feel close to, if only to avoid offending them. They may also choose to exclude people they know well but do not wish to have present on Facebook or MySpace. This category may include parents, siblings, and teachers.

Both MySpace and Facebook offer many incentives for adding people other than close friends. Many of the privacy features that were introduced during the course of our study limit non-Friends from profile viewing, leaving comments, and, in some cases, sending messages. Teens who wish to talk with peers or friends of friends are encouraged to accept requests from peers so as to open the channel of communication. Likewise, teens who use MySpace to distribute their music think it is important to accept requests from any potential fans.

Teens must determine their own boundaries concerning whom to accept and whom to reject. For many, this is not easy. In determining boundaries, there are common categories of potential Friends that most teens address in their decision-making process. The first concerns strangers. While many early adopters of MySpace gregariously welcomed anyone and everyone as Friends, the social norms quickly changed. For most teens, rejecting such requests is now the most common practice. Although teens who accept Friend requests from strangers rarely interact with these people online, let alone offline, the same concerns that keep teens from interacting with strangers online also keep them from including strangers in their lists of Friends. Yet fear is not the only reason teens choose to deny strangers.

Trevor, a seventeen-year-old working-class white boy from a suburb in northern California, says he added only people he knew in the physical world to his Friends list on MySpace because "I don't want anyone on here that I don't know" (C. J. Pascoe, Living Digital). By denying strangers, Trevor reinforces MySpace's claim that it is "a place for friends." He thinks that people who accept requests from those they do not know are trying "to seem more popular to themselves." Trevor is not alone in his criticism of those who are open with their Friends lists. Mark, a white fifteen-year-old from Seattle, complains that "there's all these people that judge [MySpace] as a popularity contest and just go around adding anyone that they barely even know just so they can have like, you know, 500,000 friends just because it's cool. I think that's stupid, personally." Those who collect large numbers of Friends on MySpace are derogatively called "MySpace whores." While this term is both gendered and sexualized in nature and those loaded references are sometimes intended, it is applied to both boys and girls and refers to attention seekers of all types, not just those seeking sexual attention.

The vast majority of those who collect large numbers of Friends are adults—musicians, politicians, corporations, and both real and wannabe celebrities. Teen musicians and activists sometimes collect Friends for the same purposes as public-facing adults—to connect with fans and develop a following. Teens also do so as a form of entertainment or competition among friends. These teens are not interested in developing friendships with those they include as Friends; they simply collect them because it is something to do. One boy said that it is fun to see which attractive women would say yes to his Friend requests. Collecting attractive women is so common that spammers started making fake profiles of attractive women to lure men.

Mass Friend collecting is just one of the practices of connecting with strangers. Teens commonly send Friend requests to bands and celebrities. Teens do not believe that such connections indicate an actual or potential friendship, but they still find value in these Friends. Bands and celebrities frequently send messages—and sometimes VIP opportunities—to the fans who are their Friends. Teens enjoy receiving these, value the occasional comment, and sometimes enjoy connecting with other fans by leaving comments themselves. Such connections serve as a public display of taste and identity (Donath and boyd 2004).

Teens also use the Friending feature to build communities based on specific affiliations. For example, Christo Sims (Rural and Urban Youth) interviewed a sixteen-year-old Haitian American girl from Brooklyn, New York, named Ono who accepts all Friend requests from people who are Haitian "because I'm from Haiti, and I want to keep all the Haitians together." By making these connections, Ono is able to inhabit a community on MySpace that is dominated by Haitians from all over the world. She may not build personal relationships with these people, but connecting to them allows her to participate in a networked public of people like her.

While most teens who connect with strangers have no expectation of building a relationship out of this performed connection, there are teens who happily add people to whom they are attracted in the hopes that one of these connections might develop into something more. This practice is often controversial, both in adult and youth worlds. Adults are concerned that this opens the door for pedophiles pretending to be teens, even though the data show that deception is virtually nonexistent on the rare occasions in which sexual solicitation occurs through these sites (Wolak et al. 2008). Also, many teens—especially girls—think talking to any stranger is risky, as it exposes them to unknown adults as well as to fellow teens who may take an unwanted interest in them.

There is little social cost to rejecting Friend requests from strangers—because these people are unknown, teens do not worry about offending them. Rejecting known individuals, on the other hand, is much more complicated. By and large, the social convention is to accept Friend requests from all known peers, including all friends, acquaintances, and classmates, regardless of the quality of the relationship. Jennifer, a white seventeen-year-old from a small town in Kansas, always accepts requests from people she knows because "I'd feel mean if I didn't." She sees such requests as a sign of niceness and an opening of potential friendship. Additionally, she thinks it is important to be nice because she would be mad if someone rejected her attempt to be nice.

As Jennifer indicates, some teens use the Friend request feature to develop acquaintances into friends. When Bob, a nineteen-year-old white male from rural California, meets someone new, he turns to Facebook to learn more about the person because "it gives you a deeper level of comfort with the person after you meet them" (Sims, Rural and Urban Youth). Using social network site profiles to research someone's tastes, style, affiliation, and social connections provides valuable conversation fodder in addition to offering signs of potential friendship compatibility. Furthermore, the online communication channels provide a low-cost and casual option for initiating conversations. As Bob explains, becoming Friends on Facebook

sets up your relationship for the next time you meet them to have them be a bigger part of your life. . . . Suddenly they go from somebody you've met once to somebody you met once but also connected with in some weird Facebook way. And now that you've connected, you have to acknowledge each other more in person sometimes.

The ritual of Friending can permit or prompt direct interaction when the teens involved see one another in school or at a group function; it lays the groundwork for building a friendship and gives reason to single the other out from the group and initiate communications. From Bob's point of view, Facebook allows teens to take a new relationship "to the next level immediately."

Bob feels comfortable sending Friend requests to people he does not know well in the hopes of future connections. Yet not all Friend requests from acquaintances are attempts to deepen the relationship. Often teens send requests to everyone they know or recognize and no additional contact is initiated after the Friend request is approved. This only adds to the awkwardness of the Friend request. As Lilly, a white sixteen-year-old from a Kansas City suburb, explains, getting Friend requests from class-mates does not even mean that they know who you are at school, making it difficult to bridge the gap between online and offline.

It's just on Facebook, you're friends. At school, you don't have to talk if you don't want to. . . . It's kind of nice, but then at the same time it's not because you know they're your Friends. . . . You don't say hi in the hall 'cause maybe they just added me because somebody else had me added and they'd be like, "I don't know who you are. Hi."

Lilly accepts requests from all classmates, even those from classmates she barely knows, but her friend Melanie prefers to mock the dynamic that this sets up. Melanie, a white fifteen-year-old, will approach classmates who send her Friend requests with comments such as "Hey Friend from Facebook," simply because she thinks it is funny. Melanie's approach to Facebook is quite unusual. Not only is she willing to call out the absurdity of being Friends online but not talking at school, but she also is willing to buck the norms by rejecting people she does not like and deleting people who annoy her. Melanie notes that Facebook "is better than real life" because while there is no simple mechanism to formally indicate disinterest in school, it is possible to say "no" on Facebook by rejecting Friend requests. Unlike Melanie, who is comfortable deleting Friends who annoy her on Facebook, most teens find deleting people discomforting and inappropriate. Penelope, a white fifteen-year-old from Nebraska, says that deleting a Friend is "rude . . . unless they're weird." Yet, while she will do it occasionally, the process of deleting someone is "scary" to Penelope; she fears that she will offend someone.

Generally, it is socially unacceptable to delete a Friend one knows. When this is done, it is primarily after a fight or breakup. In these situations, the act of deletion is spiteful and intentionally designed to hurt the other person. Teen awareness of malicious deletions adds to the general sense that deleting someone is socially inappropriate. Thus, it can be problematic when teens accidentally delete people they know. Ana-Garcia, a fifteen-year-old half-Indian, half-Guatemalan girl from Los Angeles, faced this problem when her brother decided to log in to her account and delete two pages' worth of Friends. Luckily, those she did know understood as soon as she explained what happened. Gabbie, a seventeen-year-old Chinese girl from a suburb in northern California whom C. J. Pascoe (Living Digital) interviewed, found herself on the opposite side. Her feelings were initially hurt when her friend deleted her, but she confronted him and learned that he did it by mistake. "I just asked him, I was like, 'Why did you delete me?' And he was like, 'I didn't know!' So he added me on. But he's one of my closest friends."

While deleting known people can be seen as malicious, it is socially acceptable to choose to move from an open profile to a closed one and delete strangers. In fact, this is often encouraged. Lolo, a Latina fifteen-year-old from Los Angeles, says: "At the beginning, I was just adding people just to get friends and just random boys living in New York or Texas. Then my boyfriend kinda like, 'You don't know them. You don't know them,' so I deleted them and then I had three hundred and I really knew them." Deleting strangers, like rejecting their initial Friend requests, is viewed as having no social repercussions.

The Friends feature forces teens to navigate their social lives in new ways. Although youth are in a process of actively negotiating the underlying social practices and norms for displaying friendship online, we have observed an emerging consensus about socially appropriate behavior that largely mirrors what is socially appropriate in offline contexts. The process of adding and deleting Friends is a core element of participation on social network sites. It allows teens to negotiate who can gain access to their content, but it also means that teens have to manage the social implications of their decisions. Because the peer groups that teens connect with on social network sites are the same as those they socialize with in everyday life, decisions about whom to accept and whom to reject online directly affect their offline connections. By facing decisions about how to circumscribe their Friends lists, teens are forced to consider their relationships, the dynamics of their peer group, and the ways in which their decisions may affect others. These processes make social status and friendship more explicit and public, providing a broader set of contexts for observing these informal forms of social-evaluation learning. It makes peer negotiations visible in new ways, leading to heightened stakes as well as opportunities to observe and learn about social norms from their peers.

Friendship Hierarchies

A Friend connection alone says nothing about its strength. By accepting all acquaintances as Friends, teens can avoid offending peers who might believe there to be a stronger connection. Yet an additional feature on MySpace—"Top Friends" (formerly "Top 8")—complicates matters by forcing teens to indicate whom they are closest with among their Friends. While MySpace designed this feature to allow participants to showcase their actual close friends, many teens highlight that this feature is the crux of what makes MySpace filled with social drama. These practices of displaying friendship hierarchies online are controversial and more fraught than the simple articulation of Friend connections.

Rhetoric such as "best friends forever" ("BFF") is common among children, especially young girls (Thompson, Grace, and Cohen 2001, 62). This

stems from a desire by children to understand the strength of their relationships and embedded in this is an expectation of affirmation and reciprocity. Most friendship declarations take place verbally between friends,
but girls have used symbolic accessories such as "BFF" heart charms and
friendship bracelets to formalize and display their connection. While these
practices exist, they are far more common with elementary-school children
and middle-school tweens than with teenagers. The idea of "best friends"
does not disappear in high school, but the formal symbolism fades.

In many ways, MySpace's Top Friends forces teenagers to publicly articulate their best and "bestest" Friends. This feature requires participants to
list up to twenty-four Friends' names in a grid. Designed to help participants add nuance to their Friends list, this feature quickly became a social
battleground as participants struggled over who should make the list and,
more important, who should be in the first position. Anindita, an Indian
seventeen-year-old from Los Angeles, explains:

> People will be like, "Why am I number two? You're number one on my page." I was
> like, "Well, I can't make everyone number two. That's impossible." Especially with
> boyfriends and girlfriends, get in a fight like, "Why is she before me? I'm your
> girlfriend. I should be higher than her." I'm just like, "Okay." I don't really think
> it's a big deal, the top thing. If you're friends, you shouldn't lose your friendship
> over that.

Like many teens, Anindita finds the social dynamic around Top Friends
annoying. Yet she is not immune to its effects. Even though she thinks it
should not be important, it is a topic of regular conversation among her
friends. While Anindita may see her friends' attitude as cattiness, Top
Friends surfaces insecurities by forcing teens to face where they stand in
the eyes of those around them. As Nora, a white eighteen-year-old from
Virginia, explains on her MySpace: "It's like have you noticed that you
may have someone in your Top 8 but you're not in theirs and you kinda
think to yourself that you're not as important to that person as they are
to you . . . and oh, to be in the coveted number one spot!" Many teens see
the Top Friends feature as a litmus test of their relations and this prompts
anxieties in teens about where they stand.

Reciprocity plays a central role in the negotiation of Top Friends.
Many teens expect that if they list someone as a Top Friend, that person
should list them in return. Teens worry about not being listed and
about failing to list those who list them. Jordan, a biracial Mexican-white

fifteen-year-old from Austin, Texas, says: "Oh, it's so stressful because if you're in someone else's [Top Friends] then you feel bad if they're not in yours." The struggles that teens face in constructing their Top Friends resemble those involved in choosing whom to invite for a special occasion. Nadine, a white sixteen-year-old from New Jersey, described this on her MySpace:

As a kid, you used your birthday party guest list as leverage on the playground. "If you let me play I'll invite you to my birthday party." Then, as you grew up and got your own phone, it was all about someone being on your speed dial. Well, today it's the MySpace Top 8. It's the new dangling carrot for gaining superficial acceptance. Taking someone off your Top 8 is your new passive-aggressive power play when someone pisses you off.

While there are parallels among Top Friends, speed dial, and the birthday party, there are also differences. Top Friends are persistent, publicly displayed, and easily alterable. This makes it difficult for teens to avoid the issue or make excuses such as "I forgot." When pressured to include someone, teens often oblige or attempt to ward off this interaction by listing those who list them. Catalina, a white fifteen-year-old from Austin, Texas, says: "If you're in someone else's, you have to put them in yours." Other teens avoid this struggle by listing only bands or family members. While teens may get jealous if other peers are listed, family members are exempt from the comparative urge. This is the strategy that Traviesa, a Hispanic fifteen-year-old from the Los Angeles area, takes to avoid social drama with her friends:

It's very difficult to choose a Top 8 because when you do, your friends are like, "Well, why didn't you choose me?" And this and that, and I'm like, "Well, all right fine, I'll just choose," like I choose my cousins now because I can't deal with it. Like everybody's always like, "Why didn't you put me on, why am I not on your Top 8? You're on mine."

In addition to having to decide whom to include, teens must also decide in what order those Friends are listed. Zelda, a fourteen-year-old boy from Brooklyn who was born in Trinidad, told Christo Sims (Rural and Urban Youth):

It's just your best friends; you just put them in the top whatever. If you had a girl-friend or a boyfriend, you put them first. And, then, you just go down like people that you're cool with and then people who are just normal friends. It just keeps on going down. But, it's mostly, if the people who you're really friends with, they stay at the top. And, then, sometimes, because people will be, they get mad 'cause they're like, "Oh, I'm not your friend. I'm not your best friend."

The most valuable position—the "first"—is the one in the upper left corner of the grid. This position is usually reserved for a person's "best" friend, significant other, or a close family member. While few object to a significant other's appearing first, some teens, especially girls, get jealous when other same-sex peers are listed above them on the page of the person they believe to be their closest friend. Exceptions are made for family members and it is common in some teen circles to list family first. While some teens list family to avoid conflict with friends, others do so because they see a family member as their closest friend. This is exemplified by Laura, a white seventeen-year-old with Native American roots from suburban Washington State, who said: "My sister is in position number one because she is one of my best friends and she will be there for me most likely longer than anyone else."

Although most teens find a way to manage the Top Friends feature, others prefer to avoid it altogether. Some intentionally leave Tom Anderson, the site's founder, in the first position while others find more creative solutions. One teen explained that she changed her Top Friends every month, creating themes such as "all Sagittarius Friends." After getting frustrated with the resultant social drama, Amy, a biracial black-white sixteen-year-old from Seattle, found code that allowed her to not display her Top Friends on her profile, and, thus, no one could be upset with her. While Amy's approach is uncommon, it highlights the power of this feature in shaping how teens interact with the site.

Not all teens participate in the social dramas that result from Top Friends, but it does cause tremendous consternation for many. The Top Friends feature is a good example of how structural aspects of software can force articulations that do not map well to how offline social behavior works. Top Friends suggests a single, context-free, hierarchical ranking of friends and a hard cut between "Top" friends and everyone else. This results in social drama for multiple reasons. First, teens do not necessarily think of their friends as hierarchically ranked, but the technology forces this ranking. Second, teens might feel closer to different friends in different contexts and along different dimensions. Friends from a sports team might be different from friends in geometry class. All those situational distinctions are erased in the Top Friends feature. As a result, friends from different contexts are forced into a single spot for comparison. Finally, people might feel close to some friends because they get them invited to parties and close to other friends because they help them with their homework.

Because of the ways in which Top Friends collapses the complexities of social relations and hierarchies, teens have developed a variety of social norms to govern what is and is not appropriate. While common practices ease some tensions, the Top Friends feature still causes anxieties and social pressures. Most of these stabilize through time but not without a few battle scars.

The process of articulating and ranking Friends is one of the ways in which social media take what is normally implicit and make it explicit. When teens are enmeshed in dramas about social categories, cliques, and popularity, the forced nature of Friending can be turbulent. Like the practices of accepting or rejecting Friend requests, the practices of ranking Friends translates certain forms of social connectedness into an online representation. The problem with explicit ranking, however, is that it creates or accentuates hierarchies where they did not exist offline, or were deliberately and strategically ambiguous, thus forcing a new set of social-status negotiations. The give-and-take over these forms of social ranking is an example of how social norms are being negotiated in tandem with the adoption of new technologies, and how peers give ongoing feedback to one another as part of these struggles to develop new cultural standards.

Status, Attention, and Drama

The issue of whom one is friends with, and whom one is "best friends" with, is embedded in a broader set of struggles over status among peers at school (Milner 2004). Because social media are used in a variety of friendship-driven practices, they are also home to the struggles that occur as a natural part of this process. Teens use social media to develop and maintain friendships, but they also use them to seek attention and generate drama. Often the motivation behind the latter is to relieve insecurities about popularity and friendship. While teen dramas are only one component of friendship, they often are made extremely visible by social media. The persistent and networked qualities of social media alter the ways that these dramas play out in teen life. For this reason, it is important to pay special attention to the role that social media play in the negotiation of teen status.

Teens seeking to spread rumors or engage in drama often use social media. These acts may be lightweight parts of everyday teen life or they may snowball in magnitude and become acts of bullying. Regardless of the intensity, our research shows that the acts of drama involving social media are primarily a continuation of broader dramas. Stan, a white eighteen-year-old from Iowa, said: "You'd actually be surprised how little things change. I'm guessing a lot of the drama is still the same; it's just the format is a little different. It's just changing the font and changing the background color really." While the underlying practices may be the same, Michael, a white seventeen-year-old from Seattle, pointed out that social media amplify dramas because they extend social worlds beyond the school.

MySpace is a huge drama maker, but when you stick a lot of people in one thing, then it's . . . it always causes drama. 'Cause, like . . . MySpace is, like, a really big school . . . school's filled with drama. MySpace is filled with drama. It's just when you get people together like that, that's just how life works and stuff.

Properties of social media can alter the visibility of these acts, making them more persistent and more difficult for participants to get a complete picture of what's happening or interpret the acts accurately.

Gossip and rumors have played a role in teen struggles for status and attention since well before social media entered the scene (Milner 2004). When teens gather with friends and peers, they share stories about other friends and peers. New communication channels—including mobile phones, IM, and social network sites—have all been used for the purposes of gossip. Some teens believe that the new media tend to replace the older media as a tool for gossip. Trevor, a white seventeen-year-old from a northern California suburb in C. J. Pascoe's "Living Digital" study, argued that "the Internet has taken the place of phones . . . it spreads all rumors and gossip."

While it is unclear whether or not the Internet has changed the frequency of gossip, social media certainly alter the efficiency and potential scale of interactions. Because of this, there is greater potential for gossip to spread much farther and at a faster pace, making social media a catalyst in teen drama. While teen gossip predates the Internet, some teens blame the technologies for their roles in making gossip easier and more viral. Elena, a sixteen-year-old girl from Armenia who was adopted by a Mormon family in suburban northern California, explained:

And the thing on a lot of MySpace is it brings a lot of drama. A lot of drama. Because it's like, oh, well, "Jessica said something about you." "Oh, really?" "Yeah, we heard it from this girl, Alicia." So then you click on Jessica and talk about comments that Alicia did and then you go from Alicia to her friends. It's this whole going around. And then I'm like, "I was on Alicia's email last night and she's saying this about you." It just gets really out of control, I think. And you're in everyone's business. . . . That's what happened with me and my friends. We got into a lot of drama with it and I was like, anyone can write anything. It can be fact, fiction. Most people, what they read they believe. Even if it's not true. (C. J. Pascoe, Living Digital)

Social media provide another stage on which dramas can be played out. Some of these dramas are truly dramatic, while others are mundane parts of everyday life. When content is persistent (e.g., comments on social network sites), teens can gain access to the content even when they were not present for the situation being referenced. The public nature of social network sites, in particular, makes it much easier for teens to "overhear" what is being said. Furthermore, because teens' presence as observers may not be noticeable online, social network sites can allow them to "stalk" their peers, keeping up with the gossip and lives of people they do not know well but with whom they are familiar. Penelope, a white fifteen-year-old from Nebraska, said: "If [the popular kids are] having a fight you know about it. They confront each other. They say, 'Well, if you're going to leave a comment like that on her page then you'd better send a comment to everybody because this is a war,' or something like that."

While teens can surf through their MySpace or Facebook Friends' profiles to read their comments, Facebook introduced a feature in September 2006 that made this process much easier: the News Feed. When teens log in to their Facebook, they are presented with a News Feed that lists actions taken by their Friends on the site. Some of the actions that are announced on the News Feed include when two people become Friends, when someone leaves a comment on someone else's wall, when a Friend uploads new photos, and when two people break up. Although teens can opt out of this, many of them do not, either because they do not know about the option or because the juicy updates are too alluring.

Cachi, a Puerto Rican eighteen-year-old from Iowa, finds the News Feed useful "because it helps you to see who's keeping track of who and who's talking to who." She enjoys knowing when two people break up so that she knows why someone is upset or when she should reach out to offer support. Knowing this information also prevents awkward conversations that might reference the new ex. While she loves the ability to keep up with the lives of her peers, she also realizes that this means that "everybody knows your business."

Some teens find the News Feed annoying or irrelevant. Gadil, an Indian sixteen-year-old from Los Angeles, thinks that it is impersonal, while others think it is downright creepy. For Tara, a Vietnamese sixteen-year-old from Michigan, the News Feed takes what was public and makes it more public: "Facebook's already public. I think it makes it way too like stalker-ish." Her eighteen-year-old sister, Lila, concurs and pointed out that it gets "rumors going faster." Kat, a white fourteen-year-old from Salem, Massachusetts, uses Facebook's privacy settings to hide stories from the News Feed for the sake of appearances.

As a feature that amplifies public acts, Facebook's News Feed helps rumors posted publicly to spread farther faster. Yet, according to the teens we interviewed, the vast majority of rumors spread through more private channels such as IM and text messaging. IM allows teens to converse with multiple people at once as well as copy and paste conversations to spread information. Through forwarding, text messaging can help create gossip chains. Thus, even though these channels may be more "private," information can become public through incessant sharing.

While gossip is fairly universal among teens, the rumors that are spread can be quite hurtful. Some of these escalate to the level of bullying. We are unable to assess whether or not bullying is on the rise because of social media. Other scholars have found that most teens do not experience Internet-driven harassment (Wolak, Mitchell, and Finkelhor 2007). Those who do may not fit the traditional profile of those who experience school-based bullying (Ybarra, Diener-West, and Leaf 2007), but harassment, both mediated and unmediated, is linked to a myriad of psychosocial issues that include substance use and school problems (Hinduja and Patchin 2008; Ybarra, Diener-West, and Leaf 2007).

Measuring "cyberbullying," or Internet harassment, is difficult, in part because both scholars and teens struggle to define it. The teens we

interviewed spoke regularly of "drama" or "gossip" or "rumors," but few
used the language of "bullying" or "harassment" unless we introduced
these terms. When Sasha, a white sixteen-year-old from Michigan, was
asked specifically about whether or not rumors were bullying, she said:

> I don't know, people at school, they don't realize when they are bullying a lot of
> the time nowadays because it's not so much physical anymore. It's more like you
> think you're joking around with someone in school but it's really hurting them.
> Like you think it's a funny inside joke between you two, but it's really hurtful to
> them, and you can't realize it anymore.

Sasha, like many of the teens we interviewed, saw rumors as hurtful, but
she was not sure if they were bullying. Some teens saw bullying as
being about physical harm; others saw it as premeditated, intentionally
malicious, and sustained in nature. While all acknowledged that it could
take place online, the teens we interviewed thought that most bullying
took place offline, even if they talked about how drama was happening
online.

When teens told us about being bullied, they did not focus on the
technology. They were distressed that others—often former friends—were
maliciously spreading rumors about them to others at school. For example,
Summer, a white fifteen-year-old from Michigan, described how her best
friend decided to reject her because she was not popular enough. Her
former friend began by spreading secrets, but these quickly got modified
and exaggerated as they spread. Summer did not know how the rumors
were spreading, but she knew that everyone in school knew them fast and
that many believed them. In Summer's eyes, the bullying that she experienced took place offline. Yet she also acknowledged that IM was extremely
popular among her classmates at the time. It is likely that some of the
rumors had spread through IM or phone conversations in addition to
conversations in school. For Summer, it did not matter whether it was
online or offline; the result was the same. In handling this, she did not get
offline, but she did switch schools and friend groups.

Media convergence complicates bullying dynamics. Both offline and
online elements played a role in many of the stories we heard. When teens
are harassed online, it is often by people they know offline. Cruelty that
takes place offline is often fueled by mediated rumors. Technology provides
more channels through which youth can potentially bully one another.

That said, most teens we interviewed who discussed being bullied did not focus on the use of technology and did not believe that technology is a significant factor in bullying.

While bullying exists, the teens we interviewed did not see it as commonplace. They did, though, see rumors, drama, and gossip as pervasive. The distinction may be more connected with language and conception than with practice. Bianca, a white sixteen-year-old from Michigan, sees drama as being fueled by her peers' desire to get attention and have something to talk about. She thinks the reason that people create drama is boredom. While drama can be hurtful, many teens see it simply as a part of everyday social life.

The teens we talked with were also quick to point out that most drama and gossip comes primarily from girls, not boys. As Penelope Eckert notes in her study of girls transitioning to middle school, adolescent girls take on the role of "heighteners of the social" (1996). Mark, a white fifteen-year-old from Seattle, explained that drama happens more often with girls "because they always take it more seriously." While girls are more likely to be agents in talking about drama, boys are frequently cited as the cause. A lot of drama that takes place involves crushes, jealousy, and significant others.[6] For example, girls get mad when their friends text message or IM their boyfriends or leave comments on their social network site profiles. In general, using technology to communicate with someone who is not single can be seen as an affront.

Anindita recounted the story of how she stopped speaking to her former best friend, Meghana. Anindita was dating a boy and Meghana started telling him privately to break up with her, even though the girls were supposedly friends. One day, Anindita's boyfriend showed her a text message he received from Meghana. The message read, "You're the guy I love and you don't understand." This angered Anindita and she ended the friendship. From Anindita's point of view, social media took what she saw as typical "Indian drama" and magnified it out of control. She thought that her peers enjoyed the opportunity to start a fight for no reason other than that it was possible.

Although some drama may start out of boredom or entertainment, it is situated in a context where negotiating social relations and school hierarchies is part of everyday life. Teens are dealing daily with sociability and related tensions. Lila, a Vietnamese eighteen-year-old from Michigan, sees drama as the substance of daily life while her sixteen-year-old sister, Tara, thinks that it emerges because some teens do not know how to best negotiate their feelings and the feelings of others.

danah: Do you think that drama has value?

Lila: You have something to talk about. . . . And you're like, you want to fit in, kind of thing. You know, like way back when, when you don't know who you are, kind of. Not like I know now, but you know, when you're in middle school.

Tara: You have something to do, like to be honest, to resolve. . . . You feel like you're mad at somebody and you don't know how to handle it. So you just kind of turn on them like that. So it's just like, just not like having enough experience with dealing with things.

While drama is a part of teen life and Tara and Lila are accepting of it, many teens are insecure about their friendships, unsure of whether or not friends are truly loyal and trustworthy. Social media can feed drama and complicate interactions, especially when things are already heated. At the same time, social media also can be used to try to ease tensions among friends. Teens can use the ability to publicly validate one another on social network sites to reaffirm a friendship. Social media are used also to negotiate attention. Teens use different channels to reassure their friends that they are still thinking of them. So, while drama is common, teens actually spend much more time and effort trying to preserve harmony, reassure friends, and reaffirm relationships. This spirit of reciprocity is common across a wide range of peer-based learning environments we have observed. Trying to be nice when someone else is being nice is one example of how this plays out. Penelope, a white fifteen-year-old from Nebraska, believes in responding to comments because "if someone's nice enough to say something to you then you have to be nice enough to say it back."

Others view the social script of reciprocity from a more cynical point of view, believing that teens are being selfish when they leave a comment. From this perspective, commenting is not as much about being nice as it is about relying on reciprocity for self-gain, as in this example of Christo Sims's interview with Brooklyn-based Derrick, a sixteen-year-old boy who was born in the Dominican Republic:

Christo: Why do you think people put those, the pictures and all that stuff on there?

Derrick: They just MySpace people. . . . That's what MySpace people do. They send each other comments all the time.

Christo: Do you have a sense of why do you think they're doing that, though?

Derrick: That's how they talk to each other, though. They just want to let people know that people talk to them. So if you go to their page you see that they got a lot of comments. That makes them feel like they're popular, that they're getting comments all the time by different people, even people that they don't know. So it makes them feel popular in a way. (Rural and Urban Youth)

While some teens leave comments to be nice, others hope that they will get comments in return. This can be viewed as selfish, but it also can be seen through the lens of insecurity. Many teens worry that they may appear lame if they have too few Friends or too few comments. Some opt out because they fear that these tools would simply highlight the ways in which they are not cool. Alternatively, some who view Friends and comments as markers of social worth believe that they must have many Friends so as not to be alienated from their peers. Kevin, a white fifteen-year-old from Seattle, believes that getting comments is cool "because it lets everyone who goes to your page know that you're not just a guy that has MySpace; you're a guy that has friends and a MySpace."

Successful participation is not simply about having an account on a social network site but about having one with status. Yet insecure and marginalized individuals sometimes seek the markers of cool even if they themselves are not actually perceived as cool. Teens want to be validated by their broader peer group and thus try to present themselves as cool, online and off. Even when status is not necessarily accessible to them in everyday life, there exists hope that they can resolve this through online presentations.

Two of the teens Christo Sims (Rural and Urban Youth) interviewed in Brooklyn spoke about becoming an "Internet gangster," which involves trying to act tough in your profile even if you are shy in person. Shy, a fifteen-year-old Guyanese American girl, and Loud, a seventeen-year-old Jamaican American girl, both see value in getting attention online, even when it is not available offline:

Shy: Like, when you have your MySpace account, you can portray yourself differently than you do on the street. You can picture yourself, somebody that's cool and whatnot on MySpace, and do all these other things to get all the attention that you don't really get when you're with your families or with your . . .

Loud: Or in your school.

While some teens are happy to attain status solely within the context of a social network site, most hope that if they look cool online, their peers will notice and validate them. This is often not successful. Dominic, a white sixteen-year-old from Seattle, said:

> I don't really think popularity would transfer from online to offline because you've got a bunch of random people you don't know; it's not going to make a difference in real life, you know? It's not like they're going to come visit you or hang out with you. You're not a celebrity or something.

Achieving status purely through social network site participation may not be viable, but participating and being popular online can complement offline popularity. Just as having the "right" clothes or listening to the "right" music can be an indicator of status in everyday peer groups, participating in the "right" social media in a manner that is socially recognized is often key to offline status. As with clothes and music, online participation alone is not enough to achieve status, but it is still important.

Gossip, drama, bullying, and posing are unavoidable side effects of teens' everyday negotiations over friendship and peer status. What takes place in this realm resembles much of what took place even before the Internet, but certain features of social media alter the dynamics around these processes. The public, persistent, searchable, and spreadable nature of mediated information affects the way rumors flow and how dramas play out. The explicitness surrounding the display of relationships and online communication can heighten the social stakes and intensity of status negotiation. The scale of this varies, but those who experience mediated harassment are certainly scarred by the process. Further, the ethic of reciprocity embedded in networked publics supports the development of friendships and shared norms, but it also plays into pressures toward conformity and participation in local, school-based peer networks. While there is a dark side to what takes place, teens still relish the friendship opportunities that social media provide.

Conclusion

Social media, and especially social network sites, allow teens to be more carefully attuned, in an ongoing way, to the lives of their friends and peers. Social media are integrally tied to the processes of building, performing, articulating, and developing friendships and status in teen peer networks. Teens value social media because they help them build, maintain, and develop friendships with peers. Social media also play a crucial role in teens' ability to share ideas, cultural artifacts, and emotions with one another. While social warfare and drama do exist, the value of social media rests in their ability to strengthen connections. Teens leverage social media for a variety of practices that are familiar elements of teen life: gossiping, flirting, joking around, and hanging out. Although the underlying practices are quite familiar, the networked, public nature of online communication does inflect these practices in new ways.

First, social media tend to accentuate the longer-burning trend through the past century toward teens' developing social and cultural forms that are segregated from adult society. Although some of the later chapters in this book look at countervailing trends, the mainstream, friendship-driven teen practices covered in this chapter and chapter 3 indicate how same-age cultural forms and sociability are being reinforced by always-on communication networks. Adults' efforts to regulate youth access to MySpace are the latest example of how adults are working to hold on to authority over teen socialization in the face of a gradual erosion of parental influence during the teen years. For the most part, adults participate in these practices as provisioners of infrastructures and as monitors, not as competent peers or coparticipants. Youth are developing new norms and social competencies that are specifically keyed to networked publics, such as how to articulate friendships, how to be polite to their peers, and how to create, mediate, or avoid drama. For youth who hope to succeed socially in their school-based peer networks, these kinds of new media literacies are becoming crucial to youth's participation. Given the prominence of social media in both contemporary teen and adult life, learning how to manage the unique affordances of networked sociality can help teens navigate future collegiate and professional spheres where mediated interactions are assumed.

Second, the particular properties of networked publics (e.g., persistence, searchability, replicability, and scalability [boyd 2008]) mean that certain forms of sociability are reinforced and heightened. Teens are able to keep in closer and ongoing touch with one another and to support the relationships that they are nurturing in their local peer-based networks, which most see as their primary source of identity and affiliation. They develop "always-on intimate communities" with their broader peer group. However, articulating those friendships online means that they become subject to public scrutiny in new ways; teens are able to display new dimensions of themselves but they also may have their self-representations reframed by others in a public way. This makes lessons about social life (both the failures and successes) more consequential and persistent. While these dynamics have played out through fashion, appropriating spaces and lunchrooms at school, or congregating with friends in public spaces such as the mall, social network sites make these dynamics visible in a more persistent and accessible public arena.

Social media mirror and magnify teen friendship practices. Positive interactions are enhanced through social media while negative interactions are also intensified. Teens who are growing older together with social media are coconstructing new sets of social norms with their peers and through the efforts of technology developers. The dynamics of social reciprocity and negotiations over popularity and status all are being supported by participation in publics of the networked variety as formative influences in teen life. While we see no indication that social media are changing the fundamental nature of these friendship practices, we do see differences in the intensity of engagement among peers, and conversely, in the relative alienation of parents and teachers from these social worlds. Youth continue to experience their teenage years as a time to immerse themselves in these peer-based status negotiations and to develop their social and cultural identities in ways that are independent from their parents, and they are aided now in these practices by a new suite of communication tools.

Notes

1. http://headrush.typepad.com/creating_passionate_users/2006/03/ultrafast_relea. html.

2. For an overview of social network sites and their history, see boyd and Ellison (2007).

3. We use the term "social media" to refer to the set of new media that enable social interaction between participants, often through the sharing of media. Although all media are in some ways social, the term "social media" came into common usage in 2005 as a term referencing a central component of what is frequently called "Web 2.0" (O'Reilly 2005 at http://www.oreillynet.com/pub/a/oreilly/tim/news/2005/09/30/what-is-web-20.html) or the "social web." All these terms refer to the layering of social interaction and online content. Popular genres of social media include instant messaging, blogs, social network sites, and video- and photo-sharing sites.

4. To distinguish between connections displayed on social network sites and everyday relations (boyd 2006), we capitalize "Friend" when referring to the social network site feature.

5. AOL's IM client (AIM), popular among U.S. teens, does not require this.

6. There is also a large amount of drama between significant others that plays out using social media. This is discussed in more detail in chapter 3.

3 INTIMACY

Lead Author: C. J. Pascoe

"I get out of the shower, get dressed, go to my PC, log on to MSN, and talk to Alice," said seventeen-year-old Jesse about his typical morning routine. At that time of the day, he finds it easier to instant message on MSN than to talk on the phone with seventeen-year-old Alice, his girlfriend. He has to "do my hair" in the morning. "So I go back and forth, back and forth," he said, miming his movements from the bathroom mirror to the computer in his bedroom. After logging off IM, the couple might talk on their mobile phones as they commute to school. During the school day they trade text messages about their whereabouts and plans, such as "Im in da band room."[1] After school Alice might join Jesse at his house, completing her homework while he plays his favorite video game, Final Fantasy, or they might continue to communicate by sending messages, such as "I'll be here for a while, go to sleep, I love you." The day frequently ends late, with Alice falling asleep talking on the phone to Jesse in the bedroom she shares with her two younger siblings as they watch DVDs on the bottom bunk. Though they have been dating for more than a year, Alice's parents, Chinese immigrants, do not know she and Jesse, a charming young man of mixed Anglo and African-American heritage, are a couple. Their secret relationship has been shaped and, in some ways, made possible, by the profusion of new communication technologies.

Though most teens do not carry on long-term relationships such as this one outside the purview of their parents, Alice's and Jesse's use of new media exemplifies much of what we have heard from our participants about their new media use in intimate interactions. Young people are at the forefront of developing, using, reworking, and incorporating new media into their dating practices in ways that might be unknown, unfamiliar, and sometimes scary to adults. In our interviews and observations, it has become increasingly clear that, much like in their friendship practices, teens have put new media tools to use in their courtship practices such as meeting, flirting, going out, and breaking up. This intimacy-oriented new media use exemplifies another type of friendship-driven technology practice introduced in chapter 2.

Like chapter 2, this chapter focuses on teenagers' normative new media practices. Because dating and romance are primarily teenage (as opposed to childhood) endeavors, most of the interviews are with teenagers between the ages of fourteen and nineteen and the material comes predominantly from studies that focus on friendship-driven sociability: C. J. Pascoe's study "Living Digital," danah boyd's study "Teen Sociality in Networked Publics," Christo Sims's study "Rural and Urban Youth," and Megan Finn, David Schlossberg, Judd Antin, and Paul Poling's study "Freshquest." Unless otherwise noted, the examples in this chapter come from Pascoe's study.

In this chapter we explore teens' normative and nonnormative patterns of intimacy practices and new media. In doing so we sketch out the trajectories of historic and contemporary teen courtship rituals and the ways new media have become a part of these rituals, as well as highlight themes of monitoring, privacy, and vulnerability. Looking at these themes indicates that boundary work is a central part of navigating new media in intimate relationships. These intimacy practices also show how casual, friendship-driven use of new media might be a form of informal learning through which teens develop literacy by building relationships and communicating with their intimates.

Dating, New Media, and Youth

Given that teens have been the developers and shapers of contemporary youth dating culture (Trudell 1993), it makes sense that they would quickly put new media to use in the service of their romantic pursuits. While courtship norms and practices are less formal and more varied than they were in the early and mid-twentieth century, our research on teens' new media use shows that the rituals are no less elaborate or important than those of their historical counterparts.

Dating and courtship, as enacted by contemporary American teens, is largely a twentieth-century development, as is the life stage of adolescence itself (Ben-Amos 1995). After the industrial revolution, when families declined in importance as economic units, romantic unions gradually superseded primarily economic ones as a social norm in the West. Middle- and upper-class young people courted through processes heavily moni- tored by parents, families, and communities in which young men would "call" on young women in their homes (Bogle 2008). Dating, as we now recognize it, emerged out of working-class "calling" practices, in which young ladies lacked the domestic space to entertain young men in their homes and thus the couple would go out somewhere together, a practice referred to in early slang as a "date" (Bogle 2008). As the 1920s progressed, rebellious middle-class youth emulated these working-class rituals (Bogle 2008). These imitations, along with the movement of youth from work- places to public schools, the development of school dances, and the inde- pendence afforded by the spread of automobile ownership, laid the groundwork for contemporary teen dating culture (Modell 1989). In the 1950s teen dating norms were formalized, became close to a universal custom in America, and were solidified by the practice of "going steady" (Bogle 2008; Modell 1989). Youth who "went steady" indicated to onlook- ers that they were unavailable by trading class rings, letter sweaters, ID bracelets, or by wearing matching sweater jackets—their answers, as one historian puts it, to the "wedding ring" (Bogle 2008, 17).

In the 1970s and 1980s, these types of formal dating and "going steady" practices declined as dating became "merely one form of social contact among many" (Modell 1989, 291). The decline in formality is reflected in contemporary teens' language about these types of relationships, which frequently lack a clear vocabulary to define relationship status or practices: "The terms *courtship* and even *dating* have given way to *hanging out* and *going out with someone*" (Miller and Benson 1999, 106). However, the decline in the formality and uniformity of dating practices does not mean that the centrality of romance to teenagers' lives has declined in salience. One study showed that the strongest emotion during puberty was "the specific feeling of being in love" (Miller and Benson 1999, 99), and devel- opmental psychologists consider romantic relationships an essential feature of social development in adolescence (Connolly and Goldberg 1999). Contemporary relationships among teens tend to be "casual, intense and brief" (Brown 1999, 310). They are also, for all their emphasis on privacy and exclusivity, profoundly social (Brown 1999). In adolescence "peers provide opportunities to meet and interact with romantic partners, to initiate and recover from such relationships, and to learn from one's romantic experiences" (Collins and Sroufe 1999, 126). Especially in the early flirtatious stages, "romance is a *public* behavior that provides feedback from friends and age-mates on one's image among one's peers" (Brown 1999, 308).[2] Teens learn about dating, intimacy, and romance from their friends and social circles. Further, while we usually think of these intimacy practices as individual and private, teen romance and dating rituals take place, in many ways, publicly and collectively.

Dating and romance practices and themes, so central to contemporary American teen cultures, not surprisingly are a central part of teens' new media practices (Lenhart and Madden 2007; Oksman and Turtainen 2004). Using social media, contemporary teens continue to craft and reshape dating and romance norms and rituals that are now deeply tied to the development of new media literacies. Social media technologies have provided a more extensive private sphere in which youth can communicate primarily with age-clustered friends, acquaintances, and sometimes strangers outside the purview of their parents or other authority figures. These more private channels of communication have allowed an elaboration of teens' intimacy practices, especially in forming, maintaining, and ending romantic relationships. The familial negotiations over the spheres of privacy in which these practices take place will be elaborated upon in chapter 4.

In their intimacy practices youth use three primary technologies—mobile phones (though many do still use home phones), instant messaging (IM), and social network sites. Mobile phones provide youth a way to maintain private channels of communication, maintain continual contact, and also serve as a "leash" through which teens in a relationship keep "tabs on" one another. Teens use instant-messaging technologies to maintain frequent casual contact with their intimates. As described in chapter 2, social network site profiles are key venues for representations of intimacy, providing a variety of ways to signal the intensity of a given relationship both through textual and visual representations. While most of their online relationships map closely to their offline ones, these digital spaces give teens the ability to reach beyond institutional and geographic constraints to forge romantic relationships. All these technologies allow teens to have frequent and sometimes constant (if passive) contact with one another, something Ito and Okabe call "tele-cocooning in the full-time intimate community" (Ito and Okabe 2005a, 137). Many contemporary teens maintain multiple and constant lines of communication with their intimates over mobile phones, instant-message services, and social network sites, sharing a virtual space that is accessible only by those intimates.

Surprisingly, given its centrality to teen culture, very little has been written about teens' contemporary romance and courtship practices. Researchers have directed their studies of romantic relationships toward adults (Hartup 1999) and focused on teens' sexual practices (e.g., Ashcraft 2006; Martin 1996; Medrano 1994; Moran 2000; Strunin 1994; Trudell 1993). This research orientation likely reflects an American concern with teen sexuality as out of control and dangerous (Schalet 2000). In focusing on teens' intimacy, though not necessarily sexual, practices, we take a sociology-of-youth approach, following the categories and practices important to the teens we talk to, not allowing adult anxieties to guide our research. As a result we report little about teens' sexual experiences. Given the preoccupation with youth sexual practices, not to mention current popular concerns about sex predators and youth exposure to sexual content online, it seems odd to leave sex out of a chapter on intimacy practices. However, we simply did not hear a plethora of stories about sex in our interviews, as youth tended to discuss dating, crushes, romance, and heartbreak. This omission could be due to several factors. First, such intimate details might emerge in a second or third interview, which most researchers did not conduct. Second, we conducted these interviews under constraints imposed by our universities' institutional review boards, which heavily discouraged talking to youth in general and about issues of sex and sexuality in particular. Finally, it may be that intimacy practices were simply more salient to these youth than sexual ones.[3]

So even though romance is one of the focal points of youth popular culture, because of researchers' focus on sex, we know surprisingly little about teen romance, dating, and courtship practices, apart from scattered stories on historical dating practices (Diamond, Savin-Williams, and Dube 1999). This chapter begins to remedy this problem by examining the ways teens talk about their use of new media to craft, pursue, and end intimate relationships. In the first section we trace the practices of contemporary teen courtship and its relationship to the "domestication" of technology, or the way technology defines and is defined by those communities of which it is a part (Hijazi-Omari and Ribak 2008). Teens' stories revealed a set of norms about new media use and intimate relationships. According to most of the teens we talked with, it is appropriate to meet people offline and then pursue the relationship online; if one does meet someone, one should meet that person through friends; one should proceed slowly as he or she corresponds online using the appropriate communication tool; and when breaking up, one should do so in person, or at least over the phone. In the second section we discuss some of the emergent themes about relationships and technology we see from our interviews and observations.

Youth Courtship: Meeting, Flirting, Going Out, and Breaking Up

Liz and Grady, white sixteen-year-olds, sat at the dining room table during our interview, in Liz's family's comfortable middle-class suburban tract home, explaining the role that MySpace played in the origin of their relationship. Grady said that he developed a crush on Liz during the past year, and while he had known her since freshman year, flirting with her in person felt daunting, because, as he put it, "they didn't really talk." Luckily, because they shared a mutual friend, Liz said of her MySpace, "I had him on my Friend list from freshman year . . . and that's how you can be friends, just because your friend knows this guy and you kind of hung out with them, so you're like, 'Okay, I'm going to start talking to you.'" Grady used this loose friendship on MySpace to his advantage: "When I had a crush on her, I made sure I talked to her first in class before I sent her a comment on MySpace." Grady carefully planned his first comment to be casual: "My first comment to her was 'Oh, wow, I didn't know we were Friends on MySpace,'" though of course he knew full well they were Friends. After trading flirtatious messages online, they began dating. Liz and Grady are a fairly typical example of the role new media can play in meeting, flirting, and going out. As Grady put it, it is "easier to talk to them [girls] there" than in person, because one can manage vulnerability through what Christo Sims (2007) has termed a "controlled casualness." Indeed, their process is paradigmatic of teens' contemporary meeting, flirting, and dating practices, in which they can pursue casual offline acquaintances as romantic interests online.

Teens have told us that certain technologies and certain mediated and nonmediated practices are more appropriate for certain types of relationships or relationship stages than are others (Sims 2007). As Christo Sims found in his study, "Rural and Urban Youth," in the initial getting-to-know-you part of a romantic relationship, the asynchronous nature of written communication (private messages and comments on social network sites, text messaging, and the more synchronous IM) allows for slower, more controlled intimacy exploration and development. If a given relationship intensifies (because certainly not all flirtatious relationships do), couples typically shift to phone calls, text, IM, and in-person conversations. Social network sites play an increasingly larger role as couples become solidified and become what some call "Facebook official." At this point in a relationship, teens might indicate relationship status through ordering Friends in a particular hierarchy, changing the formal statement of relationship status, giving gifts, and displaying pictures. Youth can also signal the varying intensity of intimate relationships through new media practices such as sharing passwords, adding Friends, posting bulletins, or changing headlines. When relationships end (for those that do), the public nature and digital representations of these relationships require a sort of digital housecleaning that is new to the world of teen romance, but which has historical corollaries in ridding a bedroom or wallet of an ex-intimate's pictures. In the following section we trace the different types of teen courtship practices and the role of new media in these practices.

Meeting and Flirting

As Grady and Liz's story indicates, digital communication often plays a central role in casual relationships and the early stages of serious relationships. New media have provided a variety of venues for teens to meet and/or further potential romantic interests. Instant messaging, text messages, and social network messaging functions all allow teens to proceed in a way that might feel less vulnerable than face-to-face communication. These multiple lines of communication allow teens to follow up on casual meetings or introduce themselves to someone with whom they have only loose ties, perhaps sharing a mutual friend on- or offline. At present, teens' normative practice is not necessarily meeting strangers online (though that does happen) but rather using these mediated technologies to get to know the friend of a friend or further get to know someone with whom one has had only a casual or brief meeting.

For teens interested in someone they may not know well, the plethora of publicly accessible information on a given individual provides a fresh way to "research," or get to know, those on whom they have a crush. Melanie, a white fifteen-year-old from Kansas, in danah boyd's study "Teen Sociality in Networked Publics" said that she does not "talk to people I have a crush on, but I did look up the Honduran twins in our class. We looked at their MySpace." Like Melanie, a teen can research a crush's interests, likes and dislikes, friendship circles, and online behaviors through his or her publicly available social network profiles. John, a white nineteen-year-old college freshman in Chicago, disclosed that instead of asking for a phone number he will "Facebook stalk them" to discover more, though possibly superficial, information about a girl he has met briefly but finds interesting. Much like teens may have historically researched potential love interests through their friendship networks, contemporary teens have additional new media tools for laying the groundwork for flirting and relationships.

After an initial meeting and possible research on their object of affection, teens often use a social network site or an instant-messenger program to intensify a relationship or get to know another person better. This is what adults might think of as flirting or what teens sometimes call "talking" or "talkin' to" (Bogle 2008; Pascoe 2007a). After an initial meeting, a teen might initiate this "talkin' to" by following up through digital communication. As Sam, a white seventeen-year-old from Iowa, said, "The next step, I guess, in this situation is wall posts[4] [on Facebook]—that's kind of less formal...." (boyd, Teen Sociality in Networked Publics). Sam noted that if he liked a girl he would post "stupid flirty stuff just trying to make her laugh or whatever through Facebook." At this point, teens flirt, proceeding cautiously, indicating that they like each other, trying to gauge the other's feelings while simultaneously not showing too much earnestness.

The asynchronous nature of these technologies allows teens to carefully compose messages that appear to be casual, a "controlled casualness." John, for instance, likes to flirt over IM because it is "easy to get a message across without having to phrase it perfectly" and "because I can think about things more. You can deliberate and answer however you want." Like John, many teens said they often send texts or leave messages on social network sites so they can think about what they are going to say and play off their flirtatiousness if their object of affection does not seem to reciprocate their feelings. Bob, a white nineteen-year-old[5] living in rural northern California, says he carefully edits his grammar and spelling to

give the appearance of an "off-the-cuff" comment. These kinds of deliberately casual messages are evidence of what Naomi Baron (2008) describes
as the "whatever theory of language" supported by online communication,
in which people are increasingly using more informal linguistic forms to
write and communicate. It is important, however, to recognize that these
forms of literacy are not a "dumbing down" of language but a contextually
specific literacy practice, acutely tuned to the particulars of given social
situations and cultural norms.

For example, youth use casual online language to create an intentional
ambiguity. From the outside, sometimes these comments appear so casual
that they might not be read as flirting, such as the following wall posts by
two Filipino teens—Missy and Dustin—who eventually dated quite seriously. After being introduced by mutual friends and communicating
through IM, Missy, a northern Californian sixteen-year-old, wrote on
Dustin's MySpace wall: "hey.. hm wut to say? iono lol/well i left you a
comment . . . u sud feel SPECIAL haha =)."[6] Dustin, a northern Californian
seventeen-year-old, responded a day later by writing on Missy's wall: "hello
there.. umm i dont know what to say but at least i wrote something . . . you
are so G!!!"[7] Both of these comments can be construed as friendly or flirtatious, thus protecting both of the participants should one of the parties
not be romantically drawn to the other. These particular comments took
place in public venues on the participants' walls where others could read
them, providing another layer of casualness and protection.

Generally, though not always, teens prefer to flirt with people online
that they or their friends know or have at least met offline. A minority of
teens we interviewed find meeting potential romantic interests online no
different from meeting or flirting with attractive strangers they might meet
in public, but the general sentiment was that meeting people only online
was "weird," "unnatural," "geeky," or "scary." Ellie, a first-year student at
the University of California, Berkeley, and respondent in Megan Finn and
her colleagues' "Freshquest" study, described her best friend's meeting of
her boyfriend on MySpace as weird: "It was really weird at first. She didn't
want to tell anyone because she thought it was weird too. But they had
such a strong connection that they thought they should meet. And now
they're going out." Grady, Liz's sixteen-year-old boyfriend, said something
similar about meeting girls online: "I'm not going to start a conversation
with a girl on MySpace or text messaging. I'm going to start in person first.

Then it's kind of like weird and geeky, you know?" The reasons vary as to why meeting someone online feels weird to some teens. But they all have, in some way, to do with insecurity about authenticity. Brad, a first-year student at University of California, Berkeley, said, "It doesn't seem natural, I guess. 'Cause you're not actually meeting the person face-to-face" (Finn, Freshquest). It is as if that face-to-face meeting allows one to verify who that other person is before embarking on a relationship with him or her.

If teens do meet initially online, they might use their offline friendship networks to verify the authenticity, safety, and identity of the person with whom they are corresponding. Dana, a Latina fourteen-year-old from Brooklyn, New York, met her boyfriend online through mutual friends. Her best friend's boyfriend's best friend saw her MySpace and

> he requested me 'cause he liked what he seen, and then my best friend talked to him about me and then because of MySpace we were goin' out. . . .' Cause if MySpace wasn't there, then I woulda not had him as boyfriend. We talked on AIM and then we exchange the numbers, and then I met him. I seen him before, but I got him noticed on MySpace and now we're together. (Sims, Rural and Urban Youth)

Like other teens we talked to, Dana and her boyfriend flirted online before they moved to offline communication and eventually met in person. Dana said, "He usually started getting on AIM every day, and I started talking to him from there." They communicated for two days through MySpace until they traded phone numbers and "talked like from twelve to six in the morning." Eventually they met in person in a public space—a local park— in the company of groups of friends. Dana's story is not an uncommon one. Teens regularly meet romantic interests through shared friends in online environments, using these online networks to further offline meet- ings or deepen casual ties to online friends. Teens rely on their networks to do some of the verification work in these online settings.

Though the in-person meeting went well for Dana, other respondents expressed hesitancy about moving online relationships offline for fear that people might not live up to their online personas. John, the Chicago fresh- man, asked, "What happens after you've had a great online flirtatious chat . . . and then the conversation sucks in person?" He experienced this phenomenon firsthand as he transitioned from high school to college. John had used Facebook to add as Friends "the girls you wanted to meet before school started that you thought were hot and wanted to get a head start on." However, once he reached his university in the fall, "you actually saw them and didn't say anything . . . the game was over." When asked why he didn't talk to them in person, he said, "You didn't say anything, because what are you gonna say . . . 'Hey, you're my Facebook Friend?' The key is to meet them in person . . . then Facebook them." Brad, the Berkeley freshman, expressed similar hesitancies about meeting people offline. "You don't know that's who you're meeting. It isn't a smart thing. And you'll end up idolizing the person, thinking they're just this perfect thing. But they probably aren't because no one is perfect. And it's just a big letdown." This "hyperpersonal effect" indicates that intimacy might be heightened online in a way that might not translate seamlessly into offline relationships (Walther 1996).

While most teens express hesitation about meeting people online, in the case of marginalized teens, the Internet allows them to meet other people like themselves (Holloway and Valentine 2003). This sort of digital contact provides a means for youth who didn't feel heard or who felt otherwise disenfranchised in their communities to participate in other ways (Maczewski 2002; Osgerby 2004). For example, Gabbie, a seventeen-year-old first-generation ethnically Chinese teen from California, wanted to find a Chinese boyfriend, but potential suitors were in limited supply in her immediate community. In part because of this desire, she joined the social network site Asiantown.net and struck up communication with a young man she found attractive. "Well, right now I'm talking to this guy. But he has a girlfriend. I don't know. We're just talking as like friends. It seems like he's being a little flirty, but then . . . I don't know." The boy she is talking to lives in the Central Valley, about an hour from where Gabbie lives. We rarely heard teens such as Gabbie, who lack specific offline social circles, talk about moving these relationships offline as being unnatural or weird.

In a similar way, new media also are important tools for gay teens who want to date, because "the biggest obstacle to same-sex dating among sexual minority youth is the identification of potential partners" (Diamond, Savin-Williams, and Dube 1999, 187). It allows them to meet other teens for friendship or dating and affords them a level of independence, as it does for straight teens, to carry on relationships outside the purview of their parents if need be (Hillier and Harrison 2007). The Internet can put gay teens in touch with other teens so that they can have the romantic experiences that their heterosexual counterparts presumably find more readily in offline contexts. Robert, a white seventeen-year-old at a private school in Chicago, became so frustrated about not finding other guys to date through his offline friendship circles that he wrote a Facebook "note" about his difficulties dating as a gay teen:

Every time I have a crush or something, it doesn't work out (he's not gay, not enough time, etc). I'm not a downer, but I'm just realizing that if a straight person's chance of compatibility is 1 in 100. AND only about 3 in 100 are gay, and the compatibility is still 2%, then my prospect is .03 in 100, or 3 in 10,000. That is not very encouraging!

Robert said that a friend set him up on a blind date as a direct result of the announcement he placed on Facebook: "Andrew, another gay guy at my school, and [my] friend, set me up with Matt because he saw my desperate note on Facebook!" Matt and Robert were introduced through Facebook and after the initial setup, Robert was giddy with excitement and said, "We've been texting the past few days a lot; he is really good looking, and a jock, believe it or not, but we seem to really have hit it off. I hope for the best." The two had a very sweet day picked for their first date: Valentine's Day. Much like Dana, Robert found a date through a shared friend. But unlike straight, more mainstream teens, he expressed no hesitancy about meeting in person someone he had met online.

Going Out

Technology also mediates teens' long-term, steady, and committed relationships. Teens in relationships have high expectations of contact with and availability of their significant others as well as expectations that the relationship will be publicly acknowledged through digital media. These expectations of availability are compounded by the "always on" (Baron 2008) possibilities of new media. Additionally, these media help teens reach out beyond their institutional constraints, allowing them to maintain romantic relationships their parents wouldn't necessarily approve of as well as sustain relationships that might be geographically challenging. Like Jesse and Alice, introduced at the beginning of this chapter, teens who are steadily dating frequently text or call each other, post pictures of each other on their social networking sites, rank order their Friends in a particular way, and exchange digitized tokens of affection, signaling to their significant other and their online publics that they are in a relationship.

Being in a relationship increases expectations of availability and reciprocity, which has implications for how teens use new media, given this "always on" potential. In practice this means that youth in a relationship exchange several phone calls, texts, and/or IMs a day. Teens use this intensified contact as a way to differentiate romantic relationships from other relationships—to indicate that their relationship is special or different. Zelda, a Trinidadian American fourteen-year-old from Brooklyn, New York, explained that if one is in a relationship and doesn't respond to a message, the other person will "probably get mad. If they call you and you don't pick up, they probably get mad. If they write a comment on your page you have to comment them back" (Sims, Rural and Urban Youth). He distinguishes this from interacting with friends through digital media: "It's not like it's a normal friend; it's your girlfriend or whatever. You're in a relationship; you're not supposed to just answer whenever you want." As noted in chapter 2, youth have expectations of reciprocity in online communications, and these are heightened in intimate relationships. Teens now do much of their relationship work by using new media—reciprocating in comments, differentiating their romantic attachments from less intimate friends, and giving priority to phone calls from significant others.

To signal to each other that they care and are in an intimate relationship, teens exchange small digitized symbols of affection, much like teens in the 1950s traded rings, jackets, or bracelets. Champ, a nineteen-year-old Latino who also lives in Brooklyn, explained, "Like if she's already your girlfriend, you probably send a little text message, 'Oh I'm thinking of you,' or something like that while she's working. . . . Three times out of the day, you probably send little comments" (Sims, Rural and Urban Youth). These comments are the digital interactional work that cements contemporary teen relationships. Derrick said,

You know in your head you've just got to do it. It's like she writes you a comment; write her a comment back. It's not like a friend thing. It's not like your homeboy just wrote you a comment like "oh man, this kid wrote me a comment again." Write her a comment back. (Sims, Rural and Urban Youth)

Youth do emotional work to maintain a relationship through digitized media. Rather than (though sometimes in addition to) love notes exchanged in between classes, youth demonstrate affection through private and public media channels.

These tokens are part of the interactional relationship work that happens through new media; another is the expectation of availability. Teens find that their significant others expect frequent check-ins, usually by mobile phone. Derrick said,

When you're in a relationship one thing I learned [is] always pick up the phone for your girl because she complains if you don't. . . . The thing about a cell phone when you're a teenager is if you have a cell phone and you don't pick it up you're doing something that you're not supposed to be doing. (Sims, Rural and Urban Youth)

As Christo Sims notes in his research on urban and rural teens, teenagers are expected to account for their whereabouts. They are beholden to parents in this sense but also to significant others, especially in relationships in which trust might be missing or weak. As a result it might be hard to preserve space or time for oneself outside this frequent contact. In fact, Zelda said he knows he needs to answer the phone regularly because if he doesn't, "they probably going to get mad" (Sims, Rural and Urban Youth). The phone especially acts as a sort of leash, a way to keep tabs on a significant other, much like parents keep track of their teens. Teens seemingly endure this leash because of the increased independence afforded them by the phone.

In addition to the expectations of regular, if not continual, contact, teens affirm and are expected to affirm their relationships online, both by and for their significant others and for their networked publics. Zelda underscored the importance of representing relationships online: "You gotta acknowledge on your page that you [are] like with her" (Sims, Rural and Urban Youth). They define and affirm their relationship status, give public tokens of affection, and post pictures. On Facebook, default relationship options are preset, so in addition to indicating an "official" status, teens have creatively developed ways to include nuance and detail in their relationship descriptions. The existing categories hide a variety of relationships and elide the depth or length of a given relationship, so teens sometimes remedy this by indicating the seriousness of a particular relationship through noting its duration, a particularly popular practice among youth interviewed by Christo Sims in Brooklyn, New York. According to Dana, mentioned earlier, couples write a relationship-origin date in their MySpace headline "to show that they have a relationship or something, so like that's showing more, and it shows that he's in a relationship" (Sims, Rural and Urban Youth). The statement of a relationship anniversary is both a signal of intimacy to one's significant other and a hands-off signal to other teens who might be interested in one member of a couple. Nini, a Latina thir-teen-year-old from Brooklyn, said,

If you put the relationship date, whenever you got together, the girls know that you're in a relationship and this is the date, so don't really get into it with the boyfriend, 'cause you are really falling for each other . . . they know that you're a year, so I'm not gonna mess with the boyfriend. (Sims, Rural and Urban Youth)

Nini highlights the "hands off" message, arguing that the length of time a couple has been together indicates the intensity of their relationship to potentially meddlesome outsiders.

Couples typically negotiate offline the act of putting their relationship status online, whether it be a simple "in a relationship" status on Facebook or a more nuanced relationship date on MySpace, notes Christo Sims. Teens dismiss the practice of posting these sorts of public notifications about changes in their relationships through online venues before discuss-ing it with their partner first, usually offline. Joan, a first-year student at the University of California, Berkeley, said,

Yeah, I have friends [who] have confirmed they have gone official with their boy-friends through Facebook, which is ridiculous. I have known people that are dating and they'll get a request "so and so said that you are their girlfriend." They pushed the button and they are like, "Oh my God, we're official." (Finn, Freshquest)

Teens seem to have the sense that this sort of intimate decision should be made interpersonally, not just announced digitally.

The whole of these social network profiles, not just the relationship status, are the digital embodiment of teens' relationships. When in a rela-tionship, teens rank their Friends to indicate the seriousness of their com-mitment. Derrick said that "you probably write something, have her on your Top Friends, don't put other girls, don't have girls write messages to you saying anything crazy. Just to make her feel better" (Sims, Rural and Urban Youth). When teens in a relationship do not rank their Friends in a way that reflects their relationship status—that is, they do not rank their significant other high among their Friends—conflict might result, as it did with Jesse and Alice. Jesse confessed, as he showed off his MySpace site, "Alice was actually not my original top one." Alice paused from her needle-work to jump in the conversation and said, indignantly, "I was like number twelve or something." Jesse, clearly defensive, his voice growing higher, cried, "Does it really matter? You know! Really? My number one? Really?" Alice responded a little sarcastically, rolling her eyes, "Like he's not number one on my account." Clearly, it was not the first time they had had this discussion. While for these two teens, the tension did not challenge the basic foundation of their relationship, their disagreement indicates how important these public representations of relationship intensity are. Alice's feelings were hurt by Jesse's refusal to place her above his other Friends on his list.

In addition to ranking Friends, youth in relationships need to leave public messages for and post pictures of their significant others. Doing so sends messages to their significant others about their dedication and to their digital public about the nature of the relationship. Zelda said, "Sometimes, like on MySpace, you will leave a comment, and you leave a whole bunch of stuff on there 'cause they your girlfriend and stuff, so everybody can see your name. Girls get happy for that. I don't know why. They just get happy" (Sims, Rural and Urban Youth). Zelda explained that he comments "on their pictures. Like if they got a new picture up, leave a comment 'oh, that's a nice picture you got up' or whatever." Zelda indicated the dual message contained in this sort of digital relationship work— the girlfriend is happy because this sort of work feels attentive and loving and Zelda sends a message to their community, "everybody can see your name," about his dedication to his relationship. Another form of relationship work includes posting "couple" pictures on one's social network profile. As Derrick says, "Throw a picture in there of her on your profile. Have it in your pictures like when people look at your pictures they see you and her together or something. Something that makes her say, 'Aaahhhh.' To show her that you care for her" (Sims, Rural and Urban Youth). Again, Derrick's comment shows that these tokens are both for a significant other and a teen's audience. These practices also hold members of a couple publicly accountable. Once one states that she or he is in a relationship, this insures that both members of a relationship agree on their status and are ready to make it public, thus prohibiting one member of the couple from arguing that "it wasn't official."

Breaking Up

Because of the integration of new media into their relationships, teens also experience mediated breakups. These new communication practices often require that teens take a variety of steps to sweep up the digital remnants of a given relationship and to deal with access to and the continuing digital presence of their former significant others.

The media that some youth laud as a comfortable way to meet and get to know a romantic interest are viewed as a poor way to break up with an intimate. Billy, a white seventeen-year-old from a northern California suburb, said that as he was IMing with a friend he advised his friend to break up with his girlfriend. Apparently his friend did so right then, through IM. Representative of other teens, Billy said, shaking his head, "That was bad." Grady, Liz's sixteen-year-old boyfriend, agreed that breaking up through IMs or text messages was "lame" but that only "some people do it; most people don't." In line with a theme we heard, Grady claimed that breaking up in writing either through a social network site or through a text message was "disrespectful. Because they can't say anything back or anything." Teens acknowledge that breaking up in person is preferable to using text messages, instant messages, or messaging functions on social network sites, because face-to-face interaction is more respectful. Just as teens are thankful for the ways in which they can manage vulnerability using new media in the early stages of relationships, they sense that this vulnerability should not be managed in the same way at the end of a relationship.

New media have created a public venue for digital remnants, where digital representation might outlast the relationship. For instance, Gary, a seventeen-year-old Filipino senior from northern California, had created his MySpace site with his now ex-girlfriend. He laughed sheepishly during an interview as he logged on to his profile and the site title bore both his name and that of his ex-girlfriend, reading, "Sarah will always love Gary." This passive digital residue of their history together remained long after the relationship was over.

Box 3.1 The Public Nature of Mediated Breakups
danah boyd

When I first met them, Michael and Amy, a white seventeen-year-old and a white-and-black sixteen-year-old, respectively, had been dating for a few months. Amy was one grade below Michael at the same school in Seattle, but she was much more social. Her friends had introduced her to MySpace; she, in turn, had introduced him to MySpace. Amy created Michael's first MySpace profile specifically because she wanted him to have one so that she could send him messages and comments.

As Michael learned to modify his MySpace profile, it became an homage to his three favorite things: football, his friends, and his girlfriend. Michael's profile picture showed the couple embracing. His About Me section began with "I love my girlfriend Amy." Amy's profile also reflected their relationship; she wrote about how Michael "has my heart" and included pictures. Both were in each other's Top Friends list and they performed their relationship through comments, leaving standard messages as well as sweet nothings. Their friends responded by leaving comments, teasing them about their public intimacy.

For Amy, MySpace and school were the two places where she could be with Michael. She is allowed out only on weekends and, even then, rarely. MySpace is the centerpiece of Amy's social life. As she explains, "My mom doesn't let me out of the house very often, so that's pretty much all I do is I sit on MySpace and talk to people and text and talk on the phone, 'cause my mom's always got some crazy reason to keep me in the house." Amy's lack of mobility frustrated Michael, who has much more freedom. His father is usually out with his girlfriend and thinks Michael is mature enough to take his car and do as he pleases. Michael noted that "it's almost like we're roommates more than anything." While he can do as he wishes, Amy follows her mother's rules and this means that the couple rarely saw each other except online. Being able to interact with Amy motivated Michael to log in to MySpace regularly. In some senses, the mediated performance of their relationship was their relationship.

A week after I interviewed the couple, Michael and Amy broke up. Through MySpace, I was able to watch this breakup play out. Their digitally professed love turned into a performance of animosity. The entire tone of Michael's profile changed. He changed his headline to "Michael is no longer fucking with stupid bitches." His status changed to "single." The About Me section on his profile still referenced Amy, although not by name. Rather than showcasing his love, his About Me section now proclaimed, "I hate my stupid bitch ex girlfriend." The photos were gone. Additionally, the two were no longer linked as Friends, let alone on each other's Top Friends. With the eradication of the connection, all comments also disappeared.

Amy's profile also revealed traces of the breakup. She had obliterated the relationship throughout her profile, removing all photos and textual references to Michael. He was removed from her Friend list and the list of guys she called heroes. What appeared in the place of his name was "boyfriend" with a link to a new boy: Scott. While Michael had written Amy into his bio, Scott proclaimed his love for Amy even more loudly. He had changed his name on his profile: "Scott + Amy," and his profile photo depicted the happy couple smooching. He had written two blogs: "I have fallen in love with Amy" and "Rawr! Amy is Awesome." One of these blogs contained a love poem written about Amy and the other contained a prose version of his feelings; Amy responded to these blogs with comments professing her love and other friends added approving comments. Loving messages from the new couple peppered each other's profile. Scott wrote, "I Love You" two hundred times on Amy's profile, followed by "here is the translation . . . i love you too baby. . . ."

The messages on Scott's blogs from Amy's friends made it clear that some knew that her relationship with Michael had ended, but not all appeared aware of the story behind the event. In Amy's comments, there were a handful of posts with messages such as, "what happened with you and Michael?" Amy responded to these by posting to each Friend's comment section with some variation of "alotta bullshit." Third-party references to Michael littered Amy's comments section but Michael himself was no longer present—Amy's new love had usurped him.

I was unable to ask Amy and Michael what happened, but there also was no need to. Just as they had performed their togetherness for all who were curious, so too did they perform their breakup. Michael performed the newly single angry ex-boyfriend while Amy simply replaced all references to Michael with references to Scott, erasing Michael's existence without comment. The public performance of breakups goes beyond he said/she said stories; it showcases each person's emotional reaction to the situation.

By publicly documenting their relationship and their breakup, Amy and Michael are looking for validation and support from their peers. Amy's comments are filled with supportive words from her girlfriends and she acknowledges these on their profiles. Conversely, Michael's posts about "stupid bitches" provoke his guy friends to leave comments teasing him about getting into drama with girls. When I checked back a month later, Michael had removed his picture, cleared his background and content, and deleted all his Friends. He had not deleted his profile, but his last log-in date suggests that he stopped logging in. Amy had continued to use MySpace, but every trace of Scott had been replaced with a new guy.

Even though teens say that the actual act of breaking up should not happen in a mediated way, breakups do take place online as youth sweep up the digital remainders of their relationships. Teens' breakups can be reflected passively through status changes or displayed actively through hostile public messages and announcements. Michael and Amy (see box 3.1 for their story) exemplify an actively public breakup—public animosity, angry messages directed specifically at an ex-intimate, and the seeking of public validation from their friends. Conversely, passively public breakups entail quietly removing pictures, changing one's relationship status, and reordering Friends. While these breakups also happen in public, they are tamer and perhaps more representative of the customary way teens end relationships. Trevor's most recent breakup exemplified this passively public practice. The white seventeen-year-old from suburban northern California said that he usually places the person he is dating as the top Friend on his MySpace and moves people instantaneously when they break up. But "the latest ex stayed on there for six months because I was waiting. . . . I thought I'd be in a relationship really quickly." Trevor says that his ex-girlfriends weren't upset when he removed them. "There was never drama about it. They got it. They understood. . . . I always try for that, because I really don't want to be the jerk." For teens, changing a public representation of a relationship is a normal part of these now-mediated relationships; thus, unless the couple does not agree on the status of their relationship, they are rarely surprised by this sort of alteration of an ex's profile.

After a relationship ends, teens often inhabit the same, or overlapping, networked publics. Frequently, members of a former couple can still see each others' profiles, see messages left by their ex–significant other on shared Friends' social network profiles, and receive automatic updates about their ex, should they retain him or her as a Friend. As Christo Sims's research has highlighted, these indirect communication channels mean that youth can still be in touch with and possibly monitor each other after an intimate relationship has ended. These communications can be caring, respectful, retaliatory, hurtful, or angry, or they can be ways to send messages to an ex–significant other without having to interact directly with him or her. While teens may have the sense that they should sever real-world and digital ties with their former girlfriends or boyfriends, Bob, the white nineteen-year-old from suburban northern California, said that monitoring one's ex on a social network site is

one thing that you shouldn't do but everyone does. You can go check all their stuff. Like you look at their Facebook, you look at their MySpace, you see if they take off the photos of you, you see if they changed their relationship status to something, you see if they've got a new person writing on their wall. Like you become a stalker, and a highly efficient stalker. Because all the information is already there at once.

You don't have to ask your friends or her friends if she's seeing someone new. Like you know. And then they want you to know. (Sims, Rural and Urban Youth)

Teens are aware that their exes see them in these networked publics and use the opportunity to communicate with them, though not directly. Ono, a Haitian American sixteen-year-old from Brooklyn, New York, used the opportunity provided by social network sites to communicate her anger toward her ex-boyfriend.

You want to make them feel so bad that the relationship ended. So you take out all the comments, unless, it depends, unless you are still friends with that person. Take out all the pictures. Put some other person, or maybe delete him from your Friends list, and, but you know that he's gonna look at your profile anyway, so you put other males next to you, or put pictures of another male and say how nice he looks in that outfit or whatever, or my future man, or whatever, so you could put as much anger in that person as you can, or if you guys have the same Friend, like if me and my boyfriend have you as a Friend, I'll use you to get his attention. (Sims, Rural and Urban Youth)

Ono strategized about how to use her shared public to make her ex-boyfriend feel bad by signaling that she had severed ties with him, that he was no longer her Friend, and that she was intimately connected to other boys. The same technology used to publicly affirm intimate relationships can be used to publicly demonstrate their demise and to communicate anger toward someone with whom a teen may no longer have direct contact.

Bob used the same technology to communicate to an ex-girlfriend a gentler message. He had just endured a "really rough breakup" with a girl who wanted to "get back together" with him, though he did not reciprocate her wish to reunite. He wanted to communicate to her the fact that he was not willing to reconcile, but he felt constrained because he had learned of her desire in confidence from a mutual friend. To communicate his feelings to her, he changed his relationship status on Facebook to "in a relationship," even though he was not involved with anyone. At that point his ex-girlfriend realized that "I was unavailable. I knew she would read that; I didn't tell her or anything, but I knew that she would find it. And so that ended it officially." His ex-girlfriend communicated back to him in a similarly passive way:

I go on her MySpace and there's a blog about how she can finally move on. But it's addressed to no one. Right? I know who it's talking about; she knows who it's talking about. So that was a weird instance where "I'm not telling you but I know you're going to find this." (Sims, Rural and Urban Youth)

These sorts of indirect communications can enable teens to exit relationships in a dignified way and enable them to "have their say." Instead of communicating through oral conversations, or less directly through handwritten notes or chains of friends, teens can passively communicate through their online profiles and presence.

Despite popular emphasis on the one-to-one communication opportunities provided by these technologies, youth often use them to communicate indirectly, both through the technology and through intermediaries. Christo Sims's research on the ends of relationships shows that through new media, teens can retain an indirect channel to communicate after breaking up. While teens stop engaging in continuous contact after a breakup, they still use new media to communicate indirectly with each other and their larger mediated publics. Mechanisms on social network sites for indicating status or posting to an undefined public enable teens to delegate some of the more awkward social articulation work to technology-based, mediated forms of communication.

Intimate Media: Privacy, Monitoring, and Vulnerability

Themes of privacy and vulnerability weave through teens' new media practices. The ability to monitor one another and be monitored, emotional and physical vulnerability, and tensions around privacy thread through the variety of intimacy practices in which teens engage. Digital communications allow teens a sphere of privacy, when they don't have their own spaces, to communicate with their significant others through a circumvention of geographic and institutional constraints. The ability to talk beyond the earshot of one's parents and other adults, such as teachers, is part of this circumvention. Teens told us that the ability to communicate outside of adults' view and hearing was important. For instance, Joan, the Berkeley freshman, claims that she and her first boyfriend would talk

online all the time, all the time. Like, we talked on the phone but then sometimes we talked on the phone and IMed at the same time . . . especially it's like our parents was in the room and then we would talk to them and then if there is something that you don't want your mom to hear you could type it and then you could talk about it. (Finn, Freshquest)

Similarly, youth are able to maintain relationships with people of whom their parents might not approve, much like Jesse and Alice, because of this privacy. However, given the expectations of high contact with other teens and the amount of personal information in a semipublic realm, teens also have to negotiate new boundaries and spheres of privacy in their intimate relationships (Livingstone 2008). In this sense, social media carve out a new private realm in which teens can communicate, largely outside the purview of adults, while simultaneously redrawing and often weakening boundaries around their personal spheres of privacy.

Monitoring and Boundaries

From investigating crushes, to being in contact with significant others, to enduring breakups, the aspects of digital media that let teens be constantly in touch also allow them to monitor one another more intently. This monitoring varies from researching potential love interests to using a shared password to check up on one's significant other to attempting to restrict one's significant other's communications with his or her friends. Some youth regularly check on their significant other's websites simply to see what they are up to. Gabriella, a Latina fifteen-year-old from Los Angeles, logged on to her boyfriend's profile daily as part of her routine after she logged on to her own, "just to check" (boyd, Teen Sociality in Networked Publics). Similarly, Samantha, a white eighteen-year-old from Seattle, admitted, "I have done some checking up [on my boyfriend]" (boyd, Teen Sociality in Networked Publics). This sort of "checking" behavior happens when one has a crush, when one is monitoring one's romantic partner, and sometimes after a breakup.

The importance of passwords to one's online presence is central to these monitoring practices. Sharing a password both denotes intimacy and allows a significant other to monitor the private portions and manipulate the public parts of a social network profile. For some couples, such as Clarissa and her girlfriend, Genevre, white seventeen-year-olds in northern California, sharing a password feels like a way to maintain a connection even when they are apart. In fact, as Clarissa logged on to her MySpace profile she laughed, seeing that her girlfriend had updated it and altered the background to a more attractive one. However, not all teens feel comfortable with the amount of power a significant other wields with the password. Derrick, the Dominican American sixteen-year-old living in Brooklyn, New York, argued that girls want the passwords because

they want to check up on you all the time. They want to get your MySpace password, they want to get your AIM password, they want to get your phone, your answering machine, the password. They want to get anything they . . . know that another girl can get in contact with you through. (Sims, Rural and Urban Youth)

While Champ, the Latino nineteen-year-old from Brooklyn, shares his password, he protects his privacy by changing his password regularly. "You gotta change it. . . . I'll be changing mine like every three weeks" (Sims, Rural and Urban Youth). Clarissa's, Derrick's, and Champ's varying responses to sharing a password show how this practice is both a sign of intimacy and a possible invasion of privacy. By refusing to share it, some youth attempt to set a boundary around their intimate relations, sometimes to the frustration of their significant others, usually girlfriends. This may be because some girls feel powerful when they know their boyfriend's password. Dana, the Latina fourteen-year-old living in Brooklyn, explained, "I made my boyfriend give me his password and that shows power" (Sims, Rural and Urban Youth). Given the research that documents continuing gender inequality in heterosexual adolescent dating relationships (Hillier, Harrison, and Bowditch 1999; Hird and Jackson 2001; Jackson 1998), it is not surprising that girls are strategizing ways to feel more powerful in these partnerships.

In a similar move, some of the youth we spoke with draw boundaries by altering digital footprints that might make their significant other question their commitment. These footprints may be messages, search histories, phone numbers, or texts that reveal one's intimacy practices to families, siblings, friends, or significant others. Zelda, the Trinidadian American fourteen-year-old living in Brooklyn, New York, actually deletes information on his site to get rid of evidence that might anger his girlfriend: "Sometimes I'll just go in there and I delete stuff that girls wrote me. I'll just delete it." To avoid these privacy compromises, Champ and Zelda change the names on their mobile phones. To prevent his girlfriend from scrolling through to look at his contacts and call logs, Champ records "their names different," explaining, "Yeah, if it's a girl's name, you put a boy's name that probably sounds similar to it. . . . Like, let's say the girl's name is Justine, you'll probably put Justin" (Sims, Rural and Urban Youth). While these technologies have provided a greater realm of privacy, digital footprints might compromise this privacy and thus youth are often drawing digital boundaries to protect a personal sphere.

Some of the monitoring that happens during teens' relationships veers eerily close to serious emotional control or abuse. Lolo, a fifteen-year-old Latina from Los Angeles, said that her boyfriend did not like the fact that her social network profile was public. Using the password she shared with him, "He kinda put it on private, hello. He's like, 'I don't wanna know every boy's going in there searching you'" (boyd, Teen Sociality in Networked Publics). We heard this insecurity over their claim on their romantic partners throughout our interviews with youth. Teens may intensify some of the monitoring practices we found as a way to attempt to control some of their anxiety about the stability of their relationships.

This sort of control might also intensify when economic transactions are involved. In our research, teens sometimes paid their own phone bills, but usually their parents paid. This meant that teens needed to obey their parents' rules (to the extent the parents could enforce them) about mobile phone use. Something similar happened when one's significant other paid the phone bill. Ono, the Haitian American sixteen-year-old living in Brooklyn, New York, said that her friend's boyfriend pays her friend's phone bill and as a result

he expects when he calls, even if she's not available, to just pick up and say, "I can't talk to you right now, I'll call you back." Or if he's with her, then he would be asking who else is calling if it's not her parents or something. That's what happens when he pays your bills. And yeah, he can talk to you every day, even if you're not free, because he pays for it. (Sims, Rural and Urban Youth)

Girls in this type of relationship seemingly trade one type of control, parental, for another (Hijazi-Omari and Ribak 2008). Their privacy is compromised because they do not retain economic control of their mobile phones.

Youth monitor one another in the early stages of, during, and after the ending of the relationships. This monitoring manages anxiety so central to teen relationships in which teens for the first time are crafting intimate ties with one another. The monitoring capabilities afforded by digital media seem like a way to manage such anxiety as teens seek to put to rest their fears about vulnerability and betrayal. The ability to monitor others through these new media venues both allows teens to learn about others and makes them vulnerable to surveillance and control by others.

Vulnerability

New media simultaneously increase teens' vulnerability and their control over their emotional exposure. This heightened vulnerability may allow teens to craft new and strong emotional connections with one another (e.g., see box 3.2) as well as render them more open to being victimized by their friends, acquaintances, and other adults. However, the removed and asynchronous nature of some new media also allows them to manage emotional exposure and render teens less vulnerable, especially in the early stages of a relationship.

Box 3.2   Bob Anderson's Story: "It Was Kind of a Weird Cyber Growing-Up Thing"

Christo Sims

When Bob Anderson was in middle school and the first few years of high school, social network sites such as MySpace and Facebook had not yet taken hold in the rural region of California where he lived. The popular social application was instant messaging (IM), and Bob would log on for hours a day to chat with other teens. As is emphasized in chapters 2 and 3, these online engagements typically enact and extend offline relationships and the identities associated with them. Bob's use of IM supports this observation as he primarily chatted with teenagers from school. Yet Bob, who is now in college, also tells of using IM to chat with teenagers beyond his given social worlds. While this book illustrates many instances in which interest-driven practices transcend given social worlds, Bob's story is unique as it falls within the realms of friendship and intimacy.

Bob recalls forming a friendship through IM with a teenager from the East Coast. The friendship lasted about two years but the friends never met. It took place toward the end of middle school through the early part of high school. Via conversations on IM, he and his new friend created a space where vulnerable subjects could be broached and swapped:

We kind of went back and forth on a personal level. Talking about a range of things. But we were really just going back and forth with fundamental problems and questions that people have with growing up. Going through puberty. Sexual experiences. Where you fit in society. Real friends, fake friends.

When reflecting on the experience now, he frames the experience as part of the growing-up process:

It was kind of a weird cyber growing-up thing. Just like checking in. . . . I was finding out more about myself, and trying to figure out what I was about, and trying to figure out what people were about, and trying to figure out what the world was about.

Bob makes the point that these were topics that could not be easily discussed in other contexts. As he puts it, "You can suddenly say things you would never say in person." As chapter 3 shows, new media have become integral to the ways teenagers try to manage exposure and publicity. When it comes to vulnerable subjects, there are particularly good reasons to be cautious as to what one exposes and to whom. In Bob's case, he perceives it as safer to expose vulnerabilities to a stranger than to someone he already knows. This may seem counterintuitive but actually makes a lot of sense when one considers the context of his social world. At fourteen, Bob's given social world was small and persistent. His growing up in a rural area meant that his peer-based social world was largely bound to his school. His graduating eighth-grade class had fewer than thirty students, his high-school class fewer than two hundred. Short of a major family transition, these schoolmates would make up his peer-based social world until he left for college. Within this small-world context, the consequences of embarrassing exposures and public missteps can seem global and resilient.

Through IM, Bob and his friend created a space that seemed safely distant from these given and ongoing social worlds. As such, personal vulnerabilities could be swapped without risking lasting local consequences to reputation and identity. It was a place in which private thoughts, experiences, and feelings could be voiced for the first time, an intimate sphere, confidential by means of a perceived disassociation from the given and ongoing social worlds to which he belonged.

Boys in particular, because of contemporary association of vulnerability with a lack of masculinity (Korobov and Thorne 2006), express relief about the extent to which new media allow them to control what they perceive as emotional vulnerability. They feel less exposed because they can text a girl or leave a message on her MySpace page rather than risk embarrassment by calling her and stumbling over their words or saying something embarrassing. Bob, for instance, said,

It's a lot easier to flirt digitally than it is in person 'cause there's no awkward silence. You can't say something you don't mean 'cause you could sit there at one comment on a person's profile and spend a half an hour making sure that everything is right. Like some words are lowercase on purpose. The punctuation's just the way . . . I want it to look sloppy, but it really has this, you know, acute meaning to it. (Sims, Rural and Urban Youth)

The asynchronous nature of texting and leaving messages allows boys to save face when flirting with a new girl. In this way, the controlled casualness discussed earlier is a form of emotion management and a way to control vulnerability.

The same technologies that allow youth to manage emotional exposure might also render them more vulnerable, in part because of the amount and type of information shared and the speed at which it can travel. Teens are not necessarily in control of digital representations of intimate practices or in control of the audience who sees those representations. For instance, Elena and Brett, two gregarious white sixteen- and seventeen-year-olds, respectively, from northern California, talk about how embarrassing pictures might be forwarded. Elena said, "That's a lot of drama too. They can send pics to other people." Brett continued, laughing, "People might take a picture of other people making out at a party." Elena continued, "Like so-and-so was kissing so-and-so or that so-and-so made out with so-and-so at a party. Then the next week they're like, 'Look at the picture; obviously it meant something.' Then they're with somebody else." Elena said that the picture might get "around school and you're like, 'Wait, how did you even get this picture? You weren't even at the party.' It goes further than you think sometimes." In this way, even teens' offline practices may be monitored online if people forward compromising pictures of them. This digital proof of one's intimate life may spread rapidly, outside of one's control.

The other vulnerability teens talked about is that of the stereotypical risk conveyed through fear-based narratives of the Internet, that of the stalker, the stranger, and the predator. Teens rarely mentioned these stories in our research (apart from noting that this was what adults were concerned about), but a minority of youth reported having negative interactions with predatory-type adults online. Those youth who seek out intimate communities online, such as gay teens, might be more at risk for this sort of unwanted stranger intimacy. For all the opportunities to create community for gay teens, the Internet also puts them at risk as they seek this community. Robert, the white seventeen-year-old from Chicago, told a particularly affecting story about his experience on the Internet as he was coming into his early teens.

A couple times a week, after my parents went to bed, I visited some Internet sites . . . then after a while, I found a chat room website, a gay teen chat room. I chatted with a lot of guys; eventually I started to talk to people outside of the chat room, on MSN Messenger. There were people who wanted to do things with cameras and pictures, and for a while I went along with some of it, not really doing too much. Then one day, it wasn't a teenager who sent me their pic, but an old fat man. I was disgusted, beyond words. I smashed my computer camera, deleted my MSN, and barred any memory from those times out of existence until I recollect now.

Robert was trying to explore his sexuality the best he could, as a single gay teen, but in doing so, he ended up on non-age-graded sites, which, though not inherently risky or problematic, may be dangerous for marginalized teens looking for community. Instead of getting to experiment in more public and socially acceptable ways, through structured rituals of hetero-sexuality, gay teens often must find their own way. On the one hand, the Internet is an invaluable lifeline, but on the other, it renders gay teens more vulnerable to situations such as this one.

New media allow teens to manage their vulnerability; permit them to have intensely emotional, vulnerable conversations; and render them potentially susceptible to the forwarding of information about them and vulnerable to those who wish to take advantage of them.

Conclusion: Controlled Casualness, Continuous Contact, and Passive Communication

While many adults may perceive social network sites as being simply glori-fied dating sites, this chapter, in conjunction with chapter 2, on friendship, demonstrates that teens are not one-dimensional beings interested only in prurient communications and subjects; rather they craft complex emo-tional and social worlds both publicly and privately on and offline. Academic work has rarely taken youth courtship practices seriously, but in examining the way teens talk about these practices and their emotions about them, our project demonstrates that romance practices are central to teens' social worlds, culture, and use of new media. For contemporary American teens, new media provide a new venue for their intimacy prac-tices, and render these practices simultaneously more public and more private. Teens can meet people, flirt, date, and break up beyond the earshot and eyesight of their parents and other adults while also doing these things in front of all their online friends. As chapter 2 also points out, participat-ing in these mediated relational and emotional practices is central to being a part of an offline social world. Youth are developing new kinds of social norms and literacies through these practices as well as learning to partici-pate in technology-mediated publics. These sites of peer-based learning need to be taken seriously, as they are structuring social and communicative practices that differ in some important respects from the experiences of these teens' parents, and they can become a site of intergenerational tension and misunderstanding.

When meeting and flirting, teens find online communication extremely useful. This is especially true in terms of furthering casual acquaintances. They have more freedom to get to know friends of friends or others they have met briefly at parties or other group gatherings without risking too much embarrassment. They can also use social network sites to learn about, usually unbeknownst to the other person, someone in whom they have an initial interest, be it someone they see every day in class or the person who sells them burgers at the local fast-food restaurant. While meeting people solely online is not the norm, some teens do meet and flirt that way. Others consider this brave, scary, or weird, depending on their perspective. Their messages and interactions during this time might be characterized as a "controlled casualness." Dating teens use new media often, engaging in what one might think of as "continuous contact." When in a relationship, teens frequently communicate with each other and expect their significant others to publicly acknowledge and maintain their relationship on their social network profiles. Teens' relationships also end in the presence of their networked publics. The breakups might be active or passive, but because of their shared publics, teens retain the ability to passively communicate with each other even after ending intimate ties. Their continuing indirect communication about relationship status is a way in which these sites enable intimate content to be made very public. This publicity both allows teens to exact revenge and communicate important, but indirect, messages about their emotional states to their former significant others. Because of the dearth of research on teens' intimacy practices, we lack comprehensive comparative case studies, but it seems that teens' current use of new media might be a unique moment in the recent history of teen dating practices. New media allow, and seem to encourage, teens to make relationships and relationship talk explicit. They let teens access romantic others' personal information and share versions of or information about themselves that might not be done as easily in offline circumstances. Much as friends have in the past, technology now acts as a social intermediary, enabling communication that is passive, but very important, at liminal relationship stages, such as beginnings or endings. Finally, among teens in relationships, technology allows them to maintain a passive copresence with each other and provides new ways to subvert expectations of that copresence.

As we saw in the case of friendship practices, these online tools and communication practices make peer-based interaction and pressures more consistently available to teens. Unlike more familiar forms of public space, networked publics and private communication channels such as IM and mobile phones can make it harder for parents to passively monitor their children's romantic communications (though written records of these communications often linger in digital environments should parents know how to access them). Youth call and send messages to each other directly, bypassing mediation by parents or siblings. This is part of the trend toward what Misa Matsuda (2005) has called "selective sociality," in which youth can make more intentional decisions about those with whom they affiliate. Further, some parents do not fully understand the norms that govern teens' online interactions, and the literacies they deploy in these interactions, and thus they may be tempted to resort to blanket prohibitions rather than more nuanced forms of guidance. These dynamics are explored further in chapter 4.

The snapshot of contemporary teens' intimacy practices presented in this chapter indicates that today's teens are part of a significant shift in how intimate communication and relationships are structured, expressed, and publicized. Networked publics of different sizes and scales contextualize these intimate communications and practices, allowing youth to observe the intimate interactions of others, and conversely, to display their own emotions, practices, and relationships to select publics. The new possibilities of self-expression available online, characterized by more casual and personal forms of public communication, complicate our existing norms about the boundaries between the public and the private.

Notes

1. This practice has varied with Alice's changing access to the text-message function on her mobile phone, because she depended on her parents' phone plan.

2. As with other parts of teen culture, contemporary practices of dating and romance are deeply gendered (Best 2000; Martin 1996; Pascoe 2007a). Gender difference and inequality is central to heterosexuality and thus is embedded in dating practices.

Contemporary dating practices emphasize a gender-differentiated heterosexuality in which girls frequently possess less subjectivity than boys and in which "power is naturalized through a discourse of romance" (Best 2000, 67).

3. Indeed, in spite of the current flurry of concern over what kids are doing online, the Internet and social network sites have hardly led to an explosion in teen sexual behavior. In fact, the number of teens who say they have had sex before they graduated high school has declined from 54.1 percent in 1991 to 47.8 percent in 2007 (CDC 2007).

4. A "wall" is the place on a typical social network site where someone's Friend might leave a message for him or her to read. These messages are usually visible to others, but their public nature depends on the privacy settings of a given profile.

5. Christo Sims interviewed Bob several times, such that during the course of our research Bob's age ranged from nineteen to twenty-one.

6. Like many teens, Missy wrote using typical social media shorthand. Translated, her comment would read: "Hey, hmm, what to say? I don't know. Laughing out loud. Well, I left you a comment. . . . You should feel special haha (smiley face).

7. "G" is slang for "gangsta," in this case an affectionate term for a friend.

4 FAMILIES

Lead Author: Heather A. Horst

Trudy lives in a working-class neighborhood on the outskirts of Silicon Valley, where she attends middle school (Horst, Silicon Valley Families). At the time of her interview with Heather Horst in the spring of 2006, Trudy and her other twelve-year-old friends recently had taken an interest in MySpace. During the course of the evening, Trudy decided to show Heather her MySpace page and all the things she was learning to do, which included creating a page that she thought would express her personality. She also talked about how she used MySpace to stay linked to her friends and said that in the past few months she had managed to "Friend" forty-two people. As each Friend appeared on the page, she proceeded to describe each person and the various aspects of that person's MySpace profile that she liked or disliked. Eventually she came to the profile of her friend Amanda, whose picture was an uploaded image of Hello Kitty. As Trudy talked more about Amanda's page, Heather asked if she knew why Amanda decided to place Hello Kitty as her main picture. Before Trudy could answer, Trudy's mother, who was washing dishes nearby, chimed in and reminded Trudy that one of the conditions of Amanda's participation on the site was that she agreed to use unidentifiable pictures or images; Amanda's mother did not want any real pictures of her daughter on the Internet. However, during the course of browsing through her profile, Trudy discovered that she still possessed a picture on her page "tagged" (labeled) with Amanda's name. Trudy's mom, who was still looking over her shoulder, reminded her daughter that she should delete or replace Amanda's residual picture out of respect for Amanda's parent's wishes. Annoyed with what Trudy felt represented an invasion of her privacy, she rolled her eyes and said she would take down the photo "later."

This book focuses on new media engagement, peer-based sociability, and learning from a youth perspective, with particular attention to learning with new media that takes place outside of traditional learning institutions, such as schools and families. As noted by Christo Sims in his summary of large quantitative surveys of new media use in the United States (see box 1.1), however, the vast majority of American households—89 percent—now possesses some form of access to the Internet at home. Alongside the Internet, many families throughout our study also owned mobile phones, portable music players, and gaming systems, although it is important to note that the latest gaming systems (e.g., PlayStation 3, Wii, Xbox) and devices such as BlackBerrys and iPhones remain out of economic reach for the vast majority of our study participants. In effect, a large share of young people's engagements with new media—using social network sites, instant messaging services, and gaming—occurs in the context of home and family life.

Parents, the guardians of the home and family, take seriously their role as guides and regulators of their children's participation in this new media ecology. Just as young people engage with new media based on friendship-driven and interest-driven genres of participation, parents and adults' attitudes toward new media reflect their own motivations and beliefs about parenting as well as their personal histories and interests in media. Indeed, parents often frame their purchase of new media in relation to the educational goals and broader aspirations they hold for their children. From this vantage point, computers, video cameras, and digital cameras as well as related software, education, and training become meaningful to many families because they represent an investment in their child's future, one that they hope will ensure their children's success in education, work, and income generation (Bourdieu 1984; Haddon 2004; Lally 2002; Livingstone 2002; Sefton-Green and Buckingham 1996; Seiter 2007; see also chapter 7 in this book). Parents also leverage new media as motivators or rewards for good grades and behavior; graduation or a good report card may result in a new game, mobile phone, or digital camera. While parents make efforts to embrace their kids' interest in new media, they admit that new media also incite anxiety and discomfort, which are often tied to moral panics surrounding media as well as what Ellen Seiter (1999b) has referred to as the "lay theory of media effects," or the belief that media cause children to become antisocial, violent, unproductive, and desensitized to a variety of influences, such as commercialization, sex, and violence (Alters and Clark 2004b; Cassell and Cramer 2007; Clark 2004; Lusted 1991).[1] Even the most media-immersed parents in our study described a deep ambivalence about the prominence of new media in their children's lives and their role as parents in influencing their children's participation in the media ecologies that structure their sons' and daughters' lives.

This chapter considers the home and family as an important structuring context for informal media engagement[2] and, in turn, explores how parents and other adults negotiate the incorporation of media in young people's lives. Drawing research materials from a wide range of studies—primarily Christo Sims (Rural and Urban Youth), Heather A. Horst (Silcon Valley Families), Katynka Z. Martínez (Pico Union Families), Lisa Tripp and Becky Herr-Stephenson (Teaching and Learning with Multimedia), C. J. Pascoe (Living Digital), danah boyd (Teen Sociality in Networked Publics), Patricia G. Lange (YouTube and Video Bloggers) and Dan Perkel and Sarita Yardi (Digital Photo-Elicitation with Kids)—we examine parenting strategies surrounding new media, with particular attention to the structuring and regulation of family life in the home and through new media. The three numbered boxes in this chapter illustrate the ways in which the use of new media in homes and families differ regionally. We begin by concentrating on the spatial and domestic arrangements that shape new media use in the home, such as the placement of computers. We then turn to the creation of routines and other forms of temporality, including the amount of time and the textures of kids' media use. In the final section, our analysis centers on parents' and kids' rules, and the creation, bending, and breaking of rules. We conclude by considering how parents and young people transform, negotiate, and create a sense of family identity through new media.

Parenting in the New Media Ecology

Home and family environments reflect the values, morals, and aspirations of families as well as beliefs about the importance and effects of new media for learning and communication. Writing in the moment of the first home computers, Silverstone, Hirsch, and Morley (1992) observe:

Media pose a whole host of control problems for the household, problems of regulation and boundary maintenance. These are expressed generally in the regular cycle of moral panics around new media or new media content, but on an everyday level, in individual households, they are expressed through decisions to include and exclude media content and to regulate within the household who watches what and who listens to and plays with and uses what. (20)

As research on youth and the family reveals, the anxiety surrounding the integration of new media into the home also reflects concerns about independence, separation, and autonomy that, at least in the context of Western societies, occur during the teenage years. Parents throughout our studies worried about the amount of time that kids spent online and not "with real people," as one mother described her son's "addiction" to networked gaming. Some parents lamented that they felt they had lost control, or that their kids had become too dependent upon their portable games, iPods, and mobile phones. Still other parents expressed concern over the extent to which their kids were spending "too much time" talking with their friends over instant messaging, on social network sites, or on the mobile phone. While these concerns over dependence and independence as well as control and autonomy appear to be a persistent family dynamic (Spigel 2001), Alters (2004) argues that during the past forty or fifty years there has been a shift in the nature of parenting in American family life;[3] "Since the 1960s, parents have become uneasy about how to raise children in light of increases in drug use, delinquency, pregnancy, and suicides among children and adolescents" (Alters 2004, 59) as well as broader societal changes, such as the entrée of women into the workforce and the increase in divorce rates during the past three decades. Alters further contends that parents now feel aware and accountable to themselves, and to society at large, regarding the decisions they make in the domestic sphere, a phenomenon she refers to as "reflexive parenting."

The particular *expressions* of this sense of responsibility or reflexivity—present among most, if not all, of the families we interviewed—remain closely intertwined with the cultural, social, economic, and educational capital associated with class dynamics. In a seminal ethnographic study of parenting in the United States, Annette Lareau (2003) explores parenting strategies and the implications of different approaches to parenting for children's chances in life, what she terms the "transmission of differential advantages to children."[5] Examining the ways these patterns of parenting, or the "dominant set of cultural repertoires," are traversed in everyday life, Lareau outlines two approaches to parenting that, she argues, correspond with class positioning. According to Lareau, working-class parents in the United States believe in what she terms "the accomplishment of natural growth," a parenting strategy that emphasizes informal play, often in and around the house. Lareau outlines how working-class parents, using what she considers a more hands-off approach than their middle-class counterparts, believe that kids will grow and develop naturally as they navigate the world. By contrast, middle-class parents operate with a belief that it is their responsibility to develop their children through sports, music lessons, and other activities, a practice Lareau terms "concerted cultivation." One of the main differences between the two parenting strategies revolves around the organization of children's daily lives as well as the extent to which parents think they should be involved in the inner workings of their children's activities in schools and other institutionalized settings. Lareau suggests that whereas middle-class parents tend to advocate for their children in institutionalized settings, working-class parents value respect for authority, particularly of teachers and principals, and prefer to give their children the autonomy to navigate their own relationships with peers and the outside world.

The dynamics that Lareau describes in school settings and nonmediated environments also emerge in parents' approaches toward managing media in the home. For example, Ellen Seiter's *Sold Separately* (1993) explores the role of parenting styles and attitudes toward children's media culture.[4] Conducting her research on television and the use of kids' videos and cartoons, Seiter draws connections between class, education, and aspiration in her analysis of children's media and family life in the United States. In particular, Seiter focuses on the relationship of the media industry, parents, and kids in shaping values and attitudes toward particular forms of media consumption and participation. Based on her textual analysis of children's toy advertisements and the ways in which parents interpret and attempt to control children's use of commercial television characters in their everyday play, Seiter reveals how middle- and working-class parents externalize their values through the toys and media they encourage their children to play with and ultimately demonstrates how class biases toward toys and media are reinforced (cf. Chin 2001; Livingstone and Bovill 2001; Roberts and Foehr 2008; Seiter 2005; Thorne 2008).[5]

Whereas much of the early literature on parenting and media attributed differential adoption of new media in the family to class dynamics, a recent study by Hoover, Clark, and Alters (2004) attempts to situate family modes of media incorporation in relation to the construction of family identity (see also Matsuda 2007; Spigel 1992). Building on their work based in the metropolitan areas of Colorado, the authors focus on how religious and other sociomoral beliefs, values, and worldviews (and to a lesser extent educational, social, and cultural capital associated with class dynamics) influence parenting styles and attitudes toward new media. Mirroring Silverstone and Hirsch's[6] (1992) notion of the moral economy of the household, Hoover, Clark, and Alters (2004) contend that many parents feel the pressure to restrict and control their children's use of new media due to the cultivation of their family identity, or reputation.[7] Throughout this chapter, we examine how these different discourses and parenting approaches become embedded in the strategies parents employ to regulate and maintain control over media and media uses among the family. In the following section, we describe the ways in which parents and families craft media spaces, the first of three strategies we observed being employed through the course of our research.

Crafting Media Spaces at Home

The decision to acquire new media means making decisions about where new media will fit within the current domestic ecology of media objects.[8] These decisions may revolve around the affordability of a particular medium, as well as infrastructural issues, such as the potential location of a desktop computer, laptop, or gaming system in the home (Alters 2004; Lally 2002; James, Jenks, and Prout 1998). Holloway and Valentine (2003) contend that where families place computers and other new media in the home often shapes whether they are used individually or collectively as well as how long and how frequently new media might be used. For example, when parents put a computer in their children's bedroom, kids tend to associate its presence in their bedroom with ownership. As a result, kids often take on a role as a person who can restrict the amount of time that others can access "their" digital camera, iPod, gaming machine, or other new media (Holloway and Valentine 2003; Livingstone 2003).

Public Media Spaces: Halls, Dens, Kitchens, and Recreation Rooms

Given parents' concerns over the ability to control and monitor their children's media use, many parents elect to place larger media objects, such as gaming systems and desktop computers, in the public spaces of the home. Bakardjieva (2005) finds that many Canadian families place media in the living room and construct family computer rooms as well as "wired" basements, which are designed for new media usage (Lally 2002; Livingstone 2002). Like many of the families in Bakardjieva's study, the families who participated in the Digital Youth Project situated computers in kitchens, hallways, and other spaces of the home where parents possessed the option to monitor what their kids were doing. This pattern was particularly common in many of the Los Angeles households with space constraints, as well as in Silicon Valley households, where families used kitchens and dining rooms to eat together and complete homework. Other families, such as those who live in the suburban-style developments of rural California (Sims, Rural and Urban Youth), prefer to place their computers in a shared family "den" or "study" (Bakardjieva 2005; Clarke 2001).

In some of the wealthier households in Silicon Valley (Horst, Silicon Valley Families), families designed new spaces to house new media, such as home offices, playrooms, and recreation rooms (Clarke 2004, 2007; Gutman and de Coninck-Smith 2007; James, Jenks, and Prout 1998). For example, the Chens, an Asian-American family in Silicon Valley, lived in a large five-bedroom house and were in the process of remodeling their home to integrate a recreation area as an extension on the back of their garage. The Chens planned to add a Wii to their existing media collection, which included a large-screen TV, speakers, and a PlayStation 2, so the kids could practice tennis and play other "physical" games with their friends; their dad also expressed excitement at the prospect of practicing his golf swing. Mrs. Chen hoped that this new entertainment space apart from the main house would become a gathering place for her two teenage sons and their friends. She thought that the space would enable her to know and monitor where they were as well as what they were doing. Indeed, Mrs. Chen's plan to create a house-based entertainment center for her two sons and their friends reflected the centrality of kids' engagement with games and gaming in their everyday social lives. While the full-scale recon-struction of a garage was an extreme example of household modifications to accommodate new media,[9] most families opted to modify or convert existing spaces (e.g., playrooms or family rooms) into media rooms.

Private Media Spaces: The Bedroom

While some parents prefer to place media in the public spaces of the home, the bedroom holds a special place in the imaginations of many youth. As

McRobbie and Garber ([1978] 2000) argued three decades ago, girls typically view bedrooms as important spaces where they feel relatively free to develop or express their sense of self, or identity, particularly through the decoration, organization, and appropriation of their bedroom space (Clarke 2001; Kearney 2006; Mazzarella 2005; Steele and Brown 1995). In many homes, the arrival of relatively affordable and portable media has solidified the importance of the bedroom as a space where one can use new media in these endeavors and assume individual control over one's own media world. As Livingstone and Bovill (2001) assert, "What is clear is that the media—particularly screen media—are playing an increasingly significant role within the more solitary, more peer-oriented space of the bedroom" (180–81). They further suggest that the more "media-rich" bedrooms are, the more likely it is that kids will spend time in their bedrooms using the media, away from the rest of the family and the more public spaces of the home.

As Livingstone and Bovill observe, many parents believe that when kids' bedrooms become the focal point of their activities at home, they lose the ability to monitor and guide their children's activities. For this reason, many parents fear what happens behind closed doors. Kira, a seventeen-year-old in Seattle who lives with her aunt and uncle (boyd, Teen Sociality in Networked Publics), describes the tension surrounding her bedroom and her aunt's regulation of her media usage:

[My aunt] just always wants me to be involved with the family, but then when I'm sitting out [in the living room] I get completely ignored so I don't like being out there. I mean I'll sit out there because I know she wants me to, but then once she goes to bed at 8:00 I'm in my room where I can turn on my music and watch my TV or talk on my phone or whatever. I can't pretty much even look at my cell phone in front of her because she gets mad, thinks I'm on it all the time. I'm like I just ignored five calls. How am I on it all the time? She makes me so mad.

Kira continues,

I have a TV in my room with cable and everything but my aunt flips out if I go in my room. She's like, you're always in there, you're always hibernating in there, and she thinks I'm smoking pot in my room because I light incense. Incense relaxes me; I mean I'm not stupid; I'm not going to smoke pot in my room. Like you guys aren't going to smell it?

Kira's desire to relax and be herself, what her aunt interprets as "hibernating" in her bedroom, appears to affirm Livingstone and Bovill's (2001) findings in the United Kingdom. However, Kira's own awareness that her room is not completely separated—that her family can smell what she is doing in her room—suggests that teens do understand that bedrooms are much less private in practice than they are in the popular imagination and discourse. For example, Sam, a seventeen-year-old in Cedar Rapids, Iowa, described to danah boyd (Teen Sociality in Networked Publics) that while he sees his room as relatively private, he still refrains from using media and technology while in his bedroom:

When the door is closed, but I don't . . . I don't like talking on the phone in my house at all. Just because, it's not like a two-room shack, but it's not huge, and you never know what's going to go through those walls. What's going to make them think that something . . . this is happening or whatever, so, I don't.

Similarly, fourteen-year-old Leigh in Cedar Rapids, Iowa, told danah boyd that her house does not feel private to her "just because my family is just . . . I don't know. My mom comes and looks in my room and stuff. I don't really like that" (Teen Sociality in Networked Publics). Sixteen-year-old Melissa of Marion, Iowa, complained to danah that while her room may be nominally private, her mother possesses the freedom of movement to come and go as she pleases:

Because there are a lot of things that my mom does that make me feel like it's not private. I can be taking a shower and she'll come in, go to the bathroom, and leave. She has no respect for my personal privacy. I can be sitting on the computer talking to a friend and she'll be reading over my shoulder and I don't want her to. That's not really private to me. . . . Private is kind of like a place where I can kind of go and just be by myself and not have to worry about anyone doing anything. . . . My most important thinking goes on when I'm either in bed, in the shower, or in my car. (Teen Sociality in Networked Publics)

As becomes evident, parents sometimes assert their status in the family hierarchy by moving through the home freely, even when a space is deemed to belong to their kids. In addition, sharing a room or a computer with a sibling has an impact on the sense of privacy teens feel and, in some instances, renders privacy almost impossible. Numerous teens discussed how their siblings used their computer or accessed their accounts to talk to their friends through IM or social network sites while pretending to be them. Ana-Garcia, a half-Indian, half-Guatemalan fifteen-year-old from Los Angeles (boyd, Teen Sociality in Networked Publics), described how her brother hacked onto my AIM and my MySpace, and he just started talking to people, and then the next day when I went on, they were like, "What was wrong with you yesterday? Why were you acting all mean to me?" I was like, "It was not me. It was my brother," so he does that a lot.

Ana-Garcia explained to her friends that her brother was the one responsible for being "mean." While many of her friends believed her, she worried that one day he might take his pranks too far. In an effort to make her brother stop, Ana-Garcia told her parents, but, she said that because he is the boy in the family, her brother rarely gets punished. The lack of privacy surrounding Ana-Garcia's new media usage reflects a general frustration Ana-Garcia holds about being a girl and the lack of freedom, privacy, and control this entails, at least in her family.

Alongside age and gender dynamics, the size and infrastructure of homes also contributes to the negotiation of privacy in domestic spaces. As Katynka Martínez suggests in her description of Maxwel Garcia and his family (see box 4.1), for many other low-income families who live in tight quarters, the retreat into the bedroom and the creation of bedroom culture is simply not an option.

Box 4.1   The Garcia Family: A Portrait of Urban Los Angeles
Katynka Z. Martínez

Maxwel, a fourteen-year-old seventh-grade boy, lives in a studio apartment with his mother, Lydia, and two older sisters. The tight living quarters make Bovill and Livingstone's (2001) concept of a "bedroom culture" difficult to apply to these kids' digital-media environment. Since the family's living quarters do not include traditional bedrooms, the increasing availability of media in such rooms becomes a nonissue. This is not to say that digital media are absent from the kids' home environment. Maxwel's oldest sister owns a digital camera and he owns a Game Boy. In addition, there are two television sets in the apartment, one hooked up to a VCR/DVD player and one hooked up to a Nintendo 64. Maxwel's favorite TV shows are *Yu-Gi-Oh!* and *Lilo and Stitch,* but the family makes joint decisions about what to watch on television. On the day of my interview with the family, the television was set to the local news on Spanish-language television. By the time the interview was over, Maxwel and his sister had watched Spanish-language news, the local news on an English-language television station, *X-Men 2,* and *Bend It Like Beckham.* The family does not pay for cable television or satellite service so both movies were viewed as broadcast television programming.

In the case of this family, the television set is the media object that has been used to create a shared family time-space. For example, Maxwel often watches *telenovelas,* Spanish-language soap operas, with his mother. Maxwel mentioned watching "the one at seven," and when I asked if he was referring to the telenovela *Peregrina,* Maxwel and his mom immediately answered with an enthusiastic "yes." During the interview with Lydia, she struggled to remember the title of an English-language television program that she likes. She asked her son for the title and he asked her *"el de los que siempre están fumados?"* ("The one where they're always stoned?") This description of *That '70s Show* resonated for Maxwel's mom and she said that she enjoyed watching the program because *"Cuando uno esta joven, todo se te hace fácil."* ("When one is young, everything seems easy.") Lydia does not understand the English language but says that she can follow the physical humor used in English-language sitcoms. She also enjoys watching wrestling with her kids. The theatricality and physicality of World Wrestling Entertainment make it easy for a non-English speaker to follow the television programs.

Lydia was able to attend school up to only the fourth grade in her hometown of Mexico City. She came to the United States with her then husband and two daughters. Maxwel was born in the United States a few years after the family arrived in the country. Lydia explains that while she encourages her kids' use of computers, she hasn't tried to incorporate computers into her own life. She says that she often joins her children on their trips to the public library or the local community center and looks on while they use computers. However, she does not actually use them herself because, as she explained, her first goal is to make sure that her kids have everything they need to succeed in school and her second goal is to learn English.

Lydia was unemployed at the time the interview was conducted. Her last job had been as a garment worker using embroidery machines in a factory. She brought out a hat and showed it off as both an example of the work that she did and also to draw attention to her favorite Mexican soccer team. When Lydia was employed at the sewing factory she often worked nights and did not see her children in the morning or immediately after school. When she worked into the early morning she used her mobile phone to call her kids and remind them to eat breakfast. She also expected them to call her mobile phone when they came home from school. She explained that her phone recorded the time and place from where the call was made and that she returned her kids' phone calls during her break.

Maxwel and his sisters have asked their mom for mobile phones but she does not have the funds to buy them. Her eldest daughter is twenty years old and bought her own mobile phone. This daughter sells carpet cleaner door-to-door and is the owner of the family's digital camera. When Maxwel was asked what he used the camera for, he explained that it was used "on my confirmation, or on my sisters' birthdays, or my birthdays, or my mom's birthday or special occasions." In addition to snapping pictures on these "special occasions," Maxwel had recently used the camera for a science experiment and his mother used the camera to take pictures at the march for immigrants' rights that was held in Los Angeles, and nationwide, on May 1, 2006.

The digital camera also has been used to document apartment fixtures in various stages of decay. Lydia explained that she took classes at a local community center and learned about her rights as a tenant. She also learned about how to use photographs to record landlord negligence by watching *La Corte del Pueblo* (*The People's Court*), a Spanish-language program that presents reenactments of actual court cases. She said that the ceiling in her bathroom was inflated and one day caved in, almost hitting her daughter. Another time, the kitchen ceiling caved in and released rat feces all over the room. This happened on Maxwel's birthday. Lydia said that she had made multiple pots of tamales and that she was lucky the ceiling did not fall apart while she was cooking. She took pictures of both the bathroom- and kitchen-ceiling incidents and then used these photos to argue against having to pay a full month's rent.

Lydia explained that she can barely afford to pay the rent for her apartment, let alone buy the kids video games. The family buys video games, which cost about thirty-five dollars, on credit at a local indoor swap meet. The Nintendo 64 that Maxwel owns was a gift from a friend with whom they shared an apartment. The family computer was also a gift. Maxwel's godparents gave him this computer for his birthday but it broke and his godfather took the computer to be repaired. This was a year ago. Lydia explained that Maxwel's godfather *"llevó la cabeza o el monitor, cómo se llama? Y no lo ha traída"* ("took the brain or the monitor, what is it called? And he hasn't returned it"). The godfather had taken the hard drive and Maxwel's family was left with only a printer and monitor. The monitor has been laid to rest with a plastic cover and is kept in a walk-in closet that also functions as a small bedroom (see figure 4.1). The desk that the monitor was on now serves as a table for stuffed animals and knickknacks. Although the printer is not connected to a computer, it is still kept on the bottom shelf of the desk.

While the Garcia family does not represent a media-rich household, the family was eager to share stories and artifacts related to their media practices. During the course of the interview Maxwel and his mom brought out photo albums with pictures of graduations, first communions, baptisms, and birthday parties. They also brought out their digital camera and displayed the photos of the immigrants' rights march that were still stored on the device. Lydia was asked how the immigrants' rights march compared to these family events. She explained:

Figure 4.1
The Garcia family's closet/makeshift bedroom also stores a broken computer. Photo by
Katynka Z. Martínez, 2006.

*Nosotros decidimos sacarles fotos de toda la marcha porque para mí fue un día . . . gracias a
Dios . . . especial. Todos los días son especial. Pero de ver de que si todos de nosotros estamos
apoyandonos va a cambiar todo. Entonces, para mí esos fotos me sirvieron en lo personal para
decir de que si yo apoyo a mi hijo él puede llegar más arriba. Para bien. No para cosas malas.
Pero si yo no lo apoyo, es como . . . está solo. No hay quien lo escucha, quien lo va a ayudar,
quien lo va a apoyar, quien le va a decir, "Sigue adelante." Entonces esas fotos que saqué con
tanta gente allí me hizo ver que la unión hace la fuerza para cada persona para lograr lo que
queremos para bien. No para mal.*

(We decided to take pictures of the entire march because for me it was a day that was . . . thank God . . . very special. Every day is special. But to see that if all of us support each other everything is going to change. So, for me those photos served a personal purpose because they say that if I support my son he can achieve something higher. For good. Not for bad things. But if I don't support him, it's like . . . he's alone. There is no one to listen to him, who will help him, who will support him, who will say, "Continue moving forward." So those pictures that I took with so many people there made me see that unity brings strength and makes it possible for every person to achieve what we want for good. Not for bad.)

By using the family's digital camera to collect visual evidence of landlord neglect and by also taking her kids and the camera to an immigrants' rights march, Lydia is demonstrating the multiple ways that this simple device can be used as a tool of empowerment for the whole family and even for a larger immigrant community.

Moreover, and as much of the work on domestic space and childhood reveals (Aries 1962; Clarke 2004; Miller 2001), homes and bedrooms are not static entities. Just as families upgrade media or shift the ownership of new media objects among parents and kids, homes also change through time as children grow older and families disperse. Indeed, going off to college remains an important landmark. For example, Ben, a participant in Megan Finn, David Schlossberg, Judd Antin, and Paul Poling's "Freshquest" study, described how he shared his first computer, a hand-me-down from his parents, with his brother in their bedroom after his parents bought a "new, fancy computer." Later, he managed to acquire his own computer. Ben explained:

When my sister moved out and went to college, my half sister, yeah, she went to college, then I moved into her room. And the computer was, I mean, then there were three people in different rooms and the computer was going to be in one of our rooms. And obviously we can see a lot of frictions building up, whose room is, whose room is it gonna be? Right, [my twin brother] wanted [the computer] in his room, I wanted [the computer] in my room, and finally my parents caved and just bought [a] new computer all together.

As Ben suggested, parents often expect siblings to share computers and bedrooms when they are younger. However, when Ben's sister moved out and went to college, Ben and his brother each received their own bedrooms. To resolve the conflict over where to put the shared "kids' computer," Ben's parents decided it was simply easier to buy a new computer than mediate between arguing siblings. As this example illustrates, youth are constantly struggling to gain privacy and autonomy to engage with media and online communication, and this often plays out in negotiations over the location and ownership of media in the home.

Mobility and Other Media Spaces

While homes continue to be viewed as the nexus for modern family life, families are certainly not restricted to the bounded space of the home. Parents work outside the home, and kids attend school and participate in after-school and enrichment programs. Young people also hang out at their friends' houses. These spaces provide kids with opportunities to use media not available, and sometimes not allowed, in their own home(s). As Dominic, a sixteen-year-old from Seattle, explained to danah boyd (Teen Sociality in Networked Publics) while sitting with two of his friends, "I don't play [World of Warcraft], because I don't have the money for the monthly fee, but these two do, and I . . . I'll watch them sometimes when I go over to their house, and some, maybe, occasionally I'll play with them." As Dominic suggests, many teenagers and kids learn about new media while hanging out with friends whose parents make different rules about the type and extent of media their kids can play, watch, or use. With a few exceptions, parents acknowledge that their kids do use and gain access to new media elsewhere. While they might prefer that their kids follow the same guidelines they outline at home, typically to not play first-person shooters or watch sexually explicit movies, they also recognize that what happens outside their own domestic domain remains largely out of their control. Moreover, an awareness of the potential social implications of enforcing these restrictions also play a role in parents' decisions about the extent to which they attempt to impose their own rules at other families' homes.

Young people also take advantage of opportunities to operate under a different set of rules when they visit family members whom they do not regularly live with. Andrew, a ten-year-old elementary-school student who lives in Berkeley, California, told Dan Perkel and Sarita Yardi (Digital Photo-Elicitation with Kids):

At our house we only have computer games . . . where you learn stuff. We don't have fighting games. . . . The only game that doesn't have to do with adding or subtracting or dividing or multiplying or anything that's really close to math is called Sim Theme Park, which is where you make a theme park on the computer.

While games at home are restricted to the computer and the genre of edutainment (Ito 2007; see also chapter 5), Andrew's grandmother's house is a place where "we get to watch TV, watch movies, play video games . . . once a month for a weekend. And we just came back from spring break, ten days with them." In fact, Andrew's uncle and grandparents bought a variety of game systems and games for Andrew and his brother, Nick, over the past few years, including a GameCube, Super Nintendo, Nintendo 64, PlayStation, PlayStation 2, Game Boy, Game Boy Color, Game Boy SP, Game Boy DS, and Sega Dreamcast, as birthday and holiday gifts. Out of respect for their parents, the boys' grandparents store all the games at their home, the only place where the boys can play. In fact, when their uncle buys games for Andrew and Nick on holidays and birthdays, he sends them directly to the boys' grandparents' house. Andrew and Nick's parents may not like the fact that they play games during the visits to their grandparents' house, but they also recognize that it is a different domestic space and therefore out of their control.

In this section, we outline the physical and social contexts that structure where young people access media, whether that is in the public spaces in the home, more private spaces such as the bedroom, or in the homes of friends and extended family. In all these settings, youth may desire autonomy and independence from the rules and regulations of their everyday home and family life. However, given parents' concerns about and sense of responsibility over their children's lives and activities, most parents do not grant their sons and daughters full autonomy and control over their media and communications. Rather, and as we continue to see throughout this chapter, young people's attempts to maintain privacy and ownership over their media usage and the media spaces where their engagement with new media takes place remain an ongoing struggle in their everyday lives.

Making, Taking, and Sharing Media Time

In addition to structuring the place of media in the space of the home, the family context also shapes how and when family members spend their time using new media. The temporal rhythms of the family and the household take a variety of forms, from media engagements that are shared among family members to the varied ways in which parents regulate how and what forms of media their children use. As we demonstrate, the routines that guide new media use and family life are closely intertwined with the organization of domestic space.

Spending Time Together

Almost all the families we spoke with explicitly expressed how much they valued spending time together as a family, although many teenagers noted that they still preferred to spend time with their friends, boyfriends, and girlfriends. In Heather Horst's study "Silicon Valley Families," parents who possessed disposable income noted that their family holiday represented a time to "unplug" from the mediated environment and busy-ness of everyday life (see Darrah, Freeman, and English-Lueck 2007). However, even in the families who idealized unplugging, scheduling time to watch television shows, movies, or videos together also emerged as a time to relax and take a break from the fast pace of life. Many families came to view using media as a way to facilitate communication and bonding.

Within some families, games can become the primary vehicle for parents and kids, and particularly fathers and sons, to connect (see chapter 5). Miguel, a ten-year-old who lives in the San Francisco Bay Area, described the relationship with his dad to Dan Perkel and Sarita Yardi (Digital Photo-Elicitation with Kids) that developed over playing games. Perkel and Yardi write in their fieldnotes:

The only time Miguel talked about his father during the interview (his parents were separated) was in reference to the fact that he and his dad used to play PlayStation together. He recalled a time some years prior when his dad and older cousins all played the PlayStation together and teased him for how he used the controller. These "motivators" seemed to be a powerful, good memory for him.

Miguel explained this to Dan Perkel in their interview:

Dan: Where did you learn to play all of the games on your PlayStation?
Miguel: Well, my dad, we used to play like every night . . . every Friday night, Saturday night, Sunday night, whatever.
Dan: You used to play with your dad?
Miguel: Yeah, and he would invite my cousins to come over and stuff. We'd borrow games from my uncles.
Dan: Were they all older than you?
Miguel: Yeah.
Dan: And did they teach you how to play or did you figure it out for yourself?

Miguel: They taught me how to play. Like, I used to ... you know how when you play car games the car moves to the side and stuff? I would go like this with the control [moves arms wildly from side to side simulating holding a game controller as if he were racing]. So ... they taught me how to keep still and look at the screen ... hand-eye coordination.

Dan: Hand-eye coordination? Where did you here that term from?

Miguel: TV.

Both: [laugh]

Other families view gaming as a more persistent site of family togetherness that they move in and out of fluidly (see figure 4.2). Patricia Lange (YouTube and Video Bloggers) interviewed Akmalla, a twelve-year-old white girl in Los Angeles, who regularly plays World of Warcraft with her parents.

Patricia: So for weekends you're pretty much at your computer?

Akmalla: Yeah, weekends I'm at my computer in front of the TV screen with a couple sodas in front of me. But my mom and dad play World of Warcraft as well.

Patricia: Oh, do they?

Figure 4.2
Two sisters playing games together. Photo by Heather A. Horst, 2007.

Akmalla: So we're usually just sitting in the same exact room on the same couch playing the same game going, "Oh, what are you doing?" "Oh, that's nice."

Patricia: And so do you play with, like, do you play by fighting your mom and your dad?

Akmalla: Yeah, we can play alongside each other. We can fight each other.

Patricia: Do you usually fight each other or do you band together and fight like others?

Akmalla: Well, we try to get in a group to do a quest or something and we usually end up yelling at each other because it's just a family thing. It's like we're walking and it's like, "Where are you?" "We're on the beach." "Which beach?" "This beach." "What? You ahhh!"

Rather than a forced family gathering (e.g., "family time"), the social atmosphere as well as her parents' own interest in and skills playing World of Warcraft enable the family to be together by participating in an activity that the entire family now shares as an interest.

Whereas some families spend time together hanging out playing games or watching television, other families gather around a variety of media to make websites and videos and edit digital photographs while together. In these spaces, kids are often given the opportunity to work alongside their parents (typically their father), and parents continue to support their kids' interests by buying new media for the next project. In middle-class homes, such as that of the Millers (see box 4.2), families gather around a variety of media in effort to learn about and gauge interest, practices that parents describe as taking an interest in or, in the words of many parents, "staying involved with" their kids.

In many of the studies in Los Angeles (e.g., Tripp and Herr-Stephenson, Los Angeles Middle Schools; Martínez, Pico Union Families), kids play an important role as the technology expert or broker in the family, translating websites and other forms of information for their parents. Twelve-year-old Michelle in Lisa Tripp and Becky Herr-Stephenson's study "Teaching and Learning with Multimedia" noted what she taught her mother, a single parent, from El Salvador:

How to send emails, but sometimes, I check it first, because she does it wrong. And I taught her how to like . . . sometimes, she wants to upload pictures from my camera, and I show her, but she doesn't remember, so I have to do it myself. Mostly, I have to do the picture parts. I like doing the pictures.

Box 4.2   The Miller Family: A Portrait of a Silicon Valley Family
Heather A. Horst

The Miller family lives in a leafy-green suburb of Silicon Valley in a four-bedroom home with a yard, dog, and basketball hoop at the bottom of the driveway. Like other middle-class professional parents in this study, Eli and Miriam Miller work in the technology industry and view themselves as the producers of software, code, and other systems that fuel the literal and figurative engine of the Silicon Valley economy. This close relationship with technology and the technology industry shapes the ways that families such as the Millers think about, use, and imagine the possibilities of technology and digital media.

In the construction of their family identity, the Millers decided to develop a family website that includes photographs, descriptions of family vacations (their "trip log"), as well as details about key family events, such as birthday parties, anniversaries, graduations, and bat and bar mitzvahs for their three children. The front page of the family website consists of the beaming family gathered in the water in wet suits around a dolphin after a recent visit to SeaWorld, Mom and Dad on the left of the dolphin and the three kids gathered in birth order on the right. Each family member created a funny quote typed in different-colored ink next to his or her picture—Dad typed "I think I might be touching something I shouldn't be" above his head; Iraina wrote, "I have salt watter in my mouth!!! Can we PLEESE gett this over with?"; and Jonathan commented, "I feel like a dork in this life jacket." Originally Eli Miller, who is a consultant with training and experience in engineering, created and maintained the site. Each of the kids has his or her own webpage, where Eli encourages them to express and explore their individual interests, which include information about the Darfur conflict, the youngest son's development of a podcasting site called Reality, and Iraina's recent trip to Israel.

Along with creating the family website, the Millers like to mess around with digital media. Iraina explained the use of new media when they are at home:

My brother just got a digital video camera for his birthday; it was his big present this year. And I was like . . . like I was having fun with it. I always thought it's kind of cooler in theory because for me I don't want to actually take the time and sit down and edit a whole movie because for me that's just not worth it. I love to come up with the basic concepts and then give people advice if they do it. . . . But to actually sit down and hear someone say the same words over and over and over again while you're trying to get the right cut would drive me crazy. I heard my dad trying to do it for my grandparents when we filmed them for their anniversary and we were talking about their wedding. And it was kind of a little documentary thing. Only we forgot to bring a stand that day and my mom let us kids film them. And so my dad was trying to edit it and put pictures in where it was really bouncy and stuff. . . . It would just drive me crazy because you have to hear the person say the same words over and over and over again. . . . I don't think I would ever be able to do that.

Another thing is Daddy, for his birthday, just got kind of from himself, kind of from myself—he told everyone he wanted one. He got a professional radio mic [microphone], so we've been playing around with that. And he's been tampering with it [also for Jonathan's podcasting] and you can put the sound through headphones and you can sing around with your own voice. We were just playing with it before and he got a sound board and a mic so it's really cool.

As a family, Iraina and her family's collaboration strategy involves an egalitarian-expertise model in their incorporation of digital media. Each family member—Iraina and her brother, sister, mom, or dad—develops an interest and, in turn, gets involved in the use and/or process of using the digital media that he or she enjoys. Other family members, usually their dad, then tries to develop the technical expertise that will enable everyone to experiment and play with the media objects. Eli Miller, in particular, sees the maintenance of this expertise as a way to make sure the kids extend their knowledge and interests. However, despite the relatively egalitarian ethos of creating websites and videos, there are times when the development of expertise for the family turns competitive. Jonathan suggested this when discussing his new podcasting project:

Jonathan: I think with podcasting is one of the first things I kind of gotten my dad into. He doesn't actually subscribe to podcast, but he's thinking about maybe making his own podcast or thinking of ideas even. So it's . . .
Heather: You're the one who influenced?
Jonathan: I'm the one who's doing it. So it's kind of a cool thing.

When families work together, the leadership continues to come from parents, and particularly fathers, at the beginning and end of the collaboration process, sometimes regardless of experience or expertise. But when talking to kids about the role of their parents in this process, we find that the kids who engage in these familial collaborations discover that there are opportunities to subvert the normal power dynamics in the family by becoming particularly good at or interested in a technology or practice. In such families, the proliferation of new media and technologies in the household provides kids and parents with a space to explore the possibilities of these tools. Rather than learning skills for specific educational outcomes, upwardly mobile middle-class families such as the Millers view these tools as contributing to the wider development of their kids as individuals as well as the construction of a family identity.

Similarly, Lisa Tripp talked to a mother named Rita about her motivation for spending time on the computer with her middle-school-aged son Andrew. Rita, a single parent in Los Angeles, explained:

Se me hace más para estar cerca de él, estar jugando con él, porque él es un niño muy serio. De repente es más separado. Es bien tierno, pero es de repente separado. Entonces, a él le gusta, de repente que yo esté con él. O él me dice: "mami: esto;" o "¿me ayudas a buscar palabras?" "¿Me ayudas?" Que le dejan muchas letras y le gusta buscar palabras. Y a mí como me encanta eso, y de rompecabezas también, es la forma de acercarme a él y estar más cerca con él, y que él me tenga confianza y ganas de estar siempre ahí. . . . Casi siempre me gusta estar más cerca de él, porque él tiene carácter de repente más explosivo. Y, a veces, la computadora nos sirve para quedar más en una zona de acuerdo.

(It is to be close to him, to be playing with him, because he is a very serious quiet kid. He can be very sweet, but tends to hang out by himself. Sometimes he grows distant. And then, suddenly, he'll want me to be with him but sometimes he likes me being with him. Or he might say, "Mommy, look at this," or, "Would you help me look for words?" "Would you help me?" Sometimes he gets an assigment to study several new letters at once. He gets many letters, and he likes looking for words. And since I like all that, and I also like puzzles, it is a way for me to get closer to him and be together. It is an opportunity for him to get to trust me, and continue enjoying being together for him to know that I am always there. . . . In general, I try to get I have always liked to be close to him because he has a strong temper. And sometimes the computer helps us get along better.) (Translation by Martin Lamarque and Lisa Tripp)

As Rita suggested, this give and take surrounding media is a way to become closer and feel connected; the computer mediates between the generations.

Whereas watching DVDs together on Sunday evening, helping out on the computer, or editing recordings of matches or family events structured many of our participants' use of media as a family, we also observed the importance of new media for families separated by vast geographic distances. For example, transnational families take advantage of the possibilities of new media, including cassette tapes, videocassettes, DVDs, and online media, to intensify their sense of connection and communication, such as producing videos of graduations, weddings, funerals, and other events to circulate among family members living abroad (Basch, Schiller, and Szanton-Blanc 1994; Horst 2006; Panagakos and Horst 2006; Wilding 2006). Among Silicon Valley families with transnational connections, one of the most popular ways of feeling like a family involved the exchange of emails, which were typically written by the mother in the family. Family websites and online photo albums, including photos shared through public sites such as Kodak Gallery and Shutterfly, emerged as important spaces for families to share information and pictures of one another. Families without regular or reliable Internet connections, such as in the studies of families in urban Los Angeles, viewed mobile phones and phone cards that catered to the Central American market as an important communication medium.

In addition to various forms of personal media sharing, online conversational media are increasingly used by transnational families to communicate. Transnational families with greater economic means also use new media such as Skype and webcams to enhance their sense of connection and communication.[10] Raj, a freshman who participated in Megan Finn and colleagues' "Freshquest" study, noted:

> It's pretty neat to be able to see my brother, my family twelve thousand miles away over the sea. . . . I just use Skype [Internet telephony software] for the voice capability and my webcam has some inbuilt software. . . . It's nice to be able to see each other and talk at the same time.

Voice and vision are often viewed as the ideal modes of communication because they mitigate the distances in time and space that typically plague transnational families.

Although the particular expressions of sharing media and knowledge between parents and kids vary with parents' own technical expertise, education, gender, time, and command of English, many parents expressed the desire to create spaces and times for hanging out, messing around, and, as we see in box 6.2, geeking out with their kids. Much like after-school programs that attempt to harness the passion for media in the name of learning, families may also try to leverage media in their everyday interactions. While it is promising that parents and kids can come together around interest-based practices (see chapters 5, 6, and 7), the gendered dimensions of spending time together with media—from a kids' perspective, mothers are often described by kids as "clueless" or "hopeless" outside the domain of communication technologies and fathers as being the ones who play or tinker with technology alongside their kids—suggest that new media continue to contribute to the production and reproduction of class and gender inequities in American society.

Routines and Rhythms

Although parents value the potential of new media to bring families together, they also recognize that young people's use of new media causes disruptions to school and family life. Parents attempt to counteract the possibilities for distraction from activities that they believe are more important by restricting their kids from playing games or going on IM and social network sites before schoolwork, household chores, and other productive activities are completed. In addition, they set time limits on media use, such as thirty minutes or one hour per day. Peter, a thirteen-year-old participant in Matteo Bittanti's study "Game Play," explained, "My parents let me play between four and ten p.m. during the week, but the schedule is more flexible during weekends." Twelve-year-old Akmalla notes that her parents have set controls on World of Warcraft so that she will go to bed. As she described to Patricia Lange (YouTube and Video Bloggers), "like if you try to log on after a certain hour when your parents have said no, it'll say, 'You cannot log on because your parents are controlling it.'" Such external control features are used by parents who possess a more sophisticated knowledge of computers.

In some cases, parents prohibit their kids from using new media altogether during the school and workweek, saving weekends for unstructured, nonproductive play. Nineteen-year-old Torus, who is an Indian-Italian from the Los Angeles area, discussed with Patricia Lange (YouTube and Video Bloggers) how his parents structured his time for gaming:

Before I kind of got to college and my senior year of high school, it was pretty much you play on weekends for just a couple of hours, but you have to study, all the rest of the time. So, even when we were very young . . . even when I was like eight or nine, my dad required us to study for two hours before we could play two hours of games, so it was those kind of . . . it was very clear to us that our parents thought of it as definitely a reward system, not a privilege, not a right, sort of thing. Like I could never play during the week and I hardly watched TV during the week, but on the weekend, I could usually play. Me and my brother would play.

Such routines changed seasonally. Many of the young people we interviewed noted that their parents closely monitored their use of new media during the school year, but summers and breaks remained relatively unstructured. Kids report playing games or checking social network sites up to five or six hours a day during summer breaks and other less structured times of the year. Many young people value this time because it enables them to play games that require more strategy and time investment, what is described in chapter 5 as recreational gaming.

The time allotted for media use also varies in relation to economic and other family circumstances, such as divorce or separation (Clark 2004). In Heather Horst's study "Silicon Valley Families," seventeen-year-old Archibald compared his media ecology at his mother's house with that at his father's house. Archibald spends most of his weekdays living with his mom and sister in a three-bedroom townhome on the periphery of a wealthy area of Silicon Valley; Archibald's mom works two jobs to support Archibald and his sister's attendance at a well-respected school. Archibald's father, a doctor who lives about two hours away, pays for train tickets for Archibald and his sister to visit him each weekend. They spend part of their summers with their father, at least when Archibald is not busy with soccer and volunteer activities in Latin America. Archibald described his media environment at his dad's house as well equipped with the latest computers, software, and other media, but his access is restricted since his dad always wants to spend time with Archibald and his sister during their limited time together. By contrast, at his mom's house, Archibald's media environment is more limited, but Archibald and his sister possess relatively unfettered access to the computer and other media because of his mother's busy work schedule. While the two media environments provide opportunities for accessing different media, the kids must also navigate two different series of time restrictions and rules.

One of the most striking aspects of the role of new media and technology in the home is that mothers bear most of the responsibility for upholding the morality of the family, especially in nuclear and extended families. In her study of American families, Hochschild (2003) notes that since the 1970s women carry out most of the care work within the family, a practice she terms "the second shift." Where the integration of new media and technology into the home is concerned, mothers tend to be the parent who maintains the temporal rhythms of the household, structuring what kids should be doing with their time, when kids should and should not be watching television, playing games, and going online. The exception to this rule is in single-parent families where the father is the primary caretaker and, to a lesser extent, in places such as Silicon Valley, where fathers are familiar and reasonably fond of these tools. For example, kids note that their fathers tend to be much more lenient about games, and in some cases, spend time with their sons and daughters messing around with new media. As Heather Horst discusses in box 4.2, some fathers become heavily invested in their kids' interests, such as music making and podcasting, expressing their support by buying accessories for these activities. Yet, much like the liberal fathers Hochschild describes in her study, fathers tend to restrict their control to the technological capacities of their home computer networks and tools, leaving mothers to be the enforcers of the family rules and regulations. Java, a twelve-year-old white middle-school student who lives in one of the wealthier areas of rural California, described to Christo Sims (Rural and Urban Youth) who is in charge of restricting new media: "My mom. Most of the time my mom comes up with the rules." By contrast, Java depicts her dad as a person who is into music and technology:

Well, we're basically allowed . . . that's actually my dad's thing. The music and the computers are his thing. But if they don't know the artist, the person, the CD is, then they like to listen to a few songs or they'll ask people, different people, about it before they let us buy it. But normally the mix CDs are fine. . . . Well, 'cause my dad's more into the technology and stuff. And he . . . well, he works with computers obviously so he's more into that.

In some instances fathers join forces with their kids to actively subvert the mother's rules. Kim, a participant in Megan Finn and colleagues' "Freshquest" study, described how her father bought games for her behind her mother's back:

My dad. And every time he went to Costco, he'd surprise me with just a little game without my mom knowing. My mom would get so pissed that he waste[d] money on that. "Ooh, a game." So I'd go ahead and play. After a while I think he hit upon a couple that really got me into gaming. Either it was Warcraft or something else. So he got me really into gaming and then I forced my parents to buy me a game afterwards. Like every day, I'd be, "Can we go to Computer City? Can we go [to] Electronic Boutique in the mall?"

The relatively playful nature of dads' engagements with media in domestic settings often results in negative characterizations of moms either as nagging enforcers or "hopeless," as a twenty-five-year-old AMV creator described his mother's technical skills to Mizuko Ito (Anime Fans). One Los Angeles mother named Anita (Tripp and Herr-Stephenson, Los Angeles Middle Schools) explained:

Pues . . . como le digo yo . . . casi no conozco la computadora; yo no sé usarla . . . casi yo no conozco. . . . O sea, entonces, por eso me preocupo; porque como yo a veces no sé lo que están haciendo.

(Like I said, I barely know the computer. I don't know how to use it. I don't know it. So, that is why I am worried, because sometimes I don't even know what they are doing.) (Translation by Lisa Tripp)

Although parents, particularly mothers, feel responsible for monitoring and regulating their kids' media engagements, they are often hampered in their efforts by their children's resistance to control, their own lack of technical expertise, and the subversion of their rules by other family members.

Growing Up

The rules and boundaries surrounding new media typically begin to change as kids grow up and develop judgment, "a process of critical evaluation that develops as one matures, with help from parents" (Alters 2004, 114). As Liz's mother explained to C. J. Pascoe (Living Digital):

She's going to be seventeen. She's going to graduate next year. I think she needs to be responsible. . . . Her dad would have it differently, but since I'm in control, and they're lucky that I am because I pretty much . . . I just look at them more as adults. They can figure things out. They're not doing anything against the law. They're home. She's a great student. You know?

While there is a sense of a loosening of control tied to allowing teenagers to exercise their own judgment, it is clear that parents expect their teenagers to know and, to some degree, internalize their parents' values. In the case of games, parents typically allow kids to engage with different gaming genres depending on how capable they think their children are in making these judgments. Somewhere between the ages of five and eight, kids (typically boys) tend to shift away from the edutainment genres of Leapster and other desktop computer games and upgrade to the Nintendo DS or PSP, a transition that tends to occur when the family plans a lengthier car or plane journey (see box 7.2). A few years later, in the kids' preteen and early teen years, middle-class parents "give in" (as kids describe it), or determine that their kids are mature enough to exercise judgment (see Alters 2004; Clarke 2004). As thirteen-year-old white teenager named Peter discussed with Matteo Bittanti (Game Play), "I was not allowed to play Grand Theft Auto when I was eleven because my parents felt that the content was inappropriate for me." As Peter suggested, violence and violent video games remain a particularly important preoccupation, especially first-person shooters (see box 5.2). Another gamer, twenty-two-year-old

Earendil, reflected upon his parents' boundaries concerning violent genres of video games with Mizuko Ito (Anime Fans): "Ah! But when we all hit about thirteen years old, mom didn't worry about whether we could distinguish fantasy violence with real violence and allowed more computer use!"

As has been well established in the literature on youth and mobile phones (Baron 2008; Goggin 2006; Horst and Miller 2006; Ito, Okabe, and Matsuda 2005; Katz 2006; Ling 2004, 2008; Matsuda 2005; Miyaki 2005), giving kids possession of a mobile phone also involves a determination of kids' judgment. As a general rule, few elementary-school students owned mobile phones and there was a general sentiment among parents that they should avoid buying a mobile phone for children while they are in elementary school. An exception to this rule was single parents and working-class parents who buy their kids phones in the interest of safety, since they tend to navigate independence at an earlier age (Chin 2001; Lareau 2003). Families who could not afford the cost of after-school and other enrichment programs also felt compelled to give their children mobile phones, or access to a mobile phone, while they were away from home. As CrazyMonkey, a fourteen-year-old white middle school student who lives in a single-parent household in Silicon Valley (Horst, Silicon Valley Families), recounted, "I've had a cell phone since fourth grade because I had to start figuring out my rides home . . . to and from school . . . well, just from school to home almost on a daily basis, and so my mom wanted to be able to reach me easily." But rather than owning the swanky new mobile device desired by most teenagers, CrazyMonkey had a thick, black phone that she used as her mobile to arrange for rides and check in with her mom, who could not be physically present to take her from point to point. In such cases, the mobile phone becomes a safety gap when kids take the train or bus or walk to and from school, work, or home.

In middle-class families, the decision to give kids a mobile phone typically occurs during middle or high school as teens start to invest more time in their peer worlds. As Jennifer, a white seventeen-year-old in Lawrence, Kansas, recounts to danah boyd (Teen Sociality in Networked Publics), " 'Cause junior high you start, you do more stuff, your parents let you do more stuff so they were like, well, we're not gonna know where you're at all the time, so you should have a phone just in case something happens, so their reasoning was." Kids in middle-class families tend to acquire mobile phones when they are deemed old enough or responsible enough to take on the responsibility of using or owning a phone. Parents also provide their kids with mobile phones when they obtain a driver's license or a car, in the interest of safety should they run out of gas or have car trouble. Jordan, a biracial Mexican-American fifteen-year-old in Austin, Texas, recalled (boyd, Teen Sociality in Networked Publics):

Well, I got my first phone in seventh grade so looking back, it might have been too early, but it's important now. Like starting driving, like you go out a lot more, and I think my parents feel better that I have one. Also, so I can call them at any time and if I need them, we're connected.

In these cases, the mobile phone represents a symbol of freedom, one that is used by kids to justify movement outside the home and outside the purview of their parents; when they want to go somewhere, they remind their parents that they will call and check in to let them know they are safe and parents provide their kids with the phone and freedom as an opportunity to exercise judgment.

However, it is also clear that kids do fail to exercise judgment and, when major indiscretions occur, parents place temporary restrictions on computer access, gaming, and other new media as a form of punishment. A white sixteen-year-old named Liz and her mom discussed with C. J. Pascoe why she was grounded from instant messaging (IM) (Living Digital):

Liz's mom: Well, what happened with the IMing thing is that the kids have a tendency to type things in that they normally would not verbalize to anyone. And it can get pretty vulgar and disrespectful within themselves. And it got to that point, of arguments and things happening in that aspect. So we took it away because we saw the vulgarity coming out and didn't like it. It shouldn't happen. We took it away. And then she lost interest, obviously.

Liz: No, I got it back. And then I was like, okay, I have to have it because I haven't had it in a long time. But then I started losing interest.

For Liz and her mother, being grounded was recognition of Liz's lack of judgment, her failure to meet the behavior expectations that her mother had for someone Liz's age. As Liz's mother noted, the secondary effect of being grounded helped Liz lose interest in instant messaging, a process that Liz's mother attributes to growing up. In the following section, we focus more explicitly on the negotiation of rules between kids and parents.

Making, Breaking, and Bending the Rules

As we have outlined, parents use space and time to help guide their kids' use of new media at home. Throughout many of our interviews, parents readily articulated the various rules they attempted to establish as well as how these rules reflected their beliefs about new media. Kids, by contrast, often claimed to forget rules, or stated that their parents made rules but that they were either open to negotiation or not regularly enforced. Hood et al. (2004) found in their Colorado-based study that the family discourse surrounding new media reflected the parents' *intentions* rather than actual practices. Rather than defining this discourse as failure or irony, Alters (2004) argues that rules are "part of the family's project of building and maintaining a family identity" (128) and, for this reason, parents become invested in the rules and the importance of having rules, although they acknowledge that breaking and bending the rules regularly occurs. These "media transgressions," or points at which the normal, discursive rules are bent, were pervasive among all families who struggled to uphold their own rules on a daily basis (Alters and Clark 2004a; Clark 2004). In this section, we focus on young people's engagements with mobile phones and online spaces, with particular attention to the ways in which parents and kids make, break, and bend the rules.

Plans, Minutes, and Cards

The decision to give a son or daughter a mobile phone is often motivated by the desire to maintain a sense of control over kids' movements and activities. While parents value the leash function of mobile phones, they also struggle with the day-to-day management of their kids' phone use. Typically, parents of younger kids attempt to restrict the number and types of people entered into their kids' phones. Indeed, companies such as LG, which makes the Migo, and Firefly Communications, which sells the Firefly mobile phone, have attempted to capitalize on parents' desires in their design and marketing of a phone that restricts calls to a small number of people or places (in these phones "Home" is marked as the most important number in the phones). Other parents try to control the extent to which their kids make calls on the mobile phone by providing the phone on a need-to-use basis, such as buying a "kids' phone" to be shared among siblings. This often results in conflicts, particularly if one sibling decides to assume ownership of the phone. A seventeen-year-old Mexican-American named Federico recounted the trials of sharing a phone with his sister to Dan Perkel (MySpace Profile Production): "Because my parents can't afford to pay too much money, so we have to share a phone most of the time. So she's pretty hoggy about the phone, so if I get a text message or a phone call she'll be like . . . oh, I don't know that person. Delete." Depending on the economic situation in the family, the shared-phone strategy works for only a year or two before the parents give in and buy each child his or her own phone.

By far the most effective form of parental control emerges through the selection of mobile phone plans. In many cases, parents regulate their kids' use of the phone by limiting the number of calls they can make. Nini, a thirteen-year-old Latina in Christo Sims's study of teenagers in Brooklyn, New York (Rural and Urban Youth), reflected upon her use of the mobile phone:

To call my mother, to call my father, or other important people like my grandmother to tell her to come pick me up if I need to come . . . leave out of school early, or whatever. Then I had a phone . . . like my father . . . I lost the one when I was seven, so my father didn't let me get one until I was ten, and then he gave me another one that he uses it, so I used it to call my mother. Like I only had certain friends' number, but my father says to not use my minutes, 'cause I have prepay, so he said not to use it, just put the number in if anything . . . so I always had their number in my phone. Then I lost that one, and then my father gave me one last year. I call . . . I put all my family numbers in, and then he let me put certain friends in that I really hang out with, and I could call them, but he says to make it fast so then all my minutes don't run out, and then I just got a new one because the old one . . . it got messed up, like the memory was all blurry, or whatever, so he bought me a new one.

As Nini suggests, her father imagined that the preprogrammed and prepaid phone card minutes would encourage Nini to preserve the minutes, facilitating her ability to use the phone for what he perceived to be essential calls to family. As with many other parents, over time Nini's father began to make exceptions to the rule, allowing certain friends' names and numbers to be entered into the phone.

Parents' attempts to shape kids' phone usage therefore involves a range of strategies, such as buying basic phones that come with a family plan and avoiding upgrading features, such as multimedia messaging service (MMS) or short messaging service (SMS). Middle-class families take advantage of their reliable credit history and the ease of paying bills by enrolling in family plans for their cell phones, which allow two or more phone subscribers to share a finite pool of minutes that are are billed to a single person or address. Family plans usually include phones for three to five family members and offer cheaper rates for calls within the networks and, in the United States, typically require a two-year commitment with companies such as AT&T, Verizon, Sprint, T-Mobile, and others. Many kids complained about their parents' selection of phones or plans without the latest or desired features. However, within the family plan model, parents effectively acknowledge that at least some amount of time will be used talking to friends. While there are other negatives for kids, such as parents' having access to the times, dates, and numbers that kids call, family plans make it easier and cheaper to keep in touch with family (and others on the same network) and it also guarantees that their kids will be able to call should they find themselves in difficulty.

One interesting implication of the different plans is that kids on family plans tend to have less awareness of how these plans work, or what a call or text message costs, unless their parents make them pay for certain features, such as SMS. In fact, many teens do not generally know what their parents pay each month or what the different mobile-phone plans offer until they "go over." Gabbie, a seventeen-year-old Chinese girl living in a middle-class suburb in the San Francisco Bay Area, described her experience of "going over" to C. J. Pascoe (Living Digital):

Gabbie: I have, actually. On text messages. Because we don't have that plan. And then my mom is like, "Why are we over two dollars this month?" And I was like, "Because I was text messaging."

C.J.: But only like two dollars. I've heard stories of like eight hundred or nine hundred dollars.

Gabbie: I think I've gone over fifty dollars once. And then that didn't go over very well.

C.J.: Did they make you pay for it?

Gabbie: No. They just got mad for a couple of days. After that they were fine [breathy giggle].

By contrast, many of the kids who lived in urban New York were aware of and adept with the various plans and possibilities of mobile phones. Dana, a Latina fourteen-year-old in Brooklyn, discussed with Christo Sims (Rural and Urban Youth) the way she tries to balance her mother's selection of a mobile-phone plan with her relationship with her boyfriend.

Dana: Yeah, but you . . . like when I first . . . uh-huh, when I started talkin'
to 'im, when I started with my man, I was like, "You got Sprint, right?"
[laughter] 'Cause I got worried, because then I'm the one that gets in
trouble because I don't work, you know, and I gotta be careful with my
mom, and text messages . . . they be like fifteen cents per message, and
when my mom finds out that the bill is more, I'll be like, "I don't know,
it's probably because my phone is modern," and that's my lie because my
man'll be like, "Well, I miss ya," and I'm like, "Yo, stop text messaging me,
'cause they charge," and then he'll keep on, but . . .
Christo: So you don't have a text-messaging plan?
Dana: Nah.
Christo: Who . . . your boyfriend will text message you?
Dana: Yeah, all the time, and my mom will . . . he always sends me
pictures, too, and my mom she'll be killin' me. Like she don't know it yet,
but I told her that, "Oh," I lied, "Oh, I was talkin' to my friend from
Georgia, and she sent me a text message and I had to write back to her."
"All right, don't do it again," so I haven't been using text messages.

This situation differs dramatically from that of low-income and working-
class kids such as Elena, a sixteen-year-old of Armenian descent, who is
not on her parent's family plan and therefore must maintain a continuous
cycle of credit on her own. Elena clarified her situation to C. J. Pascoe
(Living Digital):

We are all independent kind of thing because we don't have jobs kind of thing. My
sister has a job. And we won't be able to afford if there's a plan kind of thing. But
my mom and my dad have a plan. But all the kids, like me, my sister, and my
brother, have pay-as-you-go cell phones.

When she ran out of money, her phone number could not be renewed
and she lost the number. After losing her phone, which in many low-
income families can be akin to losing one's identity (see Horst and Miller
2006), Elena started negotiating with her brother to buy his old mobile
phone. In contrast to kids in middle-class families, working-class and
low-income kids such as Elena are often acutely aware of the cost of calls
(Chin 2001).

Alongside controlling and managing costs, owning a phone gives kids
and parents more freedom to control how and when they use their phones
and their private communication. One mother named Geena in Silicon
Valley mentioned that she bought a keyboard-enabled phone on which

she has learned to "type." Now that she uses SMS to communicate with her son, she thinks it is easier to keep abreast of her son's activities and movements throughout the day when she just wants to know where he is and if he is all right. She believes that the increase in communication actually improved their relationship. Geena also discovered that texting over SMS makes it easier to parent her son. As she described it, texting "takes the emotion out of" the moments when she is checking on her son's whereabouts or telling him to come home when he is out too late or somewhere she doesn't think he should be (Horst, Silicon Valley Families). By contrast, voice conversations typically lead to arguments because she can "hear" the tension in her son's voice or potentially distracting sounds on the other end of the phone. Indeed, many teens acknowledged that if their parents called when they were out late, they would answer but "I make excuses. I'm like, 'I'm at my friend Cathy's house and they really like Cathy' so they go with that [giggle]" (boyd, Teen Sociality in Networked Publics). Hearing the excuse is something Geena thinks she can avoid, or at least circumvent, via texting.

Going Online: Bandwidth, Passwords, and Privacy

Parents, guardians, and other significant adults in kids' lives spend a great deal of time managing their kids' opportunities to go online at home. At the lowest income levels, such as in urban Los Angeles, the lack of access to computers and online spaces at home, as well as the public nature of domestic life, often mitigated the issues of privacy that were available to people in better economic circumstances. Many low-income families we interviewed did not have a working computer or Internet connection in the home. In cases where computers and the basic infrastructure were present, connection speed remained a central issue. For example, Lou, a sixteen-year-old white student who lives with his grandfather and aunt in a suburb on the fringe of an upper-middle-class area in the San Francisco Bay Area, felt frustrated by his family, who refused to upgrade their dial-up connection. Lou described his Internet connection as "not even fifty-six; it's thirty-two on a good day," and he perceives his inability to obtain a quality connection at home as a severe restriction on his social life (Pascoe, Living Digital).

While in Lou's case the slow connection speed reflects apathy or lack of appreciation for the importance of going online for many teens, in other families the lack of high-quality infrastructure is intentional. Mic, a fifteen-year-old of Egyptian descent in Los Angeles, noted that his parents will not allow him to have the Internet at home: "I don't really have access to the Internet at home because my dad always hears bad things happening on MySpace and he doesn't think I'm mature enough to get the Internet at this point" (boyd, Teen Sociality in Networked Publics). The media access of one of Lisa Tripp's interviewees is restricted for similar reasons; she reported that her mother will allow her online only if she is in the same room, and that her mother often hides or takes the ethernet cable and modem with her when she leaves the house (Tripp and Herr-Stephenson, Los Angeles Middle Schools). In their "Teaching and Learning with Multimedia" study, parents consistently expressed concern about child predators' using sites such as MySpace to find kids.

Whereas concerns over child predators preoccupied some parents, others struggled with the Internet's ability to distract kids from the main work of childhood: education. Juan, a working-class Mexican immigrant supporting his two daughters as a single parent, described this dilemma:

No, no. Hay que tener Internet pero quitar esos programas. Porque muchos los quitaron, ¿verdad? Porque si no ya no se van a dedicar al estudio sino a lo demás.

(No, no. It is okay to have Internet, but you have to remove those programs [pornography and MySpace]. Many parents have removed them, right? Otherwise kids won't study, and are only going to be doing that.) (Translation by Lisa Tripp)

As Juan suggested, many parents feel compelled to be very strict about websites that are oriented to entertainment or communication with friends. Juan, and other parents like him, feels it is important to send a clear message to his kids about the value of the computer for education. Anita, a Mexican immigrant in a working-class family in Los Angeles, talked to Lisa Tripp about how she routinely argues with her thirteen-year-old daughter Nina about going online (Tripp and Herr-Stephenson, Los Angeles Middle Schools 2006).

Anita: *[Mi hija] se pone en la computadora y le digo que la computadora es para hacer tarea, no es para estar buscando cosas en la computadora. Y a veces [mis hijas] se me enojan por eso. Y les digo: "No, la computadora yo se las tengo para que hagan tarea." A veces les pregunto: "¿tienen tarea?" O: "estás haciendo tarea." Pero a veces tengo que estar lista a ver qué es lo que están haciendo. Se meten a la Internet y tantas cosas que sale salen ahí. Y se ponen a mirar*

sus amigas y eso. . . . Entonces, es lo que no le gusta a ella que yo le diga:
"¿sabes qué? La computadora no es para que andes buscando; es para lo de
la escuela."

([My daughter] sits in front of the computer and I tell her that the computer
is for doing homework, not for looking around. And sometimes [my
daughters] get mad at me because of that. And then I say, "I got this
computer so you could do your homework." Sometimes I ask, "Do you
have homework?" Or, "Are you actually doing your homework." ?" I have
to keep a close eye on them to see what is going on. They get on the
Internet, and with so many things there. They look for their girlfriends
and all. . . . They don't like me saying, "You know what? The computer is
not for you to be looking around. It is for schoolwork.")

Lisa: *¿Qué es lo que más le preocupa a usted acerca de la Internet y sus*
hijas?

(What is your main concern with the Internet and your daughters?)

Anita: *Lo que me preocupa . . . ya ve . . . es que salen muchas cosas ahí que se*
meten con niños, y a veces platican con ellos, y a veces no saben ni qué gente
es. Es lo que me preocupa, porque digo "no." Y a ver qué es lo que están mirando
ellos y uno tiene que estar siempre listo con ellos. A veces estoy que les quiero
quitar la Internet, pero a veces me dice él: "por su tarea está bien. Porque después
van a andar que 'me voy a hacer tarea,' 'que no tengo computadora,' 'que no
tengo esto.'" Pero es por lo que más peleo ahorita con ellos.

(My main concern is . . . you see . . . you hear all the time that people try
to reach kids and talk to them. Sometimes [kids] don't even know who
they are talking to. That is my concern. That is why I say, "No." I need to.
And I keep an eye on what they are looking at. One always has to stay
alert. I always need to be attentive. Sometimes I feel like canceling the
Internet, but my husband says, "It is good to keep it because of their
homework. You don't want them saying 'I need to go somewhere else to
do my homework,' or 'I don't have a computer,' or 'I don't have this.'" But
this is mostly what I fight about with them these days.) (Translation by
Martin Lamarque and Lisa Tripp)

Given the economic burdens that they take on to obtain a computer in
the first place, many parents in low-income households and in working-
class homes believe that the primary purpose of a computer and the
Internet should be educational pursuits, such as homework.

While parents may be in control of basic access, once young people go online kids assume much of the responsibility for structuring their online worlds. In much the same way that teenagers now hang out with their friends at the local Starbucks, the parking lot at In-N-Out (a popular fast-food restaurant in California), and the mall, kids define social network sites, online journals, and other online spaces as friend and peer spaces; adult participation in these spaces is problematic or "creepy." With the ability to control who can and cannot view one's profile or page with passwords, nicknames, and other tools, kids use new media to facilitate and reinforce the segmentation of their peer-driven worlds and their familial worlds (see chapters 2 and 3). Fourteen-year-old Leigh, a white teenager living in Cedar Rapids, Iowa, said, "My mom found my Xanga and she would check it every single day. I'm like, 'Uh.' I didn't like that 'cause it's invasion of privacy; I don't like people invading my privacy, so." When asked why Leigh does not want her mom to read her Xanga, Leigh responded, "I don't know, 'cause I just put stuff on there that maybe I don't want her to know" (boyd, Teen Sociality in Networked Publics).

The expressions of tensions surrounding going online varied across socioeconomic class, geographic location, and even religious background. As Christo Sims discusses in box 4.3, many rural kids who are home-schooled connect to their friends in front of their parents using sites such as Bebo. Parents in middle- and upper-middle-class families varied from parents who completely restricted their kids from going on MySpace because of the fear of, if not panic over, child predators to those who saw new media as a space to mess around and learn. Many of the parents in the latter category religiously followed the advice of parenting organizations to navigate the changing media ecology. These parents typically monitored and regulated their kids through the placement of computers and laptops in the home. Although there are a range of sites, these organizations tend to offer rules and guidelines (e.g., no more than one hour of television per day) for families to adopt. Other parents tried to educate their kids about the dangers of digital personhood. For instance, by the time many of the kids in Silicon Valley were in high school, their college applications loomed large (Horst, Silicon Valley Families). In the competitive academic environment that constitutes this particular region, many parents, teachers, and guidance counselors had successfully convinced students that a "bad" profile on MySpace or another site represented a potential threat to their record, and that this could be the difference between Stanford, Berkeley, or one of the private Claremont colleges and a less prestigious California State University school. Still other parents emphasized independence, discipline, and the need for instilling judgment. Although their particular practices differed, many of the Silicon Valley parents were quite comfortable with the role of technology in their own lives and, therefore, did not fear it in the same way as those who did not or could not use computers, mobile phones, and other new media. By contrast, many of the parents who were strict or overtly tried to ban their kids from going online often acknowledged that their own lack of familiarity with computers contributed to their anxieties.

Box 4.3   The Milvert Family: A Portrait of Rural California
Christo Sims

At first glance, Lynn Milvert's use of digital media seems to resemble the image of the wired white fifteen-year-old so often portrayed in popular culture. She spends hours each day in her music-filled bedroom, sitting in front of a computer and effortlessly switching between a social network site, multiple instant messaging applications, and even a little homework. At this level of detail her routine seems quite similar to those enacted by teenagers featured in Heather Horst's study "Coming of Age in Silicon Valley," C. J. Pascoe's study of suburban northern California teenagers (Living Digital), and many of danah boyd's teenage participants from various urban and suburban contexts (Teen Sociality in Networked Publics). What makes Lynn's case unique, however, is that she lives in a remote region of the upper foothills of California's Sierra Nevada range. And while on its surface her use of technology looks similar to that of many other youth, both the local geography and her family's unique relations to the local community—its schools, its churches, and its politics—shape the particularities of her practices with new media in quite distinct ways.

Lynn lives at the end of a meandering driveway, which branches from a single-lane private road, which, in turn, forks from a quiet two-lane county road. Homes are few and far between in this high region of the Sierra Nevada foothills. Lynn lives in a single-story three-bedroom house with her father, mother, and seventeen-year-old brother, Nate. Lynn's father grew up a quick walk down the road from where they live now. He built their current house on a part of what used to be a family ranch. Lynn's grandma, aunt, uncle, and cousins all live within walking distance. This geographic closeness affords frequent family-centered social time for Lynn. At least once a week Lynn's family tries to have meals with members of the extended family. Almost daily Lynn walks down to her grandmother's house to watch satellite TV. During the summer, she babysits her infant cousin between roughly 9:00 a.m. and 5:00 p.m. four days a week.

While most local kids attend the regional public schools, Lynn has been homeschooled since sixth grade, largely with a group of other kids from her church. Lynn's particular form of homeschooling is not conducted alone, with a parent as the tutor, but instead with a group of roughly twenty kids who share a tutor and even attend class together for three hours three times a week. Lynn's class consists of both boys and girls, ranging in age from twelve to twenty-two. She considers everyone to be friends with everyone else's friends and few people have joined or left the group since Lynn was a young child. As Lynn put it, "Most of us have known each other all our lives."

Her family's participation in the local First Baptist church reinforces the group's durable composition. While the homeschool program is administered by a separate organization, many kids in her school program also belong to the church. The church, in turn, sponsors opportunities beyond school for the homeschooled youth to get together in social settings. Every Friday the church youth group organizes a social event. Out-of-town trips are planned for roughly one weekend a month. And every Sunday afternoon the youth group holds its own session after the regular service.

At the time I visited her, Lynn's engagement with new media usually took place at home, in her room, with the door open. The computer that she and Nate share—her mom, who works from home, has her own laptop—sits on a desk with its back pressed against the wall directly across from where the bedroom door opens to the hall (see figure 4.3). Lynn's parents moved it from Nate's room after he got in trouble. Most of Lynn's practices with digital media align with her participation in, and the relations between, family, school, and church. As with many teenagers, her favorite digital technology is a social network site. But unlike most teenagers who attend the regional high school "down the hill," she chose Bebo instead of MySpace or Facebook. She perceives it as safer. And unlike some teenagers who participated in various studies for the Digital Youth Project, she doesn't use social network sites and instant messaging to build new relationships at school or to maintain weak ties across expansive networks. Instead, she uses them to participate in her existing peer group. Her friends on Bebo match her densely interconnected friends from homeschool and church almost exactly.

Contrasting with the dense composition of Lynn's social network is the geographical dispersion of homes in her neighborhood. Being an "up-the-hill" family means much greater distance between homes; in most cases, it is not possible to walk or bike to the house of a friend. This is particularly true in the snowy winters. Without a driver's license, Lynn's collocated social activity with peers either requires routine, formalized group activities—such

Figure 4.3
Lynn's bedroom with the computer she shares with her brother. Photo by Christo Sims, 2006.

as school sessions, sports practice, work, and church—or convincing a parent, or other older person, to transport them to a common location. In both scenarios, spontaneous collocated peer gatherings are difficult to achieve.

These constraints on her mobility lead Lynn to spend a good deal of time at home. As a social space largely defined by her parents, home has been a place for family, schoolwork, and, occasionally, planned socializing with friends. On the Internet, Lynn finds ways to redefine the social possibilities of time spent in the home, beyond family, beyond working alone, beyond planned sociability, and toward unplanned peer-based socializing. Yet this technological reach out of the home is not directed toward the distant, unfamiliar, and global world of the Internet; it is not even directed toward most of the other teenagers who pepper the local rural landscape. Rather, it hones toward the small, and well-established, group of friends from her homeschool and church. This dense group places each individual member in a uniquely central position, a position that contrasts with the geographic dispersion of their homes and neighborhoods, a position in relief with the group's marginal relation to the teenagers who attend the public high school down the hill. It is an inversion of geographic and social isolation, a counterpoint to their perception of living "in the middle of nowhere."

As with locked diaries and closed doors, some parents admitted they simply could not resist the temptation to see for themselves what sites such as MySpace and Facebook are all about by sneaking around online behind their kids' backs. For example, Amy, a biracial (black and white) sixteen-year old in Seattle, described to danah boyd her mom's efforts to see what was on her MySpace account (Teen Sociality in Networked Publics): "My mom made [a MySpace], just so she could look at my page, so I made it private, and I won't let her on there." James, a biracial (white and Native American) seventeen-year-old in Seattle, noted:

[Mine's private] just because of the fact that my dad made a MySpace, and there's things on there that I probably don't want my parents to see, so I set mine as private, so someone has to request me as a Friend before they can actually look at my profile. (boyd, Teen Sociality in Networked Publics)

Other parents waited until problems emerged. Gameboy, a white sixteen-year-old who participated in Heather Horst's study "Silicon Valley Families," was caught smoking pot. After Gameboy's parents found out, his dad sat down with Gameboy and went through his MySpace page to identify "the stoners," which his father claimed to identify through the pictures and images posted on Gameboy's friends' profiles, their music preferences (e.g., heavy metal), and comments on their profiles about drugs and drinking. After examining their MySpace profiles, his dad then proceeded to closely monitor Gameboy to see if he "got high" after he returned home from hanging out with the stoners.

Many kids reference similar "horror stories" of parents' breaking into their sites, pages, and profiles, acts that teenagers view as invasive and embarrassing. In some cases, parents' transgressions into their kids' media worlds are humiliating. For example, fifteen-year-old Traviesa, a Hispanic girl in Santa Monica, California, described her own horror story to danah boyd (Teen Sociality in Networked Publics):

My mom, she found out [my password] one time. I was like, "Oh, shit." And then she wrote, "Oh, I'm sorry to everybody that's on here but my daughter is fourteen years old," or she didn't even know my age, I was fifteen at the time; she was like I'm fourteen. She was like, "Oh, yeah, she's fourteen years old and she doesn't need to be talking to all you old people and this and that, and she's not going to have MySpace anymore so bye." And then she wrote that on the About Me section and I read it. I was, "what is this? Oh my God, how retarded." I think it's funny, though. Parents are stupid. I don't know, most of the time they do it for our well-being, but sometimes they just don't know what they're doing. It's really sad.

As Traviesa acknowledged, most of these parental acts are motivated by the protection of kids' "well-being" rather than harassment for the sake of harassment. However, kids view these acts as a violation of trust, much like parents' listening in on their conversations or coming into their bedrooms without knocking. They also see these online invasions as ill informed and lacking in basic social propriety. A small number of teens do share with one another what they do when they go online, such as seventeen-year-old Anindita, who told danah boyd that "[My mom] goes on [my MySpace] all the time. I even show it to her. She knows my password. I really don't care 'cause I'm not hiding anything" (Teen Sociality in Networked Publics). Yet, most families admitted that the issues of privacy and control were contentious. Teens noted that they tried not to do anything wrong, but they wanted to maintain their privacy and autonomy and felt that they possessed the skills to judge their own actions and behavior when using new media.

Conclusion

Throughout this project, we carried out research in a range of homes and communities across urban, suburban, and rural locations, revealing the ways in which the institution of the family remains a powerfully determining force in young people's new media practices. Resisting the urge to classify or evaluate families in terms of language such as "divides" and "gaps," we chronicle parental attitudes toward new media and technology as well as a broader set of beliefs about how learning and education continue to shape what becomes possible for youth of different backgrounds. The ways in which young people and their families take up new media in their everyday lives cannot be viewed as a simplistic equation between access or divisions such as "rich kids" and "poor kids." Rather, the need to balance independence and dependence, parents' values and beliefs, and parenting style shapes participation. For example, many parents worried about the allure of social network sites in their daughters' lives or the addictive power of video games for boys, but the tactics to control participation in these activities varied. While all families used time—restricting going online until kids completed their homework and giving kids more time to play on the weekend—parents who were economically well off tried to regulate their kids' participation by creating rooms specifically for playing games, homework, and socializing with their friends. By contrast, many of the less well off families in urban Los Angeles and the San Francisco Bay Area took away the power cord, deleted programs, and kept low-speed access. However, these strategies were not just a matter of economic constraints; rather, beliefs about the correlation between computer ownership and education, and parents' anxieties about their own lack of experience with media, influenced their decisions and the type of regulation parents employed. Moreover, the extent to which parents were willing to give their kids autonomy over their day-to-day media usage also revolved around the assessment of whether parents thought their kids could or, in some cases, needed to exercise judgment, as was the case with many parents who gave their kids mobile phones. Parents noted that this decision involved a consideration of their children's gender, age, as well as maturity. For example, after Heather Horst's interview with Trudy, described in this chapter's introduction, Trudy's mother explained that she needed to create different rules for Trudy and her elder brother. Because she thought Trudy was more trusting than her brother, she believed Trudy was more vulnerable to answering messages from unknown solicitors. By contrast, Trudy's parents closely monitored the completion of their son's homework and even considered placing their son in counseling for what they felt was a video game addiction when his grades dropped. For Trudy's parents and others, the ever evolving media ecology compounds the challenges of parenting kids and teenagers.

This chapter examined how families deal with media and the internal dynamics that often structure the extent to which the use of new media is encouraged, restricted, and regulated. We began with a discussion of the role parents see themselves playing in their children's (and in some cases grandchildren's) use of media, and of the relative importance of rules in shaping family life as new media take on an increasing presence in the domestic ecology. In the first section, "Crafting Media Spaces at Home," we focused on the creation of public media spaces such as recreation rooms and of private media spaces such as the bedroom. The second section examined how parents make, take, and spend time with media by focusing on the ways in which families structure time for media use during the school year and summer as well as during the weekdays and weekends. We also explored instances of families' spending time together in and around new media, a practice not commonly discussed in much of the literature on the generation gap. This sense of capturing family time is closely related to the ever present sense that kids are growing up, and that there is only a limited amount of time to spend with family and to impart family values. Whereas the first two sections analyzed the spatial and temporal dimensions of new media in family life, the third section looked at the microdynamics of rule making and rule negotiation in families in relation to the debates and practices of using mobile phones and going online.

Unlike the other chapters in this book—which discuss peer-based sociability, communication, and expression—this chapter analyzes the influence of families in shaping new media practices. We aimed to provide an important piece of the overall contextual ecology of youth new media practices; other components of this new media ecology, such as the role of commercial industries, schools, and community institutions, are touched on in relation to specific practices of interest. With our attention to the role of new media in young people's everyday lives, we believed that families, and the domestic context generally, required an extended treatment because of the powerfully determining role that parents and siblings play in shaping conditions of access. In addition, families constitute one of the primary social contexts for ongoing informal engagements with new media. In many instances in our studies, new media represented a site of conflict between parents and children, and between siblings, over issues of access and control, and much of the social negotiation around new media centered on setting boundaries and rules of various kinds. In these settings, parents are often seen as clueless or incompetent in dealing with the norms and literacies of online peer culture. However, we also chronicled many instances of parents and kids coming together around new media, even for media production. These acts became moments for cross-generational communication as well as an expression of family identity. These antagonistic and cooperative forms of parent-child dynamics appear throughout this book as structuring contexts in our descriptions of peer-based practices.

Notes

1. While acknowledging the voluminous literature on media effects (Bryant and Zilman 2002; Gunter and McAleer 1997; Singer and Singer 2001; Strasburger and Wilson 2002), our work attends to the struggles around kids' participation with new media and, in this chapter, parents' use and regulation of new media.

2. Although it is outside the scope of our work here to define "American families" or the relationship between families and the broader category of households (see Netting, Wilk, and Arnould 1984), we recognize that "family" is a mutable category that changes in relation to the social, historical, and cultural contexts (Alters 2004; Coontz 1992). The families in our study vary from the nuclear family and divorced and single-parent households to blended, extended, and transnational families.

3. Alters draws upon Mintz and Kellogg's (1988) study of parenting in American family life.

4. In the context of Europe, Sonia Livingstone (2002) argues that household income and education remain the key factors for the strategies parents take to control and manage new media in their kids' lives. For example, she argues that for individuals with high income and low levels of education, cable and satellite television, game machines, and camcorders are viewed as important. By contrast, the Internet and books are valued in homes with both high education and high income levels.

5. Recent survey work in the United States indicates that some of these dynamics may be shifting. While in the past, families with high education tended to consume less-popular media, comparisons between 1999 and 2004 indicate a changing trend. Today families with college degrees and those with less than high-school education are high media consumers and families in the middle socioeconomic brackets consume the least amount of media. Roberts and Foehr (2008) take this as evidence that economic barriers to media are no longer as salient as they once were, and that educated parents are less critical of media than in the past.

6. Silverstone and Hirsch (1992) also argue that media serve dual functions in the home, what they term "double articulation," in that media are both physical objects as well as objects that convey meaning. Lally (2002) and others have criticized Silverstone and Hirsch for attributing too much credence to the uniqueness of new media and technologies.

7. Reflecting their textual and discursive approach, Alters and Clark (2004b) use the term "public scripts" to account for the ways in which families describe how they relate to media.

8. While in the past, community and neighborhoods functioned as the locus of interaction (see Castells 1996; Lievrouw and Livingstone 2002; Low 2003, 2008; Miller 2001; Miller and Slater 2000; Morley 2000), today the home represents the primary space for family and community life and for engagement with media and public culture. We found that the home was the dominant context for youth sociability and for new media practice in almost all the regions where we were carrying out research. The one exception was the case study in Brooklyn, New York, in Christo Sims's "Rural and Urban Youth" study, in which he found that teenagers spend a great deal of time outside and on the street hanging out with friends and

traveling on the subway system. We also found this to be the case among Dilan Mahendran's Hip-Hop Music Production study participants, who took advantage of the Bay Area Rapid Transit (BART) system to move from their residential locations in the hinterlands of the San Francisco Bay Area and into the city. We did not see this mobility among our Los Angeles study participants (see Martínez's study Pico Union Families), a fact we attribute, in part, to the lack of viable public transportation in the city.

9. In other parts of the United States, basements are often converted into recreation and media rooms. Given the potential for earthquakes, basements are not common in California.

10. As Benitez (2006) has argued for Salvadoran immigrants in Washington, DC, the ability to hear and see other family members during annual teleconferencing sessions helps to counter the distance and the difficulties of travel in the wake of difficult economic circumstances and undocumented status.

5 GAMING

Lead Authors: Mizuko Ito and Matteo Bittanti

In a lengthy interview over instant messenger (IM), twenty-two-year-old Earendil described the role that gaming played in his growing up. Earendil was largely homeschooled, and though his parents had strict limits on gaming until he and his brother were in middle school, Earendil and his brother got their "gaming kicks" at the homes of their friends with game consoles. After his parents loosened restrictions on computer time when he was fifteen, his first social experiences online were in a multiplayer game based on the novel *Ender's Game* and in online chats with fellow fans of Myst and Riven. Although he did not get his first game console until he was eighteen, he considered himself an avid gamer, and when he started community college he fell in with "a group of local geeks, who like myself, enjoyed playing games, etc." Gaming was a focus of activity for him and his friends, as they engaged in forms of play and game-related production that often required high levels of gaming as well as technical expertise, including networked gaming parties and participation in a group that was developing a modification on a popular game. Throughout his late childhood and adolescence, gaming was a focus for hanging out with his local friends, for online relationships, and for developing technical expertise (Ito, Anime Fans).

Although Earendil is a more committed gamer than most of the youths we spoke to as part of our research, the diverse kinds of social experiences he gained through gaming are becoming more and more commonplace. By 1999, more than 80 percent of U.S. homes with children had a game console (Roberts, Foehr, and Rideout 2005). Between 1999 and 2004, average daily gaming time for children went from twenty-six minutes to forty-six (Roberts and Foehr 2008). Among those who responded to our background questionnaire, 90 percent reported that they currently engaged in some form of electronic gaming, and 24 percent reported that they play games daily. Gaming represents the central form of early computer experience for kids. More than two thirds of the kids we interviewed had a game console at home before the age of ten. Not only is game play time growing among U.S. youth, but forms of game play and gaming demographics are diversifying. Drawing from a survey by the NPD Group, the Entertainment Software Association (ESA) (2007) reports that 38 percent of game players are women. Women age eighteen or older represent a significantly greater share of the game playing population (30 percent) than boys age seventeen or younger (23 percent). Although the first-person shooter (FPS) game Halo 3 was the best-selling title of 2007, only 15 percent of games sold that year were rated Mature, and sales of Family games grew 110 percent over the previous year. Accessible online and casual social games have tipped the balance toward adult women, or more accurately, toward a diversified age and gender demographic.

In the past two decades, as electronic gaming has gradually become established as one of the dominant forms of entertainment of our time, there has been widespread debate over the merits of the medium. Some have accused games of promoting violence and sexism. Despite very little empirical evidence that games lead to antisocial or violent behavior, popular perception persists in painting a picture of the aggressive, isolated, compulsive gamer.[1] Unlike the image of the violent gamer, sexism in games does have some grounding in everyday practice; although in the past five years the increase has been tremendous in the number of girls and women who game, most of those gains have been made in the area of "casual" games in online and handheld platforms, and more "hard-core" and technically sophisticated forms of gaming and game modding[2] are still dominated by boys and men (Kafai et al. 2008). In contrast to these concerns, researchers have been arguing that games have important learning properties that can be mobilized for education. Research in this vein was a central part of the early games industry, and it resulted in the development of a genre of game software that came to be known as "edutainment" in the 1980s and 1990s (see Ito 2007, 2009). More recently, educational researchers have engaged with simulation and other state-of-the-art games to argue that games provide important opportunities for learning in practice (Gee 2003; Shaffer 2006; Squire 2006).

Our work speaks to these public debates by considering everyday gaming practice and how it is embedded in a broader set of media ecologies and genres of participation with new media. Rather than key our research directly in the terms of these public debates, however, we stay close to the empirical material to provide a descriptive base and set of frameworks for understanding the role of gaming in kids' lives and learning. Much of the public debate has ignored or overlooked contexts and practices of game play. The focus has been almost exclusively on what people hope or fear kids will get from their play, rather than on what they actually do on an ongoing, everyday basis. It is only recently that researchers have been moving beyond a conceptual focus on gaming representation to look at gaming practice and the broader structural contexts of gaming activity. There is still little work looking at how different genres of games intersect with different types of game play and broader structural conditions such as gender, age, and class identity. This chapter is an effort to fill in some of these gaps in the research literature by positioning game play within a broader ecology of media practices and identities.

Gaming practices are extremely diverse in nature and form; game play is a complex and multilayered phenomenon. We would like to suggest a possible framework for examining gaming as it is embedded in practice in relation to what we have learned about the other contexts of new media engagement that youth navigate. We heard about gaming practices across the different case studies in our project, though only Matteo Bittanti's study (Game Play), Arthur Law's study (Team Play), and Rachel Cody's study (Final Fantasy XI) were specifically focused on game communities. In this chapter, we draw from a wide range of different case studies, including Bittanti's, Law's, and Cody's; Judd Antin, Dan Perkel, and Christo Sims's "Social Dynamics of Media Production"; danah boyd's "Teen Sociality in Networked Publics"; Heather Horst's "Silicon Valley Families" study; Horst and Laura Robinson's "Neopets" study; Mizuko Ito's "Anime Fans" study, and Patricia Lange's "YouTube and Video Bloggers" work. In this chapter we show a diversity in terms of the ages of the participants that we describe as we transition to a discussion of interest-based practices. Unlike the contexts of family and the friendship-based peer groups we describe in earlier chapters, interest-driven practices such as gaming are not age specific, and it becomes important

to look at how youth engage with mixed-age gaming practices and discourses and also to consider the trajectories of how gaming practices
extend into adulthood. Although our focus is still on gaming in the teen
years, we quote older gamers reflecting on their practices growing up with
games or describing the cultures of gaming more generally as reflective
practitioners.

We start our discussion with a framing of the debates around gaming
and learning, suggesting how a practice- and youth-centered approach
can inform this conversation. The body of this chapter is organized in
terms of genres of gaming practice: killing time, hanging out, recreational
gaming, mobilizing and organizing, and augmented gaming. We conclude
our discussion with an analysis of the broader structural and cultural conditions of gaming that shape how the different genres of practice relate to
one another, and the ways in which individuals gain access to or are
excluded from various game play experiences.

Conceptual Framework: Gaming in Context

The dominant approach to studies of gaming and learning focus on the
relationship between the gamer and the text. This holds on both sides of
the aisle. Just as detractors assume that the violent content of the game
encourages violent behaviors (Anderson, Gentile, and Buckley 2007), proponents of games and learning generally assume that learning follows from
good game design.

Although there has been a considerable amount written on games and young
people's use of them, there has been little work done to establish an overall "ecology"
of gaming, game design, and play, in the sense of how all the various elements—
from code to rhetoric to social practices and aesthetics—cohabit and populate
the game world. . . . The language of the media is replete with references to the
devil (and heavy metal) when it comes to the ill-found virtues of video games,
while a growing movement in K–12 education casts them as the Holy Grail in the
uphill battle to keep kids learning. While many credit game play with fostering
new forms of social organization and alternative ways of thinking and interacting,
more work needs to be done to situate these forms of learning within a dynamic
media ecology that has the participatory and social nature of gaming at its core.
(Salen 2007, 2–3)

As Katie Salen, editor of *The Ecology of Games: Connecting Youth, Games, and
Learning* (2007), notes in the introduction to her book, what is largely

absent in the literature is an account of the relations among players, texts, and contexts of play. Researchers who have studied the reception of media such as books and television have argued for some time now that social context has a formative influence on reception (Buckingham 1993; Jenkins 1992; Mankekar 1999; Radway 1984). With interactive, customizable, and user-modified media such as video games, this is even more the case. Our focus is not on the relation between individual kids and game content and representation, but rather on how game play practice and activity are situated within a broader set of cultural and social engagements and contexts. The focus on activity in context means paying attention to the diversity in contexts that structure different forms of game play—the broader social and cultural ecology—rather than assuming that psychological and cognitive dispositions play the most important determining role.

Gaming occupies a complicated position in relation to structures of age, class, and gender because of its status as a technically-driven recreational activity usually associated with lowbrow, male-dominated identity and practice.[3] The moral panics over games rotting the hearts and minds of children share many of the familiar concerns voiced about television; games are frequently linked to the corrupting "bad screens" of television (and working-class culture) rather than the "good screens" of computers and middle-class culture (Seiter 1999a; 2005). Further, much like earlier forms of youth-centered popular culture, video games are a site of moral panics where intergenerational anxieties are projected onto new media (Cohen 1972). The technical sophistication of games, both as texts and practices, however, throws a unique twist into these existing cultural conflicts. While those who see gaming as an avenue into certain forms of technical expertise and learning have argued that educators and designers should work to make games attractive to girls (Cassell and Jenkins 1998; Kafai et al. 2008), others have argued that gaming reproduces sexist and consumerist logics that are often of dubious value for youth (Kline, Dyer-Witheford, and de Peuter 2003; Sheff 1993). Questions about what kids learn through games are a site of conflict among the values inflected by class, gender, and generational identity.

The controversial nature of this medium becomes explicit, for instance, in the process of establishing a set of norms about the "appropriate use" of games. Parents and kids' perspectives often collide. The nature of the clash, however, is varied. In chapter 4, we have seen these conflicts playing out in how parents from different class backgrounds regulate gaming in the home. We also have noted how certain gaming practices can function as an intergenerational wedge, where parents are shut out from certain forms of media engagement. Conflicts about how games are perceived were evident when kids talked about gender and gaming and in the larger proportion of boys who engaged in the more geeked out forms of gaming practice. In this chapter, we work to tease apart some of the specifics of how these general cultural valences play out in relation to specific game genres and genres of participation with gaming sociability and culture. Although certain core practices of recreational and geeked out gaming are strongly associated with the young, white male geek cultures that were foundational to early game practice, today we see a much more variegated palette of gaming practices. The overall statistics of an expanding gamer demographic need to be contexualized within highly differentiated forms of gaming activities. Our effort here is to specify some of these distinctions among different forms of game engagement.

When we examine gaming from the point of view of gamers and game practice, then a different set of learning issues comes into view. While we do not underestimate the relevance of the text, it is just one among a series of players in the ecological dance that results in complex social, cultural, and technical outcomes. For example, one of the most important outcomes of the practices that we call "recreational gaming" is the fact that young people develop social networks of technical expertise. The game has not *directly* and *explicitly* taught them technical skills, but game play has embedded young people in a set of practices and a cultural ecology that places a premium on technical acumen. This in turn is often tied to an identity as a technical expert that can serve a gamer in domains well beyond specific engagements with games. This is the kind of description of learning and "transfer" that a more ecological approach to gaming suggests.

We follow this approach through the body of this chapter by analyzing how gamers talk about their own investments in games in relation to the practices that they describe. In line with an ethnographic approach, we see culture and discourse as constitutive of everyday practice and vice versa. Taking gamer viewpoints and investments seriously on their own terms challenges some of the arguments that both proponents and detractors of games bring to the table. While educational proponents of gaming suggest that games provide a motivational structure that will engage kids in more academic learning tasks, gamers talk about games as killing time and a waste of time and see value in precisely those properties of games that enable a certain state of distractedness. Even in the case of games that are difficult to learn and that require sustained investments of time, gamers often enjoy the practice because it is cut off from their everyday identities. It is a space to compete in and achieve in where there will not be consequential failure in real life. The appeal lies precisely in the fact that the game outcomes do not transfer to the real-life economies of academic achievement and playing the role of the good student, daughter, or son. The real-life social ecology of a kid's life has a powerfully determining effect on what kids get out of gaming. What they learn from gaming is not necessarily what is embedded in game content, nor what parents and educators hope and fear. In the description that follows, we outline genres of gaming practice that have emerged from our research to discuss the ways in which gaming, learning, participation, and identity are intertwined in kids' everyday play.

Genres of Gaming Practice

Grounded in the previously described ecological approach to gaming, our genres of gaming are related to the genres of media participation (hanging out, messing around, and geeking out) that we outlined earlier. Rather than assume that game genres, platforms, or specific texts determine game play practice, we organize our description with different practices of play that emerged from our ethnographic material. These genres of practice correspond loosely to different genres of games, but they are *not* determined by game genre. For example, puzzle games are typical for the genre of game practice we describe as "killing time," but other games such as first-person shooters or side-scrollers on a Nintendo DS could also perform that social function. These different genres of gaming practice also are loosely correspondent with different social networks and genres of participation. Where killing time is a largely solitary activity, hanging out corresponds to our model of friendship-driven sociality. Recreational gaming is the most central practice of interest-driven peer-based gaming networks and is often a site where we see messing around genres of participation. When we move to the genres of organizing and mobilizing, and the practices of augmented game play such as modding and machinima[4] making, we are moving into the domain of geeking out. While these groups also have a peer-based structure at the core of the practice, they are more differentiated than the practices of recreational or social gaming, and there is a clear demarcation between the core production community and those who use and access their work.

Killing Time

Certain forms of gaming have long provided opportunities to fill small gaps in the day or longer stretches of waiting time. Tucking a crossword puzzle or word-search book into a commute bag, or getting out a deck of cards for solitaire, are all examples of the solitary, time-filling gaming that we are characterizing as "killing time." These are the practices in which people engage with play and gaming to procrastinate or fill gaps in the day. With video games, it happens mostly through nomadic devices such as portable consoles (Nintendo DS, Sony PSP), mobile phones, and laptops. These practices also can happen in desktop situations, such as when someone takes a break from work to play a puzzle game on Miniclip. Games are often used while waiting for relevant things to happen, as fillers between more structured events. Although we found that a wide variety of kids engaged in killing-time forms of gaming, these practices tended to skew toward either younger or less experienced gamers, or for times when more sustained gaming was not an option. For example, Christo Sims notes that students at the video-production center where he, Judd Antin, and Dan Perkel observed are keen to engage in gaming activities during the short breaks between their lessons:

The Center was largely run like a hands-on class, with an adult instructor setting an agenda and directing the students in various video production exercises and activities. The kids had unstructured time before and after class as well as during a short break in the middle of each day's session. During these free moments (maybe fifteen to twenty minutes long) many kids would get on one of the lab computers. While MySpace was a popular activity during this time, so too were casual games on sites like Miniclip as well as Flash games on websites for candy companies and other youth-targeted advertisers. (The Social Dynamics of Media Production)

The dominant discourse of this form of gaming is about boredom and filling time. Digital games are used to pass the time when traveling on a bus, car, or plane, or in other situations when there is little else to do. For instance, Nick, a sixteen-year-old black and Native American boy from Los Angeles who danah boyd (Teen Sociality in Networked Publics) interviewed, said, "If I'm bored, I play that little . . . it's a little rocket game where you shoot rocks. I play that. If I'm real bored and I really have nothing to do, that's what I do." Similarly, Natalie, an eleven-year-old white fifth grader Heather Horst interviewed as part of her study on Silicon Valley families, said: "I play with my Nintendo probably like a few times a week probably. . . . Mostly on the weekends, because sometimes my weekends are really busy, sometimes they're not, but when they're not busy, I get kinda bored, so I just play."

This genre of gaming also can be used as something to focus on in a social situation that a subject might find awkward. For instance, Monica, a Latina fourteen-year-old from Santa Rosa, California, who is part of Matteo Bittanti's "Game Play" study, said,

> Often, when I am waiting for a friend [in a public space] to show up I start playing puzzle games on my phone, not because I particularly enjoy them, but because I don't like people staring at me. . . . In a sense, I am pretending to be busy, but it's easier to fake this than, let's say, a conversation.

Portable gaming can occupy gaps in the day when one is out and about. Another teen whom boyd spoke with, Luke, a sixteen-year-old from San Francisco, said: "I always carry a [Nintendo] DS with me. It's small enough so that it can fit in [one of the pockets of] my jacket, along with one or two games." In tandem with the evolution of portable media, gaming is starting to infiltrate more and more of the little gaps in everyday life.

These examples also illustrate another key feature of gaming as killing time: its solitary nature. Even when pursued in a social context, such as at the Center (The Social Dynamics of Media Production) or when inhabiting public space, killing time by gaming involves carving out a one-on-one space with the game. We see this in an example that Rachel Cody encountered in her study of Final Fantasy XI. When members of a group are "camping," or waiting for a monster to appear in a particular place, there are often long stretches of waiting time. At these times, players would often open a new window to play a small Flash game, even while still occupying the shared social space in the multiplayer game. Although public discourse has tended to associate antisocial and solitary behavior with violent, graphically sophisticated games, we find that these forms of killing time gaming that are generally seen as "harmless" or "casual" were the ones that were most likely to be pursued as solitary activities. While we do not see these forms of gaming as sites of profound social activity or learning, they are part of the play, of the messing around with new media that are seamlessly integrated into kids' everyday life rhythms.

Box 5.1 Neopets: Same Game, Different Meanings
Laura Robinson and Heather A. Horst

Neopets[5] (see www.neopets.com) is a virtual pet website owned by Viacom that enables members to select, feed, and care for virtual pets. Reminiscent of Tamagotchi and Pokémon, Neopets's members use a virtual currency called neopoints to buy food, pets, and toys for their pets; create shops and galleries; build and decorate houses; and acquire equipment to compete with other characters or play in the Battledome. The site is also host to more than 250 casual games, varying among 3D player games, Flash and Shockwave games, PHP games, and in-world quests.

Through a variety of activities, players and participants can explore the facets of "Neopia," the virtual world where the pets live. Viacom emphasizes the creative play that can occur through these digital engagements, but in popular and academic circles (see Seiter 2005) Neopets continues to be criticized for its encouragement of capitalism, as exemplified through the salience of neopoints in facilitating participation in Neopia, the encouragement of commercial enterprises (e.g., creating shops), as well as gambling and playing the Neopets stock market (see box 7.5). Other critics focus on the immersive advertising and dislike the increasing availability of merchandise, such as "plushies" (stuffed animals resembling specific species of Neopets), as well as *Neopets* magazine, mobile-phone video games, screen savers, and breakfast cereal. Neopets's parent company, Viacom, also takes advantage of its ownership of Nickelodeon, a popular kids' television network in the United States, by marketing Neopets and its associated products to children during afternoon and Saturday-morning television shows. Although parents and others continue to be concerned about kids' lack of awareness of the immersive advertising and capitalist ethos, most kids do not differentiate between the marketing in online spaces and the marketing that occurs in everyday life on television, billboards, and the array of electronic goods in contemporary homes.

While debates over the value of consumerism in gaming marketed to kids persist, our qualitative study of Neopets players suggests that Neopets is a highly flexible gaming site that allows kids (and adults) who play to adapt their engagement to their own interests and needs. For some players, it is all about the games. For others, interest and participation in Neopets is tied to the creative possibilities inherent in sites such as Neopets. For yet others, sociality is the key draw. For example, Mike, a seven-year-old who lives in an economically well-off and highly wired household in northern California, Neopets is about the thrill of the game. Mike was passionate about playing games online—any kind of game, from Neopets to Club Penguin. When asked why he liked Neopets, he made it clear that it was all about the games—not his pet, not creating a house, not any activity except playing games. Laura Robinson asked Mike, "Are you ever worried about your pet getting hungry or having treats?" Mike quickly answered negatively, "I don't care about my pet at all. I just want to play the games!" When Mike tried to show Laura which games he liked to play, he attempted to open his site, but somehow he could not remember his password. He explained that he didn't feel any connection with the Neopets he created. Rather, he only wanted to access the games. In fact Mike repeatedly created new accounts and even played under other people's pets. His strategy was to earn points for all his friends in return for logging in at their homes.

By contrast, "newbie" Neopet player, Jackson, could not care less about playing the games. As Jackson, a nine-year-old from suburban northern California, explained, "I really like to make the pets. I even make new user names or let them die just so I can make more of them." Jackson "loves" Neopets, but not for the reasons we might expect. For Jackson, Neopets is about the creative possibilities inherent in the creative act. Creating neopets, petpets, neohomes, and any other of the virtual venues or creatures is what drew Jackson to the game. This creative orientation was not surprising when one begins to understand that Jackson comes from a highly creative family. His parents and siblings all have artistic tendencies, although they take different forms: playing the guitar, dancing, and drawing. For Jackson, Neopets becomes an extension of his home world in which creativity is honed and valued.

Yet other Neopets players value the site for social connection. Mindy, a teenage female player from California, explained why she was invested in the site during high school: "I just loved playing it with my friends." For Mindy, her Neopets experience was centered in sociality. Neopets was framed as a reason to go to a friend's house, a reason to call a friend, or a reason to chat. Mindy's introduction to the site was through a friend with whom Neopets became a conversation piece, a shared experience that further cemented their friendship. When Laura asked Mindy about her neopet, Mindy explained,

"Well, I didn't really check on my pet all that much. You know, it was more about being with other people and playing Neopets with them." As Mindy suggested, the social connection that Neopets allowed her to form with others was her key interest in the site. As with Max and Jackson, the breadth and flexibility of Neopets—be it playing games, being creative, or making and maintaining friendships—enabled Mindy to shape and customize her own engagements online.

In contrast to the genre of gaming we characterize as killing time, much of game practice centers on social activity of various forms. The genres that follow are all examples of different, more sociable forms of gaming. The first is how the hanging out genre of participation intersects with game practice.

Hanging Out

The hanging out genre of participation happens when people engage with gaming in the process of spending time together socially. It is largely a form of friendship-driven sociability; while gaming is certainly important, it is not the central focus. Video games are part of the common pool, or repertoire, of games and activities that kids and adults can engage in while enjoying time together socially. Although games are usually considered occasions to compete around clear outcomes, this orientation can often be superseded by a more conversational or relaxed mode. Played this way, games are not inherently different from traditional board games. In a sense, they represent their electronic evolution. Like board games, hanging out forms of gaming were not as strongly gendered or age specific as the more geeked out forms of gaming that we examined; though boys were more likely to talk about gaming as a social focus, the hanging out genre of gaming represents a relatively democratic and accessible form of play.

As described in chapter 4, gaming can facilitate the interaction between peers but also between youth and adults. In fact, the family is one of the most common contexts for gaming as hanging out. In their detailed studies of game play in the home, Stevens, Satwicz, and McCarthy (2007) describe the settings in the home around the game console where siblings and playmates move fluidly in and out of game engagement with one another. This family gaming increasingly includes parents as well. A study conducted by the Entertainment Software Association (ESA) states that 35 percent of American parents say they play computer and video games. Among "gamer parents," the ESA (2007) says, 80 percent report that they play video games with their children, and two thirds (66 percent) say that playing games has brought their families closer together. Hanging out genres of gaming enable people to bridge different forms of gaming expertise and to cross generational and gender divides. For instance, Steven, a twenty-one-year-old from Mountain View, California (Bittanti, Game Play), said,

At Christmas, I played this game called Scene It? for the Xbox 360 over [at] my girlfriend's house. We played with her parents as well. . . . It's a trivia game about the history of cinema and you use a big controller instead of a conventional joypad. It was fun. We got to sit down on the couch and play together, and we laughed at our mistakes and we had a really good time. I mean, I would not normally spend that much time with my girlfriend's parents, you know? [laughs]

The more casual mode of this kind of gaming sociality facilitates game play by those outside the stereotypical gamer demographic. The Nintendo Wii is in many ways the emblematic platform for hanging out as a gaming practice. This console was specifically designed to reach a broader range of players. Another example is the increasing success of music titles such as Rock Band and Guitar Hero. Games that tie into established forms of social bonding, such as music, dance, and sports, seem to invite this orientation. A fourteen-year-old white boy in Dan Perkel's study "MySpace Profile Production" described his involvement in fantasy football and basketball leagues. He plays for about five minutes a day, though many of his friends are much more involved. He said that "it is hard to stay away from it." When Dan asked for clarification, he explained: "If your friends are all talking about fantasy sports, naturally you're going to want to be in their conversation so that's basically why most people do it." Even solitary puzzle games can take on a social hanging out quality when there are others around. In his observations at a video-production center, Dan Perkel (The Social Dynamics of Media Production) frequently observed kids playing games on sites such as Miniclip in the downtime between activities. They often would invite others in the vicinity to observe their game play and move in and out of social and solitary engagement with the games.

Hanging out gaming also includes online practices such as participating in social guilds in massively multiplayer online role-playing games

(MMORPGs), where players enjoy the social affordances constructed by the games. MMORPG players spend many hours logged in to the shared space of the game, and much of that time is occupied with casual hanging out, conversation, and activities such as bartering or exploring. The time spent actively pursuing game goals is only one part of what they do online. The time and space around the more goal-directed activities of gaming becomes a site for social conversation and sharing. In Dan Perkel and Sarita Yardi's study "Digital Photo-Elicitation with Kids," they spoke to a young RuneScape player, Iris, who was ten years old and of mixed race (white and black). She enjoyed hanging out on the site because of the social environment.

I like that you can play with a lot of people at the same time. It's like you have a normal life, and you get to talk to people. And it's not only one player; it's more than one player. And it's not that you're talking to an actual robot, but you're talking to actual people playing.

She said she will play with a friend of hers in the late afternoon when they both get home, but she will also talk to others she comes across in the game. The space of the online game becomes a hangout to meet her friends both offline and online.

In Rachel Cody's study of Final Fantasy XI, the core players of the "linkshell" (player guild) she was participating in would use a voice-chat program, Ventrilo (Vent), to stay in touch with their team constantly while they were at the computer. She talks with Ryukossei,[6] a nineteen-year-old Asian-American player.

Rachel: How did you like it?
Ryukossei: I loved it. That was a great linkshell, I thought. And, like, yeah, it was pretty fun. It was good times.
Rachel: Did you make any friends?
Ryukossei: Oh, yeah. Especially the people on Vent. If I didn't have Vent, I wouldn't be playing this game, like, seriously. . . .
Rachel: Yeah, Vent made it a lot less lonely, I thought.

As noted in Cody's box 5.3, Ryukossei describes how the "24/7" connection on Vent made his teammates feel like a family. While players in an MMOG may be attracted to the game play initially, they often end up staying because of the social dimensions of the game. As described in box 5.3, players will often cite the social hanging out dimensions as one of the primary reasons to stay with the game.

In describing the more friendship-driven side of hanging out forms of game play, players often explicitly disputed public perception that games were antisocial. We found this with some of our older players, who were often reflective of their game play and more aware of the stigma (Bittanti, Game Play). Louise, a twenty-eight-year-old from Vacaville, California, said, "Playing games can be a solo act, but when you involve friends and family you become more engaged in the play. I believe this represents our human need to be connected to others in a real-world environment." Frederick, a twenty-two-year-old from San Francisco, had a similar viewpoint:

Games are shown to be social tools that, in various ways, socially connect people of the current and previous generations. It's like parents reading their children the same bedtime stories that they themselves fell asleep to as a child. I don't see how anyone could argue with that.

The practices of hanging out around games have affinities with other social games such as golf, bowling, bridge, or mah-jongg, and this is in line with our general framework of friendship-driven participation. While there are highly competitive modes of engagement with these games, the more everyday forms of engagement tend to be driven by the social activity. Just as with more long-standing forms of gaming and play, electronic games are a focus of social activity between friends and family. Although the play mechanics of the game may involve competition and representations of violence, just as in the case of sports and games more broadly, the playful conflict becomes a source of social bonding. As genres of gaming such as casual sports games, rhythm games, and social online games expand, we can expect that more and more of young people's unstructured time together will be occupied by these experiences. In their recent study of violence and video gaming, Kutner and Olson (2008) suggest that kids who do not play video games at all are more likely to be socially marginalized than those who do play. The conversations we have had with gamers also support this finding; hanging out forms of gaming have become part of the everyday and commonplace practices of social play for youth.

Recreational Gaming

While many people engage with games as a lightweight activity that fills dead time or is part of a social activity, for committed gamers competitive game play is more central to their orientation to the medium. This genre of gaming, what we call "recreational gaming," represents the core of what we think of as gaming practice: people gaming to game and getting together specifically to play games that require persistent engagement to master. If in the previous category gaming tends to be in the background, here it is in the foreground. Recreational gaming includes everyday in-home gaming, when kids are into a game, or play with friends or family. It can be both solitary and social. This form of engagement includes everyday offline gaming and dedicated services such as Xbox Live, where people enjoy playing online games such as first-person shooters and sports titles. As described in box 5.2 on first-person shooters, in recreational gaming, players can develop intense relationships to games. Unlike killing time and hanging out forms of gaming, with recreational gaming we see a stronger identification with the historically dominant gamer demographic—young males. We discuss these dimensions of gamer identity later in this chapter in the section on boundary work.

Box 5.2   First-Person Play: Subjectivity, Gamer Code, and Doom
Matteo Bittanti

Kenny is a twenty-one-year-old from San Francisco who used to play games on a daily basis when he was younger, but who then reduced his game time when he started college. He is now saving money to buy an Xbox 360 because "the love of the game is just too strong." He loves first-person shooters, a genre of game characterized by a subjective perspective that renders the virtual world from the point-of-view of the player character. According to Kenny, the "FPS embodies the quintessential traits of the medium." I decided to reproduce with minimal editing his comments on Doom, the most celebrated FPS, because they contain many interesting points. Kenny discussed the game with a specific discursive style (note the emphasis on the pronoun "I" to describe his game play experience—"My first encounter with a pinky demon scared me shitless"—that does not happen, for instance, when somebody is retelling the plot of a movie or a novel); a clear understanding of what lies beneath the formal structure of the game (to describe the experience, Kenny uses adjectives such as "exhilarating" and "dumb," [Doom] is very mechanistic and repetitive, and "simultaneously calls for civility, for rational thinking, and meticulous problem solving"); the morality code of the gamer ("I wouldn't touch the strategy guide until I beat the game"); an assumed importance of expertise in discussing games (historical contextualization); and an intense emotional investment in game practices.

Doom is my favorite video game of all time. I own all of them and have played, to some extent, all of them with the exception of Resurrection of Evil and the Master Levels of Doom. It was simultaneously a triumph of technology as well as game play, serving as, arguably, one of the most influential games of all time. The greater half of big-name titles are all, in a sense, descendents of Doom: Bioshock, F.E.A.R., Stalker, Crysis. Pretty grotesque, but Doom III simultaneously calls for civility, for rational thinking, and meticulous problem solving. The problem solving goes much deeper than switch flipping, key finding, and dashing for the exit. Every enemy you encounter is a problem that needs to be solved. Doom III is easy, but you'd never guess that based on the imagery alone: shocking, intimidating, frightening. It plays on your irrational fears, expecting you to panic, to slip, to shoot wildly at nothing, but there is a logic to the game, a code, like every game. Doom teaches one to hunt, to compose oneself as a gentleman before and after war. One must supplant, or supersede, many of the atavistic urges Doom encourages in order to truly master the game.

In Gears of War, there is nothing more satisfying than dismembering your opponent with a [chain saw]. Charging headlong into the fray, your Lancer, growling hungrily for Locust intestines, held high above your head, is exhilarating. It's also dumb. There are rules of engagement. The shotty [shotgun] trumps the [chain saw], and the sniper trumps the shotty. And I love first-person shooters! My daily gaming diet consists solely of first-person shooters! I bought Doom III the day it was released, with the strategy guide and everything. I swore, like I always do, that I wouldn't touch the strategy guide until I beat the game, for a very special reason. Doom III is *huge* on atmosphere, and I'd be hard-pressed to find a game that does a better job of creating such frighteningly gorgeous environments. The use of sound is phenomenal, and the monsters are just oozing with gory details.

My first encounter with a pinky demon scared me shitless. Hell, my first encounter with an imp left me shaking. It's scary! Well . . . at least it is until the imp is eviscerated by one shotgun blast and all that remains of that pinky demon after two well-placed shots is an incongruous pile of gore. It's this knowledge that separates one from the game. That's why I didn't touch the strategy guide. If I knew how to kill an imp before our encounter, I would have never experienced that fear.

Recreational gaming is deeply social, but unlike in the hanging out genre of gaming, the game play itself is the impetus and focus for getting together. It is interest-driven rather than friendship-driven sociality that drives gatherings in this genre of play. For example, one of our interviewees described "DS Fridays," when kids meet weekly to play specific Nintendo DS games. Annie Manion (Anime Fans), in her interviews of anime fans who lived in college dorms, found an active gaming-centered social life among some of the students. One of her interviewees, Cara, described how there was a group who would get together to play Smash Bros., and group members would develop different techniques and specialties in playing different game characters. Another example is Halo parties, where gamers gather to "frag"[7] each other. MaxPower, a white fourteen-year-old in Christo Sims's study in rural California (Rural and Urban Youth), described a Local Area Network (LAN) party, involving networking computers with sixteen kids, that he was part of. The LAN was set up with four Xboxes and four TVs. "It was for five hours straight. After the second hour, I couldn't take it anymore. I had to go out with me and my friend, Josh, just kind of went out and skateboarded a little bit while everybody was playing 'cause my eyes started to hurt." A white seventeen-year-old in Sims's study, a self-described geek, said he is part of regular LAN parties with computers, where anywhere from six to fifteen kids will get together regularly to play.

Through recreational gaming, kids build social relationships that center on game-related interests and expertise. As part of her "Silicon Valley Families" study Heather Horst interviewed an avid gamer, a white twelve-year-old who described his immersion in game play together with a good friend:

John Harker: My friend and I, we just lived [in] each other's houses alternating GameCube and PS2. Go over to a friend's, like, Xbox. . . . We have all-nighter video-game parties and so it's kind of pathetic but it's a lot of fun. . . . My friend just got, like, he's even more obsessed than I am. So he always gets games and I just go over to his house for the day and we'll make stuff, eat it, bike, and play video games. . . . Watch movies.

Heather: It's usually groups, like, how many of you can play, actually?

John Harker: I've had times when we have two TVs in the same room and we're playing joint, eight-player Halo. . . . Which is awesome. . . . Halo 2 is just an incredible game.

Heather: Okay. So you've had . . . you can do all of that.

John Harker: Yeah, and a lot of the time I just go over, "hey Joey, you want to come over to my house?" and it's just two people or something. And, well, say he's losing—he'll invite someone who's even worse than him and then he'll have someone to beat. So it just evolves like that.

As John Harker described, recreational gaming is a site of activity where more friendship-driven modes of gaming move fluidly into messing around and geeking out. As a genre of play, recreational gaming is compelling because kids can engage flexibly in these different modes of participation and learning. Like other more geeky, interest-driven pursuits, gaming differs from extracurricular activities that have higher status in mainstream teen sociality, particularly sports. At the same time, gaming is becoming a pervasive social activity among boys, so gaming virtuosity does provide some peer status as well as an important vehicle for social bonding. Gaming practices provide a focus for the development of identities of expertise, performance, and virtuosity—an arena of practice that differs from the demands placed on youth for academic performance. These are extracurricular spaces where kids can achieve in contexts that are detached from the high-stakes performance required of them in school, and where failure is not as consequential. They can frag and respawn repeatedly or change games and in-game identities if they do not like the path they have been on.

Another important dimension of recreational gaming is that the social relationships and knowledge networks that kids develop often become a pathway to other forms of technical and media-related learning. This chapter's opening discussion of Earendil is an example of how gaming became a focus of a certain trajectory of participation into different forms of media practices and literacies. Earendil's gaming interests became a focus of sociability and play in his childhood and early teen years, and in college his gamer friends introduced him to anime and to various other online activities. Gaming provided an initial focus for an interest-driven social group that became a friendship group supporting the development of technical and media-related expertise more generally. Similarly, in Katynka Martínez's "High School Computer Club" study, she noted that most of the boys associated with the club are avid gamers. After the computers in the lab became networked (in a moment they called "The Renaissance"), the boys would show up during lunch and even their fifteen-minute nutrition breaks to play Halo and Counter-Strike against one another. Again, this is an example of gaming providing a social focus for kids with broader technology- and media-related interests. As with other forms of interest-driven practice that we examine in this book, these are contexts that exhibit peer-based learning and knowledge sharing that are driven forward by the motivations of kids themselves. These dimensions of peer-based learning and the honing of expertise become even more pronounced when we turn to some of the genres to follow, such as organizing and mobilizing and augmented game play. These learning outcomes of recreational gaming call attention to the social and technological contexts of gaming practice rather than focusing exclusively on the question of the transfer of game content to behavior and cognition.

Organizing and Mobilizing

Gamers who are highly invested in their play will often become involved in more structured kinds of social arrangements, such as guilds, teams, clans, clubs, and organized social groups that revolve specifically around gaming. We refer to this as "organizing and mobilizing" practices in which the social dimensions of gaming become more formalized and structured and more identified with geeking out than with messing around. This is where we see the politicians and warlords of the gaming universe and the people they organize, collaborate with, and lead. Organized and mobilized forms of gaming are core to the practices of traditional sports as well as to games such as Dungeons & Dragons that became popular in the 1970s and 1980s. Electronic games as they became networked in the past decade have become a new site for organized forms of gaming and high-stakes competition. Gamers in networked systems can keep track of in-game skills, "gamerscores," records, reputation, experience points, and so on in international game networks. Services such as Xbox Live and the PlayStation Network are specifically designed to facilitate agonistic forms of playing in a particularly competitive environment based on a specific form of meritocracy: gaming skills. Online role-playing games enable players to organize guilds with formalized leadership and specialized roles and responsibilities. This genre of participation requires various degrees of commitment not only in terms of time and competency but also in terms of resources and economic capital, as gaming equipment (hardware, software, and services) are generally more expensive than other forms of mediated entertainment.

For the most dedicated players, competitive gaming might represent an evolution of recreational gaming, or they may engage in both genres of gaming. In a few cases, the passion for gaming can evolve into a profession. Consider, for instance, the rise of gaming as e-sport—electronic sports, or the play of video games as a professional sport—in countries such as South Korea as well as the United States and Scandinavia. "My hero is Lil Poison," said Grant, a twenty-two-year-old avid gamer from Sunnyvale, California, referring to Victor M. De Leon III, the world's youngest known professional video game player (Bittanti, Game Play). "His skills are incredible for a nine-year-old! I watch his videos online and I find them amazing." While recreational gaming is practiced by youth who have a variety of interests and hobbies, these mobilized practices are specific to a group of teenagers and young adults who often openly call themselves "gamers." The social identity fostered in tight-knit gaming groups leads to a stronger identification with gamer identity.

While gaming as hanging out or recreation takes place mostly in the private sphere (homes), mobilizing often requires dedicated spaces such as an Internet café, which can provide fast Internet connections and powerful computers. Mobilized gaming, like many other forms of geeking out, requires more specialized technical resources and social networks as well as the time and space to dedicate oneself to a serious hobby. We can see this in the difference in scale of various LAN parties, which we describe in this chapter's section on recreational gaming. Parties can vary in size from a small group of friends to large, more formal gatherings. Small parties can form spontaneously, but large ones usually require a fair amount of planning and preparation on the part of the organizer. Because of the size of these events, most require renting a conference room in a hotel or in a convention center (for a study of LAN parties, see Jansz and Martens 2005).

In his study (Team Play) of a group of middle-school boys (aged thirteen to fourteen) who were regulars in an Internet café in the San Francisco Bay Area, Arthur Law describes different social configurations among gamers. One group of boys went to the café to play the strategy game, Warcraft, on their own, and another group went to the café to play Counter-Strike as a clan. Both of these modes of game play are networked and social. Law describes how two of the Warcraft players go the café to play with a set of gamer friends with whom they keep in touch online. Patrick, in particular, is a competitive gamer who keeps close track of his ranking on Battle.net, a system that keeps track of Warcraft player statistics across the country. Law writes:

Both Patrick and Zachary organize their games outside of Warcraft. Patrick is an avid user of AOL Instant Messenger (AIM) and usually connects with his friends over AIM to get them to play a game. His contact list has more than two hundred people and about half of them play Warcraft. Patrick's family moved from Southern California a year ago and he keeps in touch with his friends online through AIM. He never gets to see any of them anymore so Warcraft is one way of hanging out with his friends online. There are a number of people from his school who were international students who have moved back to their home countries and their group routinely meets online to chat about what they're doing or just to play a few games.

The Counter-Strike players in Law's study are a group of friends who regularly come to the café together and identify as a clan. They use their clan name in their online handles. This group also has a lightweight sense of leadership in the group, where Shawn is recognized as the most experienced player. As is typical with team sports and game play, this leadership is under constant renegotiation. Law describes an instance of play when Shawn won the first round and was advising the remaining players on strategy. "Both teams ignored the advice," Law notes, and suffered as a result.

Box 5.3   Learning and Collaborating in Final Fantasy XI
Rachel Cody

Final Fantasy XI (FFXI) is a massively multiplayer online role-playing game (MMORPG) developed by Square Enix as a part of the Final Fantasy series. Although the game is not tied to the other Final Fantasy games, it shares the graphical, character, and narrative style of many of the other games, providing a major draw for players who enjoyed those games.

The game was released in Japan in 2002 and brought to North America in the fall of 2003. It can be played on four platforms: PlayStation 2, PlayStation 3, Xbox 360, and PC. In 2006, there were approximately 500,000 subscribers to FFXI (Woodard/Gamasutra 2006). FFXI offers many of the same activities of other MMORPGs. Players are able to advance themselves through levels by killing monsters for points. Players can complete missions for their nation to advance in a ranking system, and there are dozens of quests to complete in all cities and towns. Additionally, FFXI offers players a crafting system, through which they can create their own food, clothes, or weapons to use or sell. Monsters that are extremely difficult to defeat offer players a challenge, competition, and rare and valuable gear. And for those who want to try their skills against other players, Ballista offers players the chance to form teams and compete against one another in games.

Most of all, FFXI is a social game. Players often join to be with their friends and develop long-lasting relationships throughout their time in the game. For example, Kalipea, a twenty-year-old white player in Ontario, Canada, started playing the game after visiting her friend:

Well, I was at my friend's house, and she had just got it and I used to play video games all the time when I was younger. But then I never played like an online one at all. . . . It looked kind of cool and then she got the online one and said, "Oh here, go on here." And I made a trial character and I tried it out—ended up playing for like twelve hours straight.

After Kalipea got into the game, she would go over to her friend's house so they could play together, a practice that is not uncommon for players who are friends outside the game and live near each other. Kalipea described playing with her friend:

Yep. I played with her. At first we hung out a lot because she was showing me how to work around stuff. Like we both lived in the same city at the time, and I would go to her house and even like bring my computer over there and we'd have like all-night gaming sessions, just playing and hanging out. And we'd go to like the twenty-four-hour grocery store and get all this food and sit in front of the computers and eat junk food.

Within the game, players form groups, chat with one another via private messages or in-game mail, and join linkshells. The game is built to be social; from leveling to questing to crafting, people need one another to progress through the game. Scott, a twenty-five-year-old white male in Washington State, describes the necessity of playing with other people:

Because you have to rely on so many, you're just . . . you're so limited on what you can do solo that you have to rely on other people. And if you have to rely on other people you might as well do it with people you like. And I think that's . . . it's just a very interconnected play, 'cause you have to have people you know.

Players who prefer casual, individual play, and players who do not get along with others are weeded out of the game early since lengthy parties are necessary for play and reputations (and thus party invites) are dependent on a player's ability to interact successfully with party members. Communities are also a major determinant in what players do in the game, how they play, and what they desire in the game. Players learn from one another where to go in order to level, what gear to wear, and their roles within parties during leveling or killing a monster. And players who fail to align their social interactions, play style, gear, and roles to the community norms risk being cast out of a party, removed from a linkshell, or ostracized or mocked by the community at large.

Players need one another to succeed in the game, but they play with one another because that is what they enjoy. People often log in to the game looking forward to hanging out with their friends. Chat fills the lengthy downtime while players look for a party to play with, between monster fights, while waiting for monsters to spawn, and during lengthy fights. Even when the game's activities are no longer fun, people often continue playing because of their friends. Wurlpin, a twenty-six-year-old white male in San Diego who had played the game for two years, described the relationships:

You will play with these guys eight, nine, ten hours in a day sometimes, all week and in wee hours of the morning so they kind of become your family so to speak, or your group of friends that you hang out with. It is your way of hanging out with them, so, leaving that is kind of hard. And the only reason I pretty much stayed was for the people.

Ryukossei, a nineteen-year-old Asian-American in Illinois who played the game for more than two years, also commented on the strong bonds formed within the game:

Yeah, especially because we had Vent on twenty-four/seven, every time we logged on and stuff. We kind of got attached, you might say. And when someone quit, it would be really hard for them. I mean, you hung out with them. It's like a family pretty much. I mean, you're there with them the whole day and stuff like that.

Scott pointed out that the people make all the frustrations of the game worthwhile when he described an early experience in the game:

You have to go down there [to a dragon] and it takes a long time to get there, and we had like—I mean, it was the most frustrating thing we ever did. But afterwards we just couldn't help but laugh, 'cause it was this stupid dragon that killed us all, and I mean, at least five times. . . . And we were running out of time because each of us had been risen once at least, and already died, and so our timers were running out. And oh, it was just the quintessential just, us-against-the-world type of thing.

Sometimes, players spend more time with their friends within the game than they do with their friends (or even families!) outside the game. They check the websites and forums during their breaks at school to keep updated on their friends' activities and eagerly log in to the game as soon as they get home. Players often sacrifice sleep, staying up long into the night to have another adventure with their friends. The communities and relationships forged within the game extend beyond its boundaries into websites, forums, guides, instant messenger programs, emails, and even phone calls or text messages. Linkshells, especially endgame linkshells, often have dedicated websites where their members chat about in-game adventures, their homework, personal problems, or just joke around. Sites such as KillingIfrit.com and ffxi.allakhazam.com allow players to chat with one another beyond the bounds of the linkshell or their server. Forums on these sites are filled with players asking questions about crafts or quests, debating the best gear or role of different jobs, proudly telling of their most recent accomplishments, or talking about the latest drama between players or linkshells. The websites and forums become extensions of the game by providing a large community of support, advice, and socializing that players often rely on and enjoy.

Final Fantasy XI players are embedded in a rich social atmosphere where relationships and communities are forged and fostered. It is the social components of the game that often motivate players to log in and support their success. The extended communities that reach beyond the game into websites, forums, instant messenger programs, and phone calls help strengthen these relationships and influence players' experiences and success within the game. The players often play for the people.

The issue of leadership and team organization was a topic that was central to Rachel Cody's study of Final Fantasy XI (FFXI). Cody spent seven months participant-observing in a high-level "linkshell," or guild. Although many purely social linkshells do populate FFXI, Cody's linkshell was an "endgame" linkshell, meaning that the group aimed to defeat the high-level monsters in the game. The linkshell was organized in a hierarchical system, with a leader and officers who had decision-making authority, and new members needed to be approved by the officers. Often the process of joining a linkshell involved a formal application and interview. The linkshell would organize "camps" where sometimes more than 150 people would wait for a high-level monster to appear and then attack with a well-planned battle strategy. Cody writes:

One of the important things about these camps was that linkshell members behaved professionally and in line with a linkshell's expectation of conduct. Enki,[8] the head of the linkshell, was known for reprimanding or even kicking people from the linkshell for unsportsmanlike behavior during camp, spamming the linkshell during "focus" time, or making a fairly big mistake during the actual killing of an HNM.[9] Even without Enki's reprimanding people, linkshell members placed a good deal of pressure on themselves to be "perfect" at these camps and not make mistakes. They realize that their behavior is a reflection on themselves and their linkshell mates. While they had a good deal of fun between focus windows, these were high-stress times demonstrated by the constant drama that occurred.

Just as in the case of some of the practices described in chapters 6 and 7, the activities of Cody's linkshell move beyond the playful toward more serious and worklike arrangements where participants are accountable to the expectations of a team. Gaming becomes a site of organizing collective action, which can vary from the more lightweight arrangements of kids getting together to play competitively to the more formal arrangements that we see in a group such as Cody's linkshell.

In all these cases, players are engaging in a complex social organization that operates under different sets of hierarchies and politics than those that occupy them in the offline world. At the same time, the dispositions and social learning that kids pick up in gaming are not completely cut off from their real-life learning. Douglas Thomas and John Seely Brown (2007) explore this dynamic in their discussion of "Why Virtual Worlds Can Matter." They suggest the notion of "conceptual blending," in which players blend their understandings of online and offline. "The dispositions being developed in World of Warcraft are not being created in the virtual and then being moved to the physical, they are being created in both equally" (15). They conclude: "These players are learning to create new dispositions within networked worlds and environments which are well suited to effective communication, problem solving, and social interaction" (17).

Following from Thomas and Brown, we also believe that the important learning outcomes of mobilized gaming cannot be reduced to an issue of transfer of knowledge or skills. Knowledge, competence, and dispositions are developed in the contexts of intense collective social commitments. These commitments can be so strong that they compromise commitments to other social groups and activities, whether they are family, offline friends, or school. At the same time, it is important to recognize that these forms of gaming represent opportunities to experience collective action and to exercise agency and political will. This genre of game play involves jockeying for power, status, and success within competitive game play with others with whom one is deeply connected. As Thomas and Brown suggest, these forms of collective action in gaming worlds can function as training grounds for collaborative forms of work and social action.

Augmented Game Play

As games get more complex, and gaming culture gets broader and deeper, players increasingly engage with a wide range of practices that relate to knowledge seeking and cultural production through games. We call this genre of gaming "augmented game play"—engagement with the wide range of secondary productions that are part of the knowledge networks surrounding game play. These include cheats, fan sites, modifications, hacks, walk-throughs, game guides, and various websites, blogs, and wikis. In her book on cheating in video games, Mia Consalvo (2007) suggests a notion of "gaming capital" to understand the broader cultural context in which gaming knowledge and expertise are negotiated. She positions the development of various cheats and cheat codes in games as part of a much longer history in the "paratexts" surrounding gaming—texts that help gamers gain knowledge and interpret the culture of games. In our genres of game play, cheating and engaging with these paratexts is part of what we consider "augmented game play," the engagement with the peripheral and secondary texts made about and with games. Paratexts, in the form of game magazines, have been part of gaming since the early years of console gaming. As the gaming community has moved to the Internet, the volume of secondary production and information related to gaming has expanded exponentially, as has the social organization of online gaming communities. The advent of accessible video-editing tools has also created new forms of player-generated content such as machinima and video-based game walk-throughs. While it is beyond the scope of this chapter to delve into details of the world of player-generated content (see for example Hertz 2002; Lowood 2007), we would like to describe some of how young people engage in these augmented game play practices, both as creators and consumers of player-created content and knowledge.

Most players engage with augmented game play as consumers of the work of other players or of the cheats and modifications embedded in games by the developers. In our work, we did not encounter any kids who created their own cheat codes or walk-throughs, but we do have indications that access to cheats and other secondary gaming texts was common among kids. In Lisa Tripp and Becky Herr-Stephenson's study of Los Angeles immigrant families (Los Angeles Middle Schools), Herr-Stephenson had the opportunity to see how cheat codes operated in the everyday game play of Andres, a twelve-year-old Mexican American. In her field notes she writes,

Andrew picks up his controller and pulls a sheet of folded notebook paper from his pocket. On the paper are written about a half dozen cheat codes for the game. He glances at it and decides that he first needs to "get the cops off [his] back." This code he knows by heart and he enters the series of keystrokes that make his character invisible to the police officers in the game. Then, he tries the new code and is excited when his bank balance jumps up about $1,000. Then, he jumps in a car and takes off. When he crashes that car, he jumps out and quickly enters a string of keystrokes from memory. The car is instantly restored to perfect condition. I ask him how he learned the codes he has memorized and where he got the list of new codes. He tells me that there are some older kids who live in his apartment complex who give him the codes. He also has two older cousins (high-school age) who play the game and have given him some of the codes. When I ask if he thinks using the codes is cheating, he looks confused. I don't think he's ever thought about it as cheating (despite calling them "cheat codes") and instead just thinks that such codes are a normal part of game play.

What is interesting about this case is the degree to which cheat codes have been integrated as a commonsensical part of game play and have found their way into the hands of a player who does not have access to the

Internet as a way of easily accessing this kind of information. Cheat codes are a kind of gaming capital that circulates among game players in a peer-to-peer fashion and that is now an established part of the social and cultural economy of gaming.

Consalvo (2007) describes a wide range of attitudes that players have regarding what constitutes a cheat, and what an appropriate way of using cheats is. We saw similar diversity in our work. Players all realized that there were ways to work around the formal constraints of the game by using augmented and external game resources. Opinions varied as to whether players liked to use cheat codes or to what extent they should rely on strategy guides and walk-throughs. For some players, simply using strategy and hint guides constitutes "cheating" in a game. Peter, a thirteen-year-old from San Bruno, California, said,

When I play games on my PlayStation 2 I usually look for strategy hints and guides on sites like GameFAQs, but I only use them when I cannot kill a particular monster or I am stuck somewhere. I mean, I know that this is a kind of cheating, but when the game becomes too frustrating or long, I feel that I need to move on. (Bittanti, Game Play)

While Peter thought that it is a kind of cheating to look at a strategy guide when he is trying to beat a game on his own, he also enjoyed engaging with cheats as a playful activity in its own right. "Sometimes I look for cheats not to beat the game but to fool around and do funny things." Cheats are the quintessential form of messing around that has accompanied electronic gaming since the early years in the 1980s. Today, these forms of messing around are a well-established part of gaming culture for kids, and processes of subverting the official rules of a game are commonplace.

Another dimension of augmented game play is the customization and modding of games. In the early years of gaming, the ability to do player-level modifications was minimal for most games, unless one were a gamer hacker and coder, or it was a simulation game that was specifically designed for user authoring. Today, many games come with the ability to create a custom avatar and customize the game experience, and some players see these capabilities as one of the primary attractions of the game. Games such as Pokémon or Neopets are designed specifically to allow user authoring and customization of the player experience in the form of personal collections of unique pets (Ito 2008b). This kind of customization activity is an entry point into messing around with game content and parameters.

In Laura Robinson and Heather Horst's study of Neopets, one of Horst's interviewees (Asian-American twelve-year-old) described the pleasures of designing and arranging homes in Neopets and Millsberry Farms. She did not want to have to bother with playing games to accrue Neopoints to make her Neohome and instead preferred the Millsberry Farm site, where it was easier to get money to build and customize a home. "Yeah, you get points easier and get money to buy the house easily. And I like to do interior design. And so I like to arrange my house and since they have, like, all of this natural stuff, you can make a garden. They have water and you can add water in your house." Similarly, Emily, a twenty-one-year-old from San Francisco, told Bittanti (Game Play): "I played The Sims and built several Wii Miis. I like to personalize things, from my playlists to my games. The only problem is that after I build my characters I have no interest in playing them, and so I walk away from the game." Kenny, a twenty-one-year-old from San Francisco, described messing around with the editing tools in different strategy games:

I remember, when I was younger, the editing tools that came with StarCraft and all the hours I would spend crafting campaigns and single-player missions, or the multiplayer maps I would develop for Command and Conquer: Red Alert for my friends and I to play on. Oftentimes I spent more time outside of the game, crafting my own complex story lines and campaigns than playing the actual game itself.

With players such as Kenny, who are messing around with modding outside the parameters predetermined by the designers, augmented game play can turn toward more geeked out activities. Rather than working within the parameters of the game, as in the case of building a neohome or a home in The Sims, more geeked out game customization means actually hacking and rewriting the rules or creating secondary productions that are outside the sanctioned game space. These activities are tied to much more specialized forms of technical knowledge. For example, one of the participants in Patricia Lange's YouTube and video bloggers study is a white eighteen-year-old who is involved in the MUD and MUSH[10] gaming communities. Although he learned Java in high school, he says he learned C++ and C through his modding activities. Box 5.4 gives another example of a player highly committed to the creative side of augmented gaming. As described in chapter 6, these kinds of secondary productions can become intensely consumed within some circuits that rely on specialized forms of production knowledge that are outside the kind of gaming expertise centered on game play itself. Even within the technical communities of video making, machinima makers are a highly specialized lot. Not only does the making of the videos require intimate knowledge of game mechanics and video editing but also the content of the videos often references highly esoteric details. One of Dan Perkel's interviewees (MySpace Profile Production), Aaron, a fourteen-year-old white Armenian male, was involved in the production of machinima for Battlefield. He is part of a community that specializes in filming stunts in the game. Each video generally involves about twenty people. Although it used to be easier to get into the group, he says that now new applicants have to have "a talent" such as video editing or using Photoshop. Playing Battlefield and participating in his machinima got him interested in digital production and other artistic hobbies he is involved in.

Box 5.4 Machinima: From Learners to Producers
Matteo Bittanti

Tom is a twenty-year-old machinima maker who lives in the San Francisco Bay Area. His family is originally from Boston, and both his parents are educators. "I have always been fascinated by visual media," Tom said, "and machinima offered me the perfect opportunity to combine my two greatest passions, cinema and video games." He elaborated: "When I was a kid, I was blown away by *Star Wars*. It was, for me, a true epiphany. After watching it, I decided I would become a filmmaker. . . . Then I discovered video games, which I consider cinema's natural progression." "Machinima" is a term used to describe animated films created using game engines and game play footage. In 2006, Tom spent approximately eight months ("from beginning to the end") working on an ambitious film that re-created one of Julius Caesar's most famous battles. "It was much harder than I thought," confessed Tom, "also because it was my freshman year in college and I was taking many classes."

He worked an average of two hours per day, seven days a week. "I could not devote more time to 'the cause' because I was studying at the same time and I did not want to compromise my grades." School has never been a problem for Tom. "I love studying but also doing side projects that are tangentially related to the things I like." What he likes most about these "side projects" is that they allow him to be "in complete control": "I also felt that working on something so complex like a movie could have helped me to learn not only new skills but also about myself." Tom never talked about video games in terms of "gamer guilt": "I never felt that playing a game was not culturally valid: I really don't see any difference in watching a movie and playing a game. They could be both very enriching experiences." However, Tom is not a fan of television. "*That* does feel like a waste of time." In his media hierarchy, television is at the bottom because "it is so dumb" and "never really asks the viewer to do any effort." Machinima making, on the contrary, was intensely creative.

To describe the process, Tom frequently used words such as "persistence," "perseverance," and "tenacity." "I have never felt that I was going to quit, but I must admit that I underestimated the time and effort that it takes to make something *good*." Tom's biggest fear was to be perceived as "the lazy kid," "the flaky one," somebody "who cannot finish what he started." When he first announced to his friends that he was making an animated movie using video games, he felt that "bailing out of the project would have been catastrophic [laughs]." What drove him to complete his film was "a mixture of ego, stubbornness, and excitement." He added: "I kept telling myself: *Don't give up on me, don't give up on me.*" Tom is fascinated by the Roman Empire (some of his favorite books are The Decline and Fall of the Roman Empire series by Edward Gibbon) and he was surprised to see that very few games focus on this particular historical period. "There are millions of titles set in the future, like Halo, or during World War II, like the Call of Duty series, but almost nothing on the ancient Rome."

To re-create historical battles, Tom resorted to Total War, a very popular strategy game for the PC. It was an unusual choice for a machinima: in fact, most authors use first-person shooters or simulations such as The Sims. Tom wrote the script in two days, but he faced a daunting task. He needed to obtain the necessary game footage to construct a long and complex story. While most machinima last a few minutes, his intention was to create a one-hour film, quite an ambitious goal for a first-time effort. Producing a machinima requires technical and social skills: One, one has to collect the appropriate footage from a game. Such content needs to be edited. Convincing dialogues and/or proper subtitles need to be added. The implications are twofold. First, creating a machinima is much more complex than just "playing a game"—the gamer becomes a director. The implication or prerequisite is that the creator needs to understand the language and conventions of cinema as well as the inner workings of a video game. He also assumes the role of a skilled technician able to master sophisticated applications such as Final Cut Pro or Vegas. But personal expertise is not enough. Second, after he collects all the raw material, the production process becomes intensively collaborative, since each character of the game speaks with the voice of a different human being. "Machinima is a bit like puppeteering: There is always somebody pulling the strings of a doll. There is always a real person behind the simulation."

Tom recruited his roommate and his friends to play the roles of the different Roman soldiers. "Coordinating and managing ten, twelve people at the same time was not easy, but that helped me to grasp the complexity and nuances of teamwork." The result of Tom's efforts is a forty-minute animated film ("the director's cut," as he called it) that has been freely distributed online. Some loved the movie; others accused it of being "Too Hollywood-ish."

I received many letters of support, but others, inevitably, disagreed, and wrote harsh comments on my Internet Archive page. It felt I had to respond, to defend "my baby," you know. Now I would not probably do it; I would just let the film speak for itself, but discussing my intentions with other readers helped me to understand more about my creative process even though I still oppose their ideas of what machinima is or should be.

Tom thought that such creative production was "empowering" and "overall, fun." "It's the feeling of deep satisfaction that you get when you build something from scratch."

When suggested to Tom that machinima is comparable to a remix practice, a bricolage—since most of the content is already available—he disagreed: "The process required to transform game play into a coherent narrative is an act of creation in itself. It's not just a matter of reassembling what's out there. And that happens in games because game play has a potential for infinite innovation." Tom is now working on a new machinima. "I've learned so much from my previous experience," he said, "and now the production process has become smoother and faster. I started with a complete storyboard this time and created parts for specific people that I had in mind. In a sense, the infrastructure is now already in place." I asked him if he thought that the skills he had learned with machinima could be transferred to other contexts. "Absolutely! To create a machinima I had to learn editing, sculpting, but also I had to learn how to manage people and cooperate with them efficiently. It was both a personal project and a collective effort."

He quickly added: "The funny thing is that now I can't watch movies like I used to do before, you know, naively. . . . I am aware of the camera, the angles, the cuts. . . . I imagine myself reediting the films as I watch them." Tom plans to move to Hollywood after graduating to fulfill his dream of becoming a film director. "I did an internship last summer in a Los Angeles studio. It was exciting. This is the direction I intend to take after I'm done with school."

Tom's case shows how video games can become tools of production for students eager to combine the literary (the Decline and Fall of the Roman Empire series) and the visual (film). Rather than being alternative to traditional learning practices, digital games can become complementary and enriching educational experiences: the pedagogic values of such practices lie not only in the information apprehended but also (especially) in the technical, social, and personal domains that they entail. Above all, creating a machinima allowed Tom to be apprentice and producer, learner and circulator of meanings.

The activities of augmented gaming are highly varied, and in comparison to the other genres of gaming practice we describe, are a less clearly defined category of practice. We have included practices varying from game-strategy guides to secondary fan productions, cheats, and customization. What is common throughout all these practices is an orientation that points outward from the competitive practices of game play toward engaging more broadly with gaming culture. In her study of Yu-Gi-Oh! play, Mizuko Ito (2008b) suggests that these practices of personalizing games and engaging with the viral knowledge exchange surrounding games are key sites of learning and "hypersocial" exchange. Through games, kids engage in sophisticated forms of knowledge exchange in a context where they are personally invested and identified. This is not about a generic position of spectatorship but rather an active subjectivity where gamers can acquire unique, esoteric knowledge tailored to their interests, and develop their own custom content as part of this engagement. This kind of relationship to media content is a quality that has been present in fandoms surrounding traditional media, but it is much more pervasive in interactive media formats. This orientation toward remaking and customizing media is in many ways a hallmark of the digital era and a key training ground for learning critical engagement with media; it is also a pathway into various forms of creative production, which we describe in chapter 6.

Flows and Boundaries of Gaming

Gaming has been at the center of ongoing cultural debate over what are appropriate forms of media for kids. A substantial part of this debate has included discussions of age and gender appropriateness. In our discussions with parents and gamers, we have found a range of perspectives on boundaries of game play, how kids should move in and out of game engagement, and what kinds of identities kids formed with games. This is the cultural context in which the practices of game-related learning and development unfold. Before we conclude this chapter, this section reflects on the genres of gaming practice we have analyzed to consider the broader social and cultural contexts that frame game play and how kids move in and out of participation with gaming and particular game genres.

Boundary Work and Gamer Identity

Throughout our discussion of gaming genres, we touch on issues of gamer identity, particularly how gamer identities intersect with gender and geek identities. This identity work differs depending on what forms of gaming practices are at play. While killing time and hanging out forms of gaming tend to have more inclusive identity profiles, recreational gaming and more mobilized forms of gaming tend to be more exclusionary and strongly associated with male geek identity. Within the genre of practice that we have called augmented game play, the practices associated with aesthetics and design tend to be gendered female, while those relying more heavily on technical expertise tend to be gendered male. These gender dynamics are not surprising given our existing knowledge about gender and games (Cassell and Jenkins 1998; Kafai et al. 2008). To understand how these broader structural distinctions and divisions are produced, we need to understand how they emerge through different forms of play and talk.

Producing particular forms of gamer identities is a form of "boundary work." Almost without exception, kids we spoke to engaged in some forms of gaming, but they have well-developed discourses for distinguishing different kinds of game play identities. Players who engaged in the killing time and hanging out genres of gaming often described their enjoyment of games, but they do not move beyond these more casual forms of gaming. These forms of gaming are considered everyday, unremarkable activities that are part of using computers and entertainment centers, and they were the most pervasive forms of game play that we encountered. The boundary work of creating gamer identity involves constructing boundaries between gamers and nongamers, and kids who engage in killing time practices are not generally considered "gamers." Among boys, certain genres of gaming were ubiquitous and socially acceptable. These genres included sports games and FPS games such as Counter-Strike and

Halo. Among girls, the dominant social norm was that it was not socially acceptable to be identified as a gamer. In danah boyd's study (Teen Sociality in Networked Publics), she interviewed two kids who talked about some of the gender dynamics around gaming. Catalina, a white fifteen-year-old from Texas, and Jordan, a Mexican American, also fifteen and from Texas, do not really play video games, but Jordan would love to get a PlayStation 3 because she thinks that Dance Dance Revolution (DDR) looks fun.

Catalina: Occasionally, I play with my brother just like one game once a month but that's it.
danah: Does he play a lot more than you do?
Catalina: Every day.
Jordan: He's a boy.
danah: Why do you say he's a boy?
Jordan: I don't know any girls that play video games.
Catalina: I know a few that do.
Jordan: Really? Not like a lot, though.
Catalina: It's stereotypical but . . .
Jordan: Yeah, but it's kind of true.
Catalina: It's really very stereotype but it is true for the most part.
Jordan: They're all like war games, a lot of them. Like I don't care to play . . .
Catalina: Yeah, the girls I know that play video games don't play war games and stuff.
Jordan: DDR, Mario Kart, and stuff like. Like Rachel will play video games sometimes.

Catalina spells out the cultural assumptions about gaming in their friendship group. Although girls might play some of the genres of gaming associated with hanging out genres, the genres associated with recreational gaming tend to be associated with boys. These distinctions are played out in the everyday kinds of boundary work that kids are engaged in. Dan Perkel (The Social Dynamics of Media Production) spoke with Shantel, an African-American high schooler, who told him a story of how she was relegated to the margins of boy game play:

Shantel spent the weekend with her cousins, "all boys!" She said that all they do is play video games. I asked her if she got a chance to play. She told me about a trick they played on her. They gave her the controller and didn't really tell her how to play. And then when she scored they got all excited for her. But, it turns out, the other boy was playing against the computer and Shantel wasn't controlling anything. She looked mad when she told this story, or at least frustrated, because she really did want to play. I asked her if she ever did get to play and she said that she did. She said that it was hard to figure out the controls (well, she said something about all of those buttons and that she didn't know what to do). But when she said that she scored two touchdowns, she was smiling.

Gamers of the recreational and mobilized variety are often militant and vocal about their passions and can put down other players they do not see as gamers. Recreational gamers are serious hobbyists who are committed to learning and honing their game play expertise. They are engaged in both messing around and geeking out on their gaming hobbies. This can appear as obsessiveness to nongamers, and their practices can be exclusionary to "noobs" (beginners, short for "newbies"). This is not the more open and accessible mode of gaming that we see in the hanging out genre. One young woman Matteo Bittanti interviewed for his Game Play study, Lynn, a twenty-one-year-old from Santa Rosa, California, described her younger brother's game play with a popular online first-person shooter, Counter-Strike, as a space that was highly social and that he was invested in in a way that was inaccessible to her.

My younger brother has been playing Counter-Strike on our home computer since late 2003. . . . There have been times where I have just sat in our computer room and watched him, so I've seen these player interactions for myself. . . . For example, Luke's screen name is NubMuffin, because, well, he likes muffins, and thinks it sounds good (whether there is another reason, he chose not to tell me). It really has become a second name for him, because even when playing under a different moniker, his friends still refer to him as "muffin," even on Xfire (where his name is currently "Saddam got pwned!"). And then his Counter-Strike clan was called "teh_noobz." Both are examples of insider language, and both are interesting as they present a false identity to other players. "Nub" and "noobz" pokes fun at how new players are targeted, and partially disguises the ability of the players. I believe this secondary identity is one of the primary reasons he returns to Counter-Strike again and again. I can see the attraction in improving your standings, taking advantage of environment glitches, or using the surfing or Warcraft mods. However, his online identity exists apart from the physical, and he has built [it] up outside of his local friends and family. When I watch my brother yell, laugh, and react to his friends through the game talk, teamspeak, and Xfire, it's not the brother I deal with day-to-day. He's a much gruffer person.

The kind of game play that Lynn described here contrasts with hanging out modes of game play that are more accessible; here her brother is relying on insider knowledge and expertise and a social network that is primarily interest-driven rather than being grounded in the local given relations of family and local friends. NubMuffin is a gruffer, more masculine identity than the one Lynn interacts with every day, and he goes online to find a peer group that supports this more specialized form of practice and expertise. Just as a stadium or auditorium provides a space where a kid might develop an alternative identity as an athlete or a performer, game spaces provide contexts bracketed from their primary, everyday contexts and identities. Lynn's discussion also indicates the role of the spectator in these performances as well as the gendered nature of the spectator role in gaming. Bittanti (Game Play) finds similar dynamics at work in another interview with nineteen-year-old Mary, who also watches her brother play.

I never really understood what was so great about Counter-Strike. Watching my brother play obsessively might have caused me to turn away from the game because it felt overrated and typical boy genre (and the graphics weren't that appealing at the time either). Typical as in aimlessly hunting down other people, shoot and kill, rake in the points, et cetera. When Counter-Strike's popularity reached its peak, I watched my brother play this game a couple times and he explained to me the basic rules and goals and such. After a couple rounds, I noticed how the players were chatting to each other and I had no idea what some of the words meant, like "lag," "owned," "pawned," et cetera. Eventually, I got pulled into the game as my brother got popped by the same guy a couple times in a row and he was desperately trying to get revenge, ha ha.

In this example, Mary positioned herself as an outsider to her brother's practice, not understanding "what was so great about Counter-Strike" and describing it as a "typical boy genre." At the same time, she was interested enough to play a spectator role, and she got drawn in as a support person to her brother's play. This dynamic has much in common with the stereotypical role that girls have played in relation to more masculine forms of sports, that of the spectator and cheerleader (Adams and Bettis 2003; Shakib 2003).

For the boys who do engage in the more geeked out forms of game play, relationships that kids build through recreational gaming provide a space for socializing that is an alternative to the mainstream status regimes that boys navigate in their everyday lives. One white thirteen-year-old, an avid gamer in Heather Horst's Silicon Valley families study, noted: "Well, as far as sports and music go, I'm not that big of a person on those. I am,

I think, by definition, a geek. The main things I actually normally do are either homework-related or video games or hanging out with friends." Similarly, a fifteen-year-old of Egyptian descent in danah boyd's study (Teen Sociality in Networked Publics) described sport-related identity and gaming identity as distinct from each other. "I'm not really much of a sports person. So it's pretty much the games and systems and that's pretty much it, although I don't really own any systems right now." As a genre of practice, engagement with recreational gaming parallels much of the social activity and identity play that young men have historically developed through sports, but there is an important difference in how these activities are culturally identified. Like gaming, sports are interest-based activities that are strongly gendered and focused on competition and performance; the difference is that the identities and reputation cultivated in sports translate to status in the mainstream friendship-driven popularity negotiations in a way that gaming identities do not (Edley and Wetherell 1997). Although we found that it was socially acceptable for mainstream boys who were popular within their local friendship-driven networks to engage in recreational gaming, kids who were more deeply involved in recreational gaming tended to self-identify as "geeks" rather than boys who are into sports.

Among recreational gamers, those who identify with the first-person shooter genres, which have been demonized by the mass media, see their interests in oppositional terms with those of mainstream culture. Players tend to reject some forms of gaming considered "too mainstream," such as the so-called casual games typical of killing time practices. Matteo Bittanti (Game Play) spoke to one player of these games, twenty-one-year-old Steven, who was particularly articulate about these oppositional stances.

Society as a whole looks down on video-game culture because they see it as a collective of geeks or geeky guys who live their lives through virtual reality. They judge video gamers on the basis that they could be doing something more productive or essentially more creative with their lives. Ninety percent of these people have never picked up a controller for themselves and [need to] just let go of stereotypes. They haven't allowed themselves to be submerged into a culture about pushing boundaries and storytelling and character development and scenery exploration. They don't allow themselves to be a part of the creative genius or problem solving. They are only a part of the judgmental side of society.

Young adult players such as Steven are part of the definition of the sub-cultures of forms of gaming, ones based in a certain kind of gamer pride and defined against mainstream norms. Players of first-person shooters are demonized by the mainstream because of the violent content of the games. By contrast, MMORPG players are often stigmatized as being socially marginalized. Although FPS and sports games were fairly ubiquitous among boys, it was rare for us to find MMORPG gamers in the mainstream teen demographic. This is partially due to the cost involved, but there are also important cultural distinctions between gamers. Here the discourse revolves around commitment of time and energy to the online world, and both those on the inside and outside of these practices often describe them in these terms. For example, in her YouTube and video bloggers study, Patricia Lange interviewed an eighteen-year-old who was an avid gamer but who says he does not play World of Warcraft because role-playing games "suck up too much time."

In an interview with Katynka Martínez (High School Computer Club), Altimit (an eighteen-year-old Filipino American) and Mac Man (a seventeen-year-old Filipino American) distanced themselves from the "real, dead, hard-core" MMORPG player in a discussion of the World of Warcraft (WoW) *South Park* episode.

Altimit: Because there's a couple of kinds of gamers. There's me, I'm hard-core semi.

Katynka: Hard-core semi.

Altimit: Then there's the real, dead, hard-core ones, which I can't even kill. I know them, trust me.

Mac Man: And then the casual one.

Altimit: The casual ones, and medium ones. The hard, the ultra-hard-core ones are like those in WoW, the one that we saw in the *South Park*.

Mac Man: Yeah, the . . .

Altimit: The guy.

Mac Man: Yeah, the guy.

Altimit: No life. He has everything. He goes to buy, he has Dungeons and Dragons. Stuff, food. He's like all day . . .

Mac Man: He has all this sodas and stuff around.

Altimit: Yeah, he has in a single room.

Mac Man: It's like a beast in there.

Altimit: Yeah, he doesn't go anywhere. He just stays there. Everything's just there.

Katynka: Do you guys know any of these people, like in real life or do you just know that they exist?

Mac Man: I know them in . . .

Altimit: I don't know that they exist . . .

Mac Man: Yeah, they never get out of their house. Yes. They stay there all day.

Although the boys refered to ultra-hard-core gamers who have "no life" and are "like a beast," Altimit admired "real dead hard-core" players who are highly skilled at shooters. He suggested there is a difference between gamers who let games control their lives and those who use their skills to acquire money and status. A player who is able to balance game play with other dimensions of life and still succeed is "normal" in his view, compared to the guy who sees the online game as his whole life. The former is the kind of gamer with whom Altimit would like to identify. As described in box 7.3, Altimit admires a professional first-person shooter player he described as "the best gamer in the world." Unlike MMORPGs, first-person shooters have subcultural capital as a form of gaming that relies on masculine performance and virtuosity that provides high status among most teenage boys.

A final form of boundary work deserves mention—the issue of generational differences in understanding of games. As described in chapter 4, we saw some instances of hanging out gaming that would involve parents, but for the most part, gaming was the province of kids. Even when gamers talked about playing with their parents, it was almost always in the genre of hanging out, not the more geeked out forms of game play that rely on mutual respect and expertise. We can expect, as members of the current gaming generation start raising their own children, that these dynamics will start to change. For example, one participant in Mizuko Ito's Anime fans study was a serious gamer, even competing in major tournaments, and acted as a gaming mentor and hero for his son. Further, with the popularity of platforms such as the Wii and the Nintendo DS, we can expect more intergenerational sharing around gaming. At the same time, the rapid rate of technology change with regard to gaming is likely to continue to produce a generation gap in gaming experience, even for parents who are avid gamers. The processes of distinction that core gamers engage in, defining their practices in opposition to mainstream culture, are likely to continue to produce an elite geeked out gaming culture that will be out of reach to most of the older generation.

The different forms of boundary work, of making distinctions between different kinds of gaming identities and between the world of gaming and mainstream culture, demonstrate how varied kids' game-play experiences are. When considering how games contribute to learning (of both the celebrated or demonized variety), we need to be specific about which forms of gaming and gaming identity we are referring to. Gaming practice is articulated in relation to the broader cultural and social dynamics of youth culture. Some of the most important outcomes of geeking out on games are experiences of mastery that translate into identity and status within peer groups that care about gaming and technical expertise. When one considers these dimensions, gender becomes important not only in terms of gender representation in video games but also in terms of participation in certain social, cultural, and technical worlds. As gaming becomes increasingly central to young people's socialization into networks of technology expertise and learning, the persistent gender gap in recreational gaming is problematic. Although we are seeing a broadening base of participation in the killing time and hanging out genres of gaming, recreational gaming is still a male-dominated sphere.

Transitions

Our descriptions of genres of gaming practice and identity provide us with a vocabulary for discussing trajectories of learning and participation with games. As we discuss in chapter 4, parents often make determinations about what is age appropriate when making decisions about game access. Recreational and mobilized forms of gaming generally peak in the early teen years, when parental prohibitions have been relaxed but before kids are fully transitioned into a focus on dating and peer-status negotiations that characterize the later teen years. When a teenager starts to transition to adulthood, or starts college, video games are often left behind (Bittanti, Game Play). Mary, a nineteen-year-old from Alameda, California, said: "I guess when I went to college [I gave up gaming]. I did not have enough time to socialize and still play games and most of my friends were into MySpace and Facebook and so I stopped playing altogether." For others, such as Chris, a twenty-nine-year-old from San Francisco, quitting gaming was work related. "When I joined my business firm, I did not have to play anymore to 'feel powerful,' you know. I had 'real' responsibilities and goals. Also, my free time decreased dramatically and spending hours in front of a screen just felt wrong." Although many gamers persist in their hobby despite the crush of real-world accountabilities, many gamers also report moving out of engagement when they thought that it was no longer productive or that it was interfering with other responsibilities.

In retrospective discourses of game play, the more geeked out forms of gaming are associated with a period in one's life when one has time to waste. Dave, a white seventeen-year-old from rural California Christo Sims interviewed (Rural and Urban Youth), reflected on an earlier game "addiction" from when he was in seventh grade to distance himself from that moment in his life. He described how he was highly involved in The Sims, and that it was "bad" and "addicting." He says of the game that "it's kind of creepy now that I think about it."

Dave Cody: I played it for hours every day; that's actually the only thing my parents have ever taken away from me.
Christo: Oh, really?
Dave Cody: Yeah.
Christo: And why?
Dave Cody: I was just like a zombie. I was just logged on to it and I'd be there for hours, hours on end and it was horrible. I couldn't walk away from it . . . .
Christo: Uh-huh, and what was the, you said you sort of had a system for it or something?
Dave Cody: Uh, yeah, that was weird, I just had a, like certain points where people would sleep and stuff like that. I don't know how to put it, like certain people would make breakfast for people in the morning and stuff like that. I got way into it. It was, no it was gross. I wish I'd never got that far into it, but I just had way too much time on my hands.
Christo: Uh-huh. Why do you think it's gross, though?
Dave Cody: Just the fact that you get so far into someone else, like a person who's not even real, like you try to control their life, like playing God almost, you know? It's like, I don't know. . . . It's not normal, I don't think.

When Christo interviewed him, Dave Cody was a starting football player at his high school, and though he played sports games, he distanced himself from the more feminine forms of recreational gaming he had been involved in earlier. Playing with The Sims was not a genre of gaming that was a suitable transition to the more mainstream forms of male sociability and identity of his later teenage years.

In the case of Dave Cody, his earlier forms of game play were out of alignment with the social identity he wanted to maintain in high school. MMORPG players, particularly those who are involved in competitive guilds, need to make hard decisions about whether their lives and identities outside or inside the game take priority. Commitments to competitive guilds are highly demanding of players' time and attention. Ryukossei, the ninteen-year-old Asian-American in Rachel Cody's study of Final Fantasy XI, described how he had to quit the game to deal with real-life commitments:

I quit because, I get very emotional as I talk about this. Nah, I'm just playing, I'm just playing. I quit because of school, pretty much. It was right when I was about to take that break and I was, like, right when the semester was going to end, I was, like, I know my parents would never let me play any games ever because they would probably know that it would be the game's fault that pretty much did it. And it was . . . And the majority was the game that got me to drop out. But I'm not going to blame it all on that 'cause it was my fault too. So that's pretty much why I quit.

Another player Cody interviewed, twenty-year-old Kalipea,[11] reflected on the time in her life when she was immersed in game play.

Like when I played, I played. That's all I did. I would go to school, I would come home, I would eat while playing, and then I would go to sleep, I'd wake up, I'd check my fricken auction house, go to school, go home, eat while playing, play for all night, and that was it. I wouldn't go out with friends, I wouldn't have friends over, and I wouldn't hang out with my roommates, which they hated last year and this year until I quit. I would once in a while, but in general if they were like, "Oh, do you wanna go out to the bar; go out drinking?" I'm like, "No, I wanna play." Or, "I don't feel good" and then stay home and play. I would always make up something. . . . I was really addicted.

She went on to describe how she eventually left the game as well as most of the relationships she had fostered online. This discourse of addiction and "recovery" is a theme that emerges among players who were formerly immersed in gaming. Their earlier social context, in which gaming was dominant, is framed as unnatural and compulsive; they have switched frames to a more mainstream notion of social health.

Players who have left the game have difficulty reconciling whether that time spent playing was time wasted or simply a moment in their lives that they were investing in a different set of relationships and commitments. Another one of Cody's interviewees (twenty-six-year-old white male) reflected:

Wurlpin:[12]   Yeah, it is a lot of lost time. Well, let me rephrase that. It is a lot of time dedicated. I could never say it is lost time because there was a lot of memories and it took me to a lot of places and I am very happy with how it all went, but it is also, it is a lot of time.
Rachel: It is a lot of time when you think about what you devoted. Imagine if you spent that much time in school?
Wurlpin: Exactly [laughs]. And that is exactly the case. It is kind of like you start to think to yourself, "Well, what else could I be doing? Yes, I am making memories, but how else can I be more productive or how else can I do something better for myself?" So like I said, it wasn't lost, but it was definitely, um, invested.

It is clear that we are entering an era in which gaming is not an activity confined to a particular life stage. At the same time, our interviews with gamers of different ages demonstrate that there are clear ebbs and flows to gaming activity, and players may move in and out of more intensive forms of gaming practice. As a focus of hanging out social activity, gaming becomes a way of moving into practices of messing around and geeking out with new media. As youth move away from more home-centered sociability of early childhood to a moment when the peer group starts to take over, and youth become interested in romantic relationships, there is an initial shift away from recreational gaming practices. In a similar move, older players may move away from their intense interest-driven forms of gaming practice when the demands of adult responsibility set in. This is particularly true for gamers who are engaged in the more organized forms of gaming that entail a high degree of social and time commitment. Although killing time forms of gaming are easy to maintain in the margins of other life responsibilities, the more geeked out forms of recreational gaming, organizing and mobilizing, are more difficult to maintain. Regardless of whether kids sustain a strong gaming interest or interest-driven peer groups around gaming, when kids pass through more geeked out gaming practices, they have picked up certain dispositions toward technology and interest-driven learning that are not characteristic of hanging out and killing time genres of gaming.

At the same time, we have seen many interest-driven gamers who are sustaining their hobbies into adulthood and who are able to balance real-life and gaming commitments. We have seen instances in which hard-core gamers will move to a different form of interest-driven activity, transferring their passionate engagements into other hobbies. They talk about not having time to game during times in their lives when they have other pressing responsibilities, or are engaged in a different hobby, but they plan to return to gaming at some point. Gamers will bring their interest-driven and geeked out dispositions to other kinds of media engagements. Many of the anime fans Mizuko Ito interviewed were active gamers and described how they divide their interests between their hobbies or decide at certain times in their lives that they will focus on one or another. Much like traditional hobbyists will decide to focus on a project intensely for certain periods, recreational gamers will move in and out of intense engagement depending on game releases, their social gaming activity, or the other rhythms of their lives.

Conclusion

This chapter describes different genres of gaming practices and the discourses that create boundaries between various forms of game play, and we analyze them in terms of issues of learning and development. Our goal in this discussion is to begin to tease apart the diversity of practices and identities that often get lumped under the gaming label. This chapter is more suggestive than conclusive with regard to the learning outcomes of engagement with a wide range of gaming practices. We can, however, venture some initial conclusions with regard to the general findings of our work.

Our work is not focused on issues of gaming representation and content learning, but we focus on the broader social and cultural ecology that contextualizes game practice. We emphasize the importance of cultural genres of game play and how they intersect with identity formations such as geek and gender identity. Where we find some potential issues of concern are not in issues of game addiction and alienation but rather in the inverse—the issue of exclusion from certain forms of gaming. In line with research on gender and games, we found that there is a persistent gender gap with regard to participation in forms of gaming that are tied to technology-related learning and certain forms of interest-driven participation. Although girls are participating at high levels in killing time, hanging out, and the less technical forms of modding and customizing, the core practices of recreational and mobilized gaming are culturally coded as male. Similarly, although we found that the more accessible forms of gaming were pervasive across different socioeconomic divides, access to mobilized and augmented forms of gaming were limited to those with high-end gaming resources, both technical and social.

Geeked out gaming activities of recreational, mobilized, and augmented game play are those activities that are most likely to be pathways into technical expertise and other forms of interest-driven learning. Gaming provides an accessible entry point into geek identities and practices that are tied to technical expertise and media literacy, but clearly this entry point is more accessible to some. In line with recent research in this area, we also believe that lack of access to game-centered sociability is of greater concern than the fears about game addiction (Beck and Wade 2004; Kutner and Olson 2008). Gaming is quickly becoming a lingua franca for participation in the digital age.

Finally, our ecological view of gaming suggests a different frame for the questions surrounding learning and transfer of game-related knowledge and skills. Rather than focus on the issue of content and knowledge transfer (of either the desirable or the undesirable variety), our focus on gaming practice suggests that learning outcomes of gaming are neither direct nor obvious. Few of us believe, for example, that the most valuable lessons that kids learn from sports are the game rules or the competitive and often aggressive "content" of the sport. Rather, we might emphasize sportsmanship and teamwork in addition to the more obvious physical benefits of sports. We understand that sports are embedded in a broader social ecology that is worthwhile for kids to participate in. Here we make a similar argument for games—that the most important benefits of gaming, if they are to be had, lie in a healthy social ecology of participation, an ecology that includes parents, siblings, and peers. Recasting the debate over games and learning in this more ecological frame is an important corrective to many of the dominant discourses of gaming that have focused on game content and design.

Notes

1. For a review of the literature on gaming, violence, and aggression, see Kutner and Olson (2008). Although there are some indications that high levels of play with Mature-rated video games is correlated with aggression, there is no conclusive evidence that there is a causative relation or that game play has any correlation with violent crime. After completing an extensive study of video games and violence, Kutner and Olson (2008, 8) conclude: "The strong link between video game violence and real world violence, and the conclusion that video games lead to social isolation and poor interpersonal skills, are drawn from bad or irrelevant research, muddle-headed thinking and unfounded, simplistic news reports." In this chapter, we do not engage directly with the empirical material on video games, violence, and aggression, but rather we focus on actual social practices of gaming and what game players describe as meaningful outcomes of their play.

2. "Modding" involves players and users making modifications to technology. This can involve modifying game chips or designing new elements of games such as cheats, interface elements, or game levels.

3. Although there has been almost no work that takes a critical look at how class and racial identity intersects with gaming, survey work indicates that in contrast to personal-computer adoption, game-console adoption is not biased toward white and higher socioeconomic status families. In fact, through the 1990s black families adopted consoles at higher rates than white families, and even now families who are high school–educated adopt consoles at higher rates than those with higher educational backgrounds (Roberts, Foehr, and Rideout 2005). Ellen Seiter (2005) has noted how in her fieldwork with youth from diverse backgrounds that working-class boys were generally more familiar with gaming consoles than computers, though they would often search for gaming culture when they had access to the PCs at the center where she was observing.

4. "Machinima" is a contraction of "machine" and "cinema," and it refers to the practice of making videos using a game engine.

5. Originally created in November 1999, Neopets is widely recognized as one of the "stickiest" sites on the Internet. In July 2007, Viacom announced that by the end of 2008, "Neopets (www.neopets.com) will be transformed into Neostudios, which will focus on developing new virtual world gaming experiences online, while continuing to grow and evolve the existing ones."

6. "Ryukossei" is a real character name.

7. In gaming jargon, "frag" is roughly equivalent to "kill," with the main difference being the player can respawn and play again.

8. "Enki" is a real character name.

9. High-level notorious monsters—these are the most difficult monsters in the game.

10. MUDs and MUSHs are text-based online games.

11. "Kalipea" is a real character name.

12. "Wurlpin" is a real character name.

6 CREATIVE PRODUCTION

Lead Authors: Patricia G. Lange and Mizuko Ito

Two fourteen-year-old boys from the Washington, DC, area have an account on YouTube in which they post videos made by their own video-production company. Their videos often sport a personalized introduction in the form of their logo, written in LEGO building blocks, set ablaze by a lighter. One of the boys, Max, hopes to be a director or filmmaker and thought it was important to have a production company, since some of his favorite filmmakers, such as Steven Spielberg and George Lucas, have production companies too. Max also has a number of friends who pitch in by acting in his videos, which are often put together quickly and spontaneously in the context of social activities. For instance, the boys became bored at a slumber party and felt inspired to make a horror film that was well received after they posted it on YouTube. In another instance, a simple outing with Max's mother at the beach turned into a YouTube sensation when he recorded her singing along to the Boyz II Men song playing through her headphones. She was unaware that people around her could hear her and had started to laugh. Max posted the video on YouTube and it attracted the attention of ABC's *Good Morning America*, on which the video eventually aired. In the two years since it was posted, the video has received more than 2 million views and more than 5,000 text comments, many of them expressing support. Max's work has also attracted attention from another media company, which approached him about the possibility of buying another of his videos for an online advertisement. He regularly receives fan mail and comments on his videos. This example illustrates the new possibilities that the Internet offers for kids to receive feedback not only from peers but also from media companies. The advent of this socially based, digital milieu means they can connect with large numbers of dispersed others and test wider reaction to their work.

Digital and online media are opening new avenues for young people to create and share media. Surveys conducted by the Pew Internet & American Life project indicate a rapid growth in what it describes as online "content creation," particularly among youth (Lenhart et al. 2007). The growing availability of digital media-production tools, combined with sites where young people can post and discuss media works, has created a new media ecology that supports everyday media creation and sharing for kids engaged in creative production. Social network sites such as MySpace and Facebook, blogs, online journals, and media-sharing sites such as YouTube, deviant-ART, and FanFiction.net are all examples of sites that enable youth to post or repost content in the context of ongoing personal communication. Media educators are beginning to consider this new media ecology's potential to reshape the conditions under which young people engage with media and culture, moving youth from positions as media consumers to more active media producers. In what Henry Jenkins (2006) and his colleagues have described as "participatory culture," budding creators can develop their voices and identities as media creators through ongoing interaction with engaged peers and audiences (Jenkins 1992; Jenkins et al. 2006). Conversely, researchers also are concerned that the blurring of the boundaries between social communication and media production could degrade the standards of the latter. For example, Naomi Baron (2008, 6) asks, "Could it be that the more we write online, the *worse* writers we become?"

Drawing from a range of case studies, this chapter describes different modes of new media production that young people engage in, analyzing these practices in relation to learning and the development of skills and identities as media producers. We draw primarily from our case studies on youth media production by Dan Perkel (MySpace Profile Production), Dilan Mahendran (Hip-Hop Music Production), Patricia G. Lange (YouTube and Video Bloggers), Sonja Baumer (Self-Production through YouTube), Mizuko Ito (Anime Fans), and Becky Herr-Stephenson (Harry Potter Fandom). Discussion of game-related production is largely covered in chapter 5. The focus of this chapter is on the social processes of interest-driven genres of participation, but we also describe how kids get involved in messing around with new media through their more friendship-driven practices, and we draw from studies on the friendship-driven side to describe some of these dynamics. The interest-driven groups that are the focus of this chapter tend not to be segregated by age, though all have strong youth participation. As with chapter 5, we include accounts by young adults who participate in these groups, and we draw on retrospective accounts of how they got involved in creative production. The chapter is organized as a progression from these messing around genres of participation toward deepening immersion in geeked out participation centered on creative production. We are not assuming that kids necessarily move in a linear fashion from hanging out, to messing around, to geeking out. In fact, kids will often move fluidly back and forth between these genres. Rather, we use this as an organizing heuristic to present the different genres of participation available to youth that involve digital media production.

After introducing our conceptual framework for production, new media, and learning, we begin our description with practices of everyday, personal media production—the creation and sharing of personal photos, videos, and online profiles. After describing a range of practices of media creation and sharing, we turn to a consideration of how young people transition to practices that they self-identify as "media production" and the creation of works that are circulated beyond personal networks. How do young people get started on practices such as video production and editing, web comics, or machinima? From there the chapter describes how young people improve on their craft in the context of digital media production and online exchange. What kind of creative communities and collaborations do youth engage in through the course of producing new media? What are the mechanisms they describe for how they improved their craft? And finally, how do they gain audiences and receive recognition and fame for their work? In the conclusion, we discuss the implications of our ethnographic findings for media education.

Creative Production in the Digital Age

What constitutes "creative work" is contested by scholars. The term traditionally has been used to describe "imaginative" or "expressive" work, where "expressive" refers to sharing aspects of the self (Sefton-Green 2000, 8). Our understanding of what constitutes creative production includes imaginative and expressive forms that are also shaped by kids' individual choices and available media. The influx of digital media into everyday life is reshaping these understandings, particularly our assumptions about the relation between media production and consumption. Media theorists have argued for decades that media "consumption" is not a passive act and that viewers and readers actively shape cultural meanings (Buckingham 2000; Dyson 1997; Eco 1979; Jenkins 1992; Kinder 1999; Radway 1984; Seiter 1999b). Contemporary interactive and networked media make this perspective difficult to ignore. Developments in the technology sector in the past decade have pushed this understanding into common parlance and consciousness. "Web 2.0," "user-generated content," "modding," "prosumer,"[1] "pro-am,"[2] "remix culture"—these buzz words are all indicators of how creative production at the "consumer" layer is increasingly seen as a generative site of culture and knowledge. A decade ago, creating a personal webpage was considered an act of technical and creative virtuosity; today, the comparable practice of creating a MySpace profile is an unremarkable achievement for the majority of U.S. teens. As sites such as YouTube, Photobucket, and Flickr become established as fixtures of our media-viewing landscape, it is becoming commonplace for people to both post and view personal and amateur videos and photos online as part of their everyday media practice. In turn, these practices are reshaping our processes for self-expression, learning, and sociality.

In the case of young people, new media production is framed by ongoing debates about the appropriate role of media in young people's lives. Our discourse about media and creativity is framed by a set of cultural distinctions between an active/creative or a passive/derivative mode of engaging with imagination and fantasy. Generally, practices that involve local production—creative writing, drawing, and performance—are considered more creative, agentive, and imaginative than practices that involve consumption of professionally or mass-produced media—watching television, playing video games, or even reading a book. In addition, we commonly make a distinction between active and passive media forms. One familiar argument is that visual media, in contrast to oral and print media, stifle creativity, because they do not require imaginative and intellectual work. Popular media, particularly television, have been blamed for the stifling of childhood imagination and initiative; in contrast to media such as music or drawing, television has often been demonized as a commercially driven, purely consumptive, and passive media form for children and youth.

Media educators have argued for critical engagement with television and other forms of commercial media, developing programs that teach youth about the conditions under which media are created and revealing the ideological dimensions of popular media. In his review of media-education efforts, David Buckingham has described how media education has been turning more and more to programs that emphasize media production rather than relying exclusively on the "inoculation" approach to media education (Buckingham 2003). In the older inoculation approach, media education focused mostly on teaching kids to deconstruct texts so that they would not be adversely affected by violence or manipulated by deceptive commercial content (Bazalgette 1997; Hobbs 1998). In contrast, emerging youth media programs have been motivated by the belief that engaging in media production should be the cornerstone of media education and lead to youth empowerment through the development of self-expression (e.g., Chávez and Soep 2005; Goodman 2003; Hobbs 1998; Morrell and Duncan-Andrade 2004). These educators believe that shifting youth identity from that of a media consumer to a media producer is an important vehicle for developing youth voice, creativity, agency, and new forms of literacy in a media-saturated era. Compared to programs that focus on critical engagement, production-oriented programs are still relatively sparse in media education. In at least some contexts, however, there seems to be a growing recognition of their importance (Buckingham, Fraser, and Sefton-Green 2000).

Today, these long-standing debates about media, kids, and creativity are being reframed by the proliferation of new forms of digital media production and social media. What is unique about the current media ecology is that photos, videos, and music are closer at hand and more amenable to modification, remix, and circulation through online networks. In the past few years, it has become common for personal computers to ship with a basic kit of digital production tools that enable youth to manipulate music, photos, and video. In addition to the new genres of creative production that are being afforded by digital media-creation tools, we see networked publics as affording a fundamental shift in the context of how new media are created and shared; media works are now embedded in a public social ecology of ongoing communication (Russell et al. 2008). As is common when new media capabilities are introduced, it takes some time for literacy capacity to build and for people to come together around new genres of media and media participation that make use of these capabilities. Given that multimedia production tools have become mainstream as consumer technologies only in the past decade, we are now at a transitional moment of interpretive flexibility with regard to literacy and genres associated with the creation of digital music, photos, and video. The practices that we describe in this chapter need to be situated as part of this transitional moment, when youth are experimenting with new digital cultural forms and, in interaction with adult mentors and parental guidance, are developing new forms of media literacy.

Judged by the standards of traditional media production, many new genres of digitally remixed derivative works would be considered inferior to original creations that did not rely on appropriation of content produced by others. Contrary to this view, Marsha Kinder points out the historical specificity of contemporary notions of creativity and originality. She suggests that children take up popular media in ways that were recognized as creative in other historical eras. "A child's reworking of material from mass media can be seen as a form of parody (in the eighteenth-century sense), or as a postmodernist form of pastiche, or as a form of Bakhtinian reenvoicement mediating between imitation and creativity" (1991, 60). In a similar vein, Anne Haas Dyson (1997) examines how elementary-school children mobilize mass-media characters within creative-writing exercises. Like Ellen Seiter (1999a), Dyson argues that commercial media provide the "common story material" for contemporary childhood, and that educators should acknowledge the mobilization of these materials as a form of literacy. These theorists point to the more socially embedded and relational dimensions of creative production that are in line with much of what we see proliferating on the Internet today.

Renee Hobbs (1998) describes how one of the central debates in the field today is the question of how central popular cultural texts should be used in media education. Although educational institutions have traditionally devalued popular culture, Buckingham, Fraser, and Sefton-Green (2000, 151) argue that students tend to learn a "great deal more from reworking forms with which they have greater familiarity and a personal engagement already." They argue that the most successful school-based media-production programs enable students to manipulate genres with which they are most familiar, to receive regular and frequent interaction with audiences (and knowledgeable peers), and to redraft and iterate their media production multiple times (Buckingham, Fraser, and Sefton-Green 2000, 151). In a similar vein, the New Media Literacy project, headed by Henry Jenkins at MIT, is one example of a project that is building frameworks for incorporating popular cultural practices and the aesthetics of remix into media-production programs.

These approaches are in line with a New Literacy Studies approach as described in the introduction to this book, seeing creativity as a process of not only creating original works but of recontexualizing and reinterpreting works in ways that are personally meaningful or meaningful in different social and cultural contexts. These approaches are efforts to bridge the more recreational practices and media literacy that kids are developing outside school with more formal and reflective educational efforts that center on media production. As with all efforts to bridge the boundaries between instructional programs and everyday peer-based youth culture, these translations are fraught with challenges. Even in educational programs that recognize the importance of new media literacy, educators struggle to develop frameworks for assessing and giving appropriate feedback on student work. Teachers tend to assume the media are "doing the work" when kids engage in critical, remix, and parodic forms of production that use elements from other media (Sefton-Green 2000). Teachers are also wary of media work that appears to be "too polished" or "suspiciously flashy," particularly those genres with which kids are more familiar than teachers (Buckingham, Fraser, and Sefton-Green 2000).

These difficulties in translating recreational media engagement into school-based forms point to persistent tensions between peer-based learning dynamics and genres and those embedded in formal education. Educators have examined a wide range of topics relating to the tension between in-school and out-of-school forms of literacy (Bekerman, Burbules, and Silberman-Keller 2006; Hull and Shultz 2002a; Mahiri 2004; Nunes, Schliemann, and Carraher 1993); media literacy is somewhat unusual in that we are dealing with both an intergenerational tension (between adult authority and youth autonomy) and a tension between educational and entertainment content (Ito 2007). This chapter, to inform educational efforts in media education, is an effort to describe the kind of new media literacies and creative production practices that youth are developing in their peer-based social and cultural ecologies. Any effort to translate popular and recreational social and cultural forms into educational efforts needs to be informed by these youth-centered frames of reference. The peer-based learning genres we see in youth online participation differ in some fundamental, structural ways from the social arrangements that kids find in schools. Simply mimicking genres or sharing an assessment dynamic is not sufficient to promote the forms of learning that youth are developing when they are given authority over their own learning and literacy in these domains.

In the sections to follow, we describe how young people are engaging in the production of digital music, images, and videos, and how these activities are contextualized in their everyday life-worlds. Digital media production is on its way to becoming a part of our everyday communication and online socialization, as well as an integral part of a diverse range of more geeked out forms of media engagement. Throughout our description, our goal is to describe the social, cultural, and technical contexts that motivate youth engagement with creative production and the networks of learning resources that help them improve their craft. These networks can vary from the more mainstream friendship-driven networks that support learning how to create a MySpace profile to the more specialized communities of interest centered on video production and remix. Although structured educational efforts can help fill this gap, successful youth producers in highly technical areas are generally driven by an ethic of being "self-taught" (Lange 2007b). They structure their learning as an integral part of their own individual passions for creating media, and they draw from a network of offline and online human resources and artifacts on an as-needed basis. We have found that in less technically driven areas, kids learn from peers through observation and informal questions situated within the context of social activities (such as making videos while on an outing or making a profile page while hanging out). Even among youth who are more technical and espouse an ethic of being self-taught, narratives of how they get started contain many references to peers, family, and other adult mentors who provided advice and encouragement in their media-production efforts.

When we turn to the geeked out production processes that youth are involved in, we see networked publics supporting interest-driven social relationships that are centered on creative production. We describe some of the cases we have seen of young people's engaging in production collaborations, where, similar to what Becker (1982) observed in his study of art worlds, different participants develop specializations to contribute to the shared enterprise. Networked media add to the creative production process by providing opportunities to circulate work to different publics and audiences and to receive feedback and recognition from these audiences. As we discuss in chapter 7, youth have been largely shut out from the skilled labor market, and this includes domains of creative production. Further, access to different kinds of public spaces and venues is also restricted for youth. These structural conditions are one reason why youth access to networked publics is potentially so transformative but also deeply challenging to our established modes of regulating and protecting youth. The current concern over how youth are circulating personal videos and photos in social network sites is inextricably tied to the more celebrated examples of how youth creative talent is flourishing online. By describing youth creative production in terms of the underlying dynamics of online participation in networked publics, we hope to provide a broader framework for these debates over youth expression and media engagement.

Everyday Media Production

We begin our description of different practices of creative production with a discussion of some of the most pervasive and everyday forms of media production that we have observed in our studies. Certain forms of digital-media creation, such as digital photography and online profile creation, are now commonplace among young people. Although youth who engage with these forms of media creation do not necessarily see themselves as "media producers," they are often engaged in sophisticated forms of media creation. As described in chapter 2, the period from 2004 to 2007 saw widespread adoption of social media by teens, particularly social network sites such as MySpace and Facebook. Although the focus of participation on these sites has been the practices of friendship and intimacy described earlier in this book, one side effect of these friendship-driven practices is that many youth become involved in the production and social sharing of digital media. This involves the creation and customization of online profiles as well as the production and circulation of personal media such as photos and videos. Although home movies and personal photos have been part of youth culture for some time, possibilities for online sharing mean that these media have a new kind of social life within networked publics.

Personal Photos and Videos

The vast majority of photographs and videos are produced not from a creative impulse but to capture personally meaningful events and relationships. While the increasing availability of digital recording devices is a precondition for these forms of everyday media production, they are not themselves the driver of these practices. Digital photography and videotaping grow out of existing practices of self-archiving (such as journaling, scrapbooking, and keeping photograph albums) and are propelled by the growth in avenues to share these media with friends and family. Although our study did not focus on these forms of media production and sharing, we have seen indicators of the growth of digital photography and videotaping and its circulation online. In this sense our work supports the conclusions of other research in this area, which describes how the spread of digital cameras and camera phones has led to more ubiquitous forms of image capture and sharing (Koskinen, Kurvinen, and Lohtonen 2002; Ling and Julsrud 2005; Okabe and Ito 2006; Van House et al. 2005).

Interviews with youth who are active online are often peppered with references to digital photos they have taken and shared. In box 2.1, Katynka Martínez describes two sixteen-year-old girls and their practices of taking photos together and sharing photos through Photobucket. Many teens also view new media as "something to do" while they are hanging out with their friends. Flutestr, a white sixteen-year-old participant in Heather Horst's study (Silicon Valley Families), described how she likes to kill time looking at pictures on her camera phone when she is hanging out with friends:

So I took pictures . . . I went to Vegas and I didn't bring my camera because it runs out of batteries really quickly and it has no memory. We have to buy a memory card for it, but I kind of forgot it. So I had my cell phone so I took pictures of, like, the resorts and the casinos and stuff. And then that was really cool so I had them on there. And I have pictures of all my friends. Like, if I'm bored I'll take out my camera and, like, try and play with it. So I use a camera phone a lot.

In another example, Alison, an eighteen-year-old video creator from Florida (who is of white and Asian descent) in Sonja Baumer's study (Self-Production through YouTube), aspires to be a moviemaker. At the same time, she sees her videos as personal media.

I like watching my own videos after I've made them. I am the kind of person that likes to look back on memories and these videos are memories for me. They show me the fun times I've had with my friends or the certain emotions I was feeling at that time. Watching my videos makes me feel happy because I like looking back on the past.

These forms of casual, personal media creation can lead to more sophisticated and engaged forms of media production. For example, Inertia,[3] a twenty-four-year-old white male from England and an accomplished anime music video (AMV)[4] creator, described in an IM interview how he first got involved in editing:

Inertia: Straight after finishing university I made a dozen little projects and music videos to camcorder footage, sometimes with anime music, but hadn't really tried with actual AMV footage still. . . . I used to just film everything . . . like a real first year photography student or something . . . anything funny or memorable I'd try to film it.

Rachel: So how did you learn to edit then?

Inertia: It was bad, sooo bad most of it should never see the light of day . . . but i still edited it into music videos to remind us of the fun we'd had over the years. i learnt by trying really . . . first time I picked up an editor was just before I got into anime, but I couldn't do much. I literally would just take home movie stuff, put it together, cut out bad bits, and save. (Ito, Anime Fans)

These cases demonstrate how the increasing availability of digital media-creation tools opens avenues for young people to pick up media production as part of their everyday creative activity. Although the practices of everyday photo and video making are familiar, the ties to digital distribution and more sophisticated forms of editing and modification open up a new set of possibilities for youth creative production. Digital media help scaffold a transition from hanging out genres to messing around with more creative dimensions of photo and video creation (and vice versa).

Sharing Personal Media

One of the primary drivers of personal media creation is sharing this media with others. Chapters 2 and 3 describe the ways in which the traffic in media and practices such as profile creation is embedded within a social ecology, where the creation and sharing of media is a friendship-driven set of practices. Online sites for storing and circulating personal media are facilitating a growing set of options for sharing. Youth do not need to carry around photo albums to share photos with their friends and family; a MySpace profile or a camera phone will do the trick. Consider the following two observations by Dan Perkel (The Social Dynamics of Media Production) in an after-school computer center:

Many of the kids had started to arrive early every day and would use the computers and hang out with each other. While some kids were playing games or doing other things, Shantel and Tiffany (two apparently African-American female teenagers roughly fifteen to sixteen years old from a low-income district in San Francisco) were sitting at two computers, separated by a third one between them that no one was using. They were both on MySpace. I heard Shantel talking out loud about looking at pictures of her baby nephew on MySpace. I am fairly sure she was showing these pictures to Tiffany. Then, she pulled out her phone and called her sister and started talking about the pictures.

This scene Perkel describes is an example of the role that photos archived on sites such as MySpace play in the everyday lives of youth. Shantel can pull up her photos from any Internet-connected computer to share casually with her friends, much as researchers have documented that youth do with camera phones (Okabe and Ito 2006). The fact that photos about one's life are readily available in social contexts means that visual media become more deeply embedded in the everyday communication of young people. In this next example from Perkel's study (MySpace Profile Production), we get a glimpse into how young people take and modify photos with this social sharing in mind.

I sat down next to Janice (a teenager roughly fourteen to fifteen years old who appeared to be African-American), who was on one of the computers at [the center]. I saw her on Yahoo! Mail dragging photos from her email to the desktop of one of the [center's] computers. She told me that she had been to [the] Stonestown mall in San Francisco with her cousin and had taken pictures. One of them was over her mock kissing a mirror and later I would see this picture as her profile picture on MySpace. Another picture had some special effects. She told me that she had done this at the Apple Store. Then, she proceeded to upload them to her MySpace account, though I noticed that it took her several attempts. The story here is that she took the photographs in one location, used Yahoo! as a way to move her pictures around from different locations, took advantage of the Apple Store to do some creative editing to at least one of the pictures, and then finally used [the center] as a place to upload them to her MySpace profile.

The case that Perkel describes is particularly notable in how Janice mobilizes multiple infrastructures to create photos to share on MySpace: taking digital photos at an outing with a cousin, modding photos at an Apple Store, and finally use of the community center to upload photos to the Internet using Yahoo! and MySpace. In one of her studies (Pico Union Community Center), Katynka Martínez also documents how youth see online photo-sharing sites as a way to share photographic records of their everyday lives, and how they often develop highly sophisticated strategies for authoring and sharing. Martínez conducted diary studies in which kids documented their everyday media use. Stephanie, a sixteen-year-old Latina of Colombian and Irish descent, said that one of her best friends takes her camera to school every day. "Sometimes we'll be like . . . she will tell me or I'll tell her, 'Straighten your hair,' or I'll tell her, 'Straighten your hair.' So we'll straighten our hair and then we'll be like, 'Okay. We're gonna take pictures tomorrow for MySpace.'" Stephanie shared her Photobucket account with Martínez, showed her hundreds of photos that she has saved, and explained that she will do searches on media and topics that interest her and save the photos she likes. Her close friends share their Photobucket passwords, and they go on to each other's accounts to view photos they've found online as well as photos they've taken. This case is described in more detail in box 2.1.

These stories from our case studies provide a window onto how digital media are reshaping long-standing practices of personal photography. Young people take photographs with opportunities for near-term social sharing in mind. Then they mobilize a suite of different technologies to modify and circulate those photos, creating new opportunities for this visual media to enter the stream of everyday conversation and sharing.

Profiles

Just as the sharing of photos and videos online is blurring the boundary between personal communication and creative production, online profile creation also lies in the boundary between hanging out and messing around genres of participation. Profile modification is most pervasive on MySpace; other sites such as Facebook, Blogger, LiveJournal, deviantART, or YouTube also enable members to create custom profile pages. As teens create their profiles, and post and link on their own profiles and their friends' profiles, they are engaged in acts of social communication and everyday media sharing and "consumption" that also entail creating their own digital media. In several of our studies, we have had a chance to both watch the process of profile creation, sometimes through the course of several weeks or months, and also discuss the profile-creation process with teenagers, many of whom created profiles on MySpace. These observations provide a window onto how youth engage with profile creation as a form of creative production embedded in their everyday social relations.

Perkel (2008) describes the importance of copying and pasting code in the process of MySpace profile creation, a practice in which youth will appropriate media and code from other sites to create their individual profiles. He characterizes MySpace profile creation as a process of "copy and paste" literacy, in which youth appropriate media and code from other sites to create their individual profiles. Although this form of creative production may appear purely derivative, young people see their profiles as expressions of their personal identities. Some youth described how one of the main draws of MySpace was not just that this was the site that their friends were already using, but that the site seemed to allow a great deal more customization than other sites. Carlos, a seventeen-year-old Latino high-school senior from a low-income neighborhood in northern California, for example, described how his cousins sold him on the site because it was a site where he could put up "all your pictures, change the background, and customize it and do all that." This chance to not just go online and be social but also to make something excited Jacob, a seventeen-year-old African-American high-school senior, who noted, "It was tight. I was like—this is real. It's the only website where you can actually come up with your own stuff" (Perkel, MySpace Profile Production).

This ability to customize gives youth freedom in defining layout, media, colors, music, and the like, but this also involves a certain amount of technical complexity. For most of the cases that we documented, at least one other person was almost always directly involved in creating kids' profiles. When asked about this, common responses were that a sibling, a cousin, or a friend showed them how to do it. In their research, Judd Antin, Christo Sims, and Dan Perkel (The Social Dynamics of Media Production) watched in one after-school program as people would call out asking for help and others would come around doing it for them (literally taking the mouse and pushing the buttons) or guiding them through the process. In an interview at a different site, Carlos told Perkel that he had initially found the whole profile-making process "confusing" and that he had used some free time in a Saturday program at school to ask different people to help him. Then later, when he knew what he was doing, he had shown his cousin how to add backgrounds. He said he had explained to her that "you can just look around here and pick whichever you want and just tell me when you're finished and I'll get it for you." The story about Jacob in box 6.1 provides an example of how he "got one of his girls to do it."

Box 6.1 "MySpace Is Universal": Creative Production in a Trajectory of Participation

Dan Perkel

I interviewed Jacob, a seventeen-year-old African-American, on a Saturday morning at a technology program run out of a school in the East Bay city of Richmond, California. A community leader at the site said that the school was in an area of town where all of the "drive-bys" and the "shootings" happened. As the community leader surveyed the room of twenty to thirty highschoolers sitting at rows of computers, some typing, some browsing the web, others talking loudly, he compared the program to that of a flower struggling to grow out of a concrete wall. While there were a lot of "brown versus black" problems in the community, he said, everyone in the program was working together.

Jacob, unlike most of the other kids in the Saturday program, did not attend that high school. A woman from another community center, whom he called his "job-finder lady," had suggested that he come to the program to see if there was a job for him or some way to get paid to be there. That had not yet panned out. But at the time of the interview, it was his fourth week at the program, which he said "just looks like a club or something." While he still felt a bit of an outsider, he was having fun and had even stayed there until late in the evening the week before. Among other things, being there gave him another opportunity to work with computers, something that was becoming more and more important to his life and career aspirations.

Like many kids we talked to for this project, Jacob had a MySpace profile that served as a communications hub. He used it to keep in touch with the friends he left in Atlanta when he moved to the East Bay the summer before. He also used it to talk to other friends he did not see every day and to follow up with girls he met at parties. But unlike many of the kids who were observed elsewhere or interviewed at that site, Jacob was trying to design MySpace layouts, using the HTML and CSS (cascading style sheets) code that translates into the background colors, borders, fonts, and other design elements of a MySpace profile.

Jacob traced his own interest in web design to when he had been introduced to digital media creation for a fifth-grade class assignment. He learned how to use the software in ways that the others students did not. By the time he was in high school he surprised another teacher who, after giving a Power Point assignment, had assumed that he would not do very well (he admits to having frequently fallen asleep in class because he was often bored). Jacob described this initial experience of mastery, and the recognition of it by his teacher and others, as an important part of his background in engaging with web design and MySpace.

Besides attending the Saturday program, Jacob also had been participating in a program at his own high school. Because that program was "too cramped," he was part of a group that stayed after school, but this did not seem to bother him as he had access to computers and software that he did not have anywhere else. The list of "nice programs" he mentioned included Photoshop, Flash, Dreamweaver, Fireworks, and others. He had access to all the things "you need to [build] any kind of website, or any kind of project or picture." The current session of the technology program had moved on from doing web design, but the teachers still let him hang out after school and work on his projects: "By my own will, not because somebody is telling me."

By the time he had started web design at his after-school program, he had been introduced to MySpace. He was excited about the site, especially because it gave him an opportunity to customize it: "It was tight. I was like—this is real. It's the only website where you can actually come up with your own stuff." At first, he "didn't know nothing about HTML" and had to get help making his profile. Jacob said that at first, like other kids described in this chapter, he did not know how to copy and paste the code to change the background or how to add videos and music. He would call up the girl who introduced him to MySpace. She would call friends of hers, and they would guide him through the process or sometimes log in to his account and change things for him. But once he moved to California and realized that they could still get into his account, he changed the password. He did not change his layout because he still did not know how to do it. In some ways, Jacob's depiction of himself at this point is of someone more dependent on the help of others than were other people we talked to about their use of MySpace. But Jacob eventually realized that even others who have learned how to change a profile do not know how to modify the code or the layouts they get from other people, a point expressed in other discussions with teenagers: "And that's what they do, just take it. All these websites . . . even the girls, they don't understand HTML. But they know how to get it from somebody else."

It was his after-school program in high school that led him to make the connections between web design and the bits of code that people use to change MySpace layouts. During his interview, Jacob said that he was in the middle of working on his first one. The layout on his site, though, came from another source of inspiration, a girl he had found on MySpace who made layouts for others to use. He said he does not talk to her, but he uses her layouts and now knows how to change them and modify them if there is something he does not like:

I'm just taking her designs and editing it my own self, putting my own little two cents in. The design itself is good, but I might want it a different color. And once I get it that different color that I want then I'm going to post it on there.

While acknowledging that other people know how to get pieces of code, he also set himself apart by noting that he knew something else that may lead to other opportunities. For example, the girl whose MySpace layouts he had been using and modifying is advertising her layouts on her profile. He speculated that she was probably making money on her activity.

It seemed that Jacob was considering if and how he could get paid to do the same thing. In fact, he had a job interview for UPS back in Atlanta doing web design and was considering it, but he felt conflicted about the location:

It's the UPS headquarters and the man said you have a job here if you do this design and finish school. And I was like, "I can do that. But . . . but how long . . . I'm young." "Don't matter; I'm the boss, I can do whatever I want. If I want to employ you, it don't matter how young you are as long as you pass high school. It don't matter." So I was like, "Cool, cool; I might just do that."

In his view, MySpace provided him another option, especially if he could follow the model of the girl who designed his layouts. He saw MySpace as a site where he could do the kind of creative work he wanted to do and reach anyone he might want to:

It's connected to almost everybody. . . . I mean, anybody around California you can probably get connected through MySpace. It's like not one person in the United States you can't get connected to, unless they don't have one. But now? Everybody has one. From the oldest people to the youngest people have a MySpace now. They might not use it, but they got one. So that's the point. MySpace is universal.

MySpace, to Jacob, is a universal connector to friends in the area and across the country. But it is also a universal space to display his emerging creative design skills, which he sees as an opportunity for the future.

Although most youth did rely on others for help in creating their profiles, we did find some youth who were able to figure things out without the help of a more knowledgeable peer. For example, Federico, a seventeen-year-old Mexican American high-school senior from the East Bay area of northern California (Perkel, MySpace Profile Production), stated:

I was like going through websites trying to look for backgrounds and stuff. Oh, how do I put this, and where do I put it? And how do I copy it and stuff? Because I pretty much didn't know nothing about computers. But then after that I was like . . . I started clicking buttons and looking at stuff. I'm like—okay, remember this place and site. And then I keep messing up and it looks all weird. Then after that days went by and then I started learning little by little how to do it. But it was hard.

What people ended up putting on their profiles was usually not the result of planning and careful consideration, but of whatever they happened to see while making or revisiting their profiles. In many cases, teens may initially work on a profile and largely leave it as is except for some minor modifications later. For instance, danah boyd (Teen Sociality in Networked Publics) spoke with Shean, a seventeen-year-old black male from Los Angeles, who said, "I'm not a big fan of changing my background and all that. I would change mine probably every four months or three months. As long as I keep in touch with my friends or whatever, I don't really care about how it looks as long as it's, like, there." In some cases, boyd also observed, teens created a MySpace profile as a way to relieve boredom.

This approach toward tinkering and messing around is typical of the process through which profiles are made and modified. Some of the people we interviewed talked about just putting up on their profiles material that was humorous. Carlos described his profile almost in terms of a collection of images and things he had found. Pointing at the various images on the screen, he noted: "I got Six Flags and the fat little kid. Got this dude and that girl. Got Itchy and Scratch [from *The Simpsons*]." When Dan Perkel (MySpace Profile Production) asked him to describe his process of finding things and deciding to put them on his profile, Carlos corrected him:

I pretty much don't . . . I just go to a certain website and if it looks like it has a lot of funny stuff I just go through that whole page and if I find something I like I just copy, paste it, and put it there. And I won't save it or nothing; I'll just keep on going through the website and copy and paste until I got anything I want. And from there I just save it. And if it looks good, it looks good. If it doesn't, I still keep it the same.

For youth who saw online profiles primarily as personal social spaces, this casual approach to their profiles was typical, and they tended not to update them with much frequency, or only when they grew tired of one. Nick, a sixteen-year-old male from Los Angeles who is of black and Native American descent, told danah boyd (Teen Sociality in Networked Publics),

That's the main time I have fun—when I'm just putting new pictures and new backgrounds on my page. I do that once every couple of months because sometimes it gets real boring. I'll be on one page. I'll log on to my profile and see the same picture every time. I'm, man, I'm gonna do something new.

For most youth, profile creation is a casual activity in defining a personal webpage and graphic identity, pieced together with found materials on the Internet. This is a form of messing around that can provide some initial introductions to how to manipulate online digital media.

We found that personal media creation was often a starting point for broadening media production into other forms, a transition between hanging out, friendship-driven genres of participation and messing around and geeking out. As kids shared personal media such as posting videos or sharing fanfic they often connected with others in ways that encouraged them to increase production and broaden participation in communities of interest, both online and off. By creating profiles and creating, modifying, and sharing visual media, youth are developing visual and media literacy in ways that are driven by their desire to participate in friendship-driven practices. We now turn to a discussion of how kids transition from messing around with new media production to more geeked out modes, describing cases that illustrate the broad range of engagement kids have with making media, developing skills, and making social connections.

Getting Started

Personal media creation and sharing can be understood as a jumping-off point for entry into more challenging forms of creative production. Just as casual tinkering with videos or photos can lead to a more abiding interest in digital media production, social network sites can be a vehicle for youth to experiment with having public profiles as creative producers. In our interviews with young media creators, we have collected many accounts of how they got started in media production. These narratives often begin with a story of how they were "playing around" with media devices that were available to them, and then they move on to a story about how they picked up more advanced skills in media production. Often they reference being inspired by a particular media work or creator in deciding to pursue their own productions. One eighteen-year-old Brazilian editor, Gepetto,[5] describes this trajectory, beginning with the first time he saw an AMV. His friend had given him a CD with some anime episodes, and there was an AMV on it as filler. "I was amazed at the idea that such a pretty little video clip was made by a fan just like me. . . . I was really affected by the video. I put it on loop and watched it several times in a row." He went on to make his own video soon after seeing this first AMV. "My first video took about two and a half hours to make and it turned out extremely horrible. But I loved it." The key here is that beginning editors see AMVs as inspiring and impressive but also something that they can aspire to, something made by "a fan just like me" (Ito, Anime Fans). Amateur media provide a more accessible model than professional media do, and a community of available peers to start kids off in creative production.

Unlike those in many other forms of specialized practice, experts in information technology often emphasize that they picked up their skills outside of formal training and instruction. Members of technical hierarchies pride themselves on being self-taught—learning how to manipulate code, technical devices, and networked forms of distribution on their own (Lange 2003, 2007b). The media creators we interviewed often reflected these values by describing how they were largely self-taught, even though they might also describe the help they received from online and offline resources, peers, parents, and even teachers. Portelli (1991) notes that exploring these tensions is particularly useful because they represent the realm of desire and what interviewees wish to convey in terms of identities of expertise and appropriate participation in technical, social groups. For example, one successful web comics writer, SnafuDave[6], whom Mizuko Ito (Anime Fans) interviewed said: "Basically, I had to self-teach myself, even though I was going to school for digital media . . . school's more valuable for me to have . . . a time frame where I could learn on my own" (see box 7.1). Despite his adoption of "self-taught" discourse, SnafuDave nonetheless described learning to use Photoshop, Flash, and Illustrator by making use of online tutorials and a network of graphic artists he met online. When makers describe themselves as self-taught, they are generally referring to the fact that they did not receive formal instruction, and they will acknowledge various sources of help they used to get started.

Adults are not simply bystanders to their children's expert technical creative endeavors; we found a number of cases in which parents and educators played an important role in influencing their children's involvement with media, either by providing resources; introducing kids to genres, software, or sites; or by working in collaboration with kids. One group of successful young YouTube video makers talked about how their uncle had a cable television show, which they eventually inherited. The boys described themselves as able to figure out technical aspects of video making on their own, but they acknowledge that they helped each other out and originally learned from their dad. A sixteen-year-old white girl from New York named Ashley, who wishes to be a filmmaker, noted, "I learned to use the camera just by playing around with it, and I used an editing program on my mom's iMac computer." As described in box 6.2, Ashley also revealed a number of ways in which her mother helped her learn how to make good videos. Many youth also described how school projects in video making provided the impetus for them to get started in video production. After-school programs and community centers also provide spaces where kids could mess around and learn about creative production with knowledgeable adults and peers. Despite the centrality of self-directed learning in young people's stories of how they got started in video production, successful entry into production is enabled by a wide range of social and technical resources that support as-needed help and learning. What self-motivated youth require to pursue these interests is not so much a formal instructional setting as access to wide-ranging sources of expertise.

Box 6.2   All in the Family
Patricia G. Lange

A mother and daughter named Lola and Ashley have a series of shows on YouTube. Ashley is a sixteen-year-old white girl who characterizes herself as a "future filmmaker" on her YouTube page. From New York, the mother-daughter team summarizes and provides commentary about current reality-TV shows. They first learned about YouTube through a television show that reviews and comments on television. Contrary to the idea that YouTube replaces television, the mother-daughter team's discussions and critiques heavily draw on material from shows such as *Survivor, Big Brother, Beauty and the Geek,* and *Top Chef,* to name just a few of their preferred topics. Their body of work is impressive; they've made more than two hundred videos together since they first established their account in the fall of 2006. To achieve a kind of recognizable branding, their videos share certain consistent features. For instance, they always sit in front of a graphic with the name of the show they are discussing in their video. Lola sits on the viewer's left-hand side and Ashley on the right (see figures 6.1, 6.2, and 6.3). Their banter is unique to them yet comfortably familiar to many people who may recognize the sense of fun and friendship their videos convey.

The following transcript is excerpted from their recap of the season premiere of *Top Chef Chicago,* which is a television show in which there is a weekly challenge or contest to make a certain dish. The challenge for that week's episode required each chef-contestant to make a deep-dish pizza. Lola and Ashley provided some personal observations on the episode:

Lola: They need to each re-create a signature deep-dish pizza.

Ashley: And they have ninety minutes to do it.

Lola: Poor Stephanie; she cut her finger in the first thirty seconds.

Ashley: [rolls her eyes] The girl is a bundle of nerves!

Lola: I know.

Ashley: She was like [grabs her finger and shudders], "Aaah!"

Lola: Well, they were all workin' pretty damn hard on their pizzas.

Ashley: [nods]

Lola: At the end, there was a lot of pizzas getting stuck in those pans.

Ashley: Some broke-down pizzas!

Lola: Yeah, and I think Richard actually used two; was it Richard who used two pans?

Ashley: I think so.

Lola: Andrew had to use a cast-iron skillet to make his.

Ashley: They only had enough pans so that each person could have one.

Lola: Well, there was a lot of doughy, bready pizzas because if you're not used to working with deep-dish pizzas you don't really realize how much that shizz is going to rise.

Ashley: It was like that pizza-bread stuff that you get from . . .

Lola: Focaccia.

Ashley: Yeah. That stuff is good! That's not what they were asking for!

Lola got started making videos on YouTube because her daughter expressed an interest in going to film school and pursuing a career in film or communications. Ashley persuaded Lola to help her and be a part of her video-making experiences. Reluctant at first, Lola agreed to help her daughter pursue her goals. Lola is now a key part of a video-making production team

Figure 6.1
Lola and Ashley recap *Top Chef*. Screen capture by Patricia G. Lange, 2008.

Figure 6.2
Lola and Ashley review *Beauty and the Geek*. Screen capture by Patricia G. Lange, 2008.

Figure 6.3
Lola and Ashley discuss *Big Brother*. Screen capture by Patricia G. Lange, 2008.

that is gaining popularity on YouTube. Between the time of their interview in May 2007 and August 2008, they gained more than 1,900 subscribers and more than 140,000 views of their YouTube channel page (which is YouTube's equivalent of the MySpace profile page). One of their fans has even done compilations of their reviews and commentary. For Lola, the purpose of their YouTube presence is to help her daughter build a portfolio that will help her pursue her career goals and enhance her already impressive scholastic record. They hope Ashley will obtain a scholarship to a prestigious college that they could otherwise not afford. Notably, Lola sees their joint video making not only as a means to an end but also as a way to stay close to her daughter and be involved in her life. Setting very few limits on her computer time, Lola stays regularly involved in Ashley's online activities. Ashley is comfortable sharing the account with her mother. As she put it, "I would never put anything up that my mom doesn't approve of and I have nothing to hide anyway." Lola emphasizes her interest in having a close relationship with her daughter through video making. Lola said:

I wanted to be involved with my kids but I think it's more important that the kids want the parents to be involved. [Because] I've seen other parents . . . my mom doesn't understand the type of relationship I have with my kids and she's like, "I'm gonna get you some software so that you can spy on your kids when they're on the Internet." And I'm like, "I don't have to do that. I know exactly where they are because I'm with them."

[So] I feel sorry for the people who have to have a relationship like that with their kids where the kids feel like they have to sneak around behind the parents' back and they don't know what's going on. So I thought that was important. So that's probably another reason why I wanted to do the videos with [my daughter] because, you know, I wanted to stay involved.

Ashley characterizes both of her parents as very technical, with formal educational training in computers and related technical subjects. She also has close relatives who have degrees in film. Ashley's home environment is filled with computers; each child has his or her own and there are some to spare. Ashley reports that their living room alone has four computers. The house is also well networked. Ashley has learned a lot about computers and video by playing around with cameras and editing software. She also describes how her mom teaches her good video-making techniques, such as keeping things short and avoiding too many transitions. They characterize themselves as "best friends," and Ashley trusts her mother's advice and is grateful to have a second opinion. They often watch their videos and those of other video makers on YouTube to improve their technique.

Lola and Ashley like watching television. Their process involves watching shows they like and taking notes about what they would like to say. After Ashley finishes her homework they set up the camera, even if it is after 11:00 p.m. because they think consistency is important. Lola can do the editing quickly, but she often encourages Ashley to practice so she can gain more experience. Lola now characterizes Ashley as more proficient, having picked up computer-related skills quickly. Their goal is generally to put up a video once per week, although quality is more important than meeting a weekly deadline. Ashley said, "I would like to post at least once a week but I'd rather have fewer good-quality videos than a lot of bad ones which were hastily made."

After the video is posted, Ashley works to promote the videos by networking with other people on YouTube, posting bulletins on MySpace, and alerting friends via instant messaging. She often subscribes to popular YouTubers so that other people will see her channel icon and potentially check out her work. Ashley is on YouTube every few hours during the day, although she does not watch videos at school. When she logs on to YouTube, she checks the number of views on her videos and then checks for comments and new subscribers or Friend requests. As a rule, she agrees to automatically accept Friend requests because her major purpose on YouTube is to network to promote her work. After checking for new Friend requests, she then looks to see if the people she has subscribed to have new videos, and then she examines favorite categories, such as "animation" and "pets and animals." Lola also finds herself frequently checking their account. She spends hours a day on the site. She characterized herself as "hooked" and she joked that her daughter tells her she has a "sickness."

Unlike many other YouTubers, Ashley has very little interest in meeting other YouTubers in person. For her the site is more of a way to achieve future goals. Yet, like many YouTubers, she also enjoys watching videos. Her most memorable moments involve encountering enjoyable videos from some of her favorite video makers. For Ashley, "The best thing about YouTube is that there is always something to watch no matter what you like. The worst thing is that you will never be able to see everything you want to see because it's just too immense." Some of Lola's most memorable moments also revolve around favorite videos—and she appreciates that YouTube is widely accessible. For Lola, one of YouTube's biggest weaknesses and strengths is that it is free and therefore available to anyone. The site contains a lot of "any old garbage" as well as fun videos. More important, it gives a range of video makers such as themselves—or as they say, "average people" with "real opinions"—an opportunity to express themselves and promote their work.

Specialization and Collaboration

As young people begin to develop their expertise in creative production, they often also work to develop a unique voice and specialty. Unlike schools, which might ask young people to perform to more standardized forms of achievement, recreational settings provide opportunities for youth to develop more targeted expertise and delve into esoteric and niche domains of knowledge. For example, Gepetto, introduced earlier, turned to the online community of AMV editors for more specialized knowledge of editing. Although he managed to interest a few of his local friends in AMV making, none of them took to it to the extent that he did. He relies heavily on the networked community of editors as sources of knowledge and expertise and for models to aspire to. In fact, in his local community he is now known as a video expert by both his peers and adults. After seeing his AMV work, one of his high-school teachers asked him to teach a video workshop to younger students. He joked that "even though I know nothing, [to my local community] I am the Greater God of video editing." The development of his identity and competence as a video editor would not have been fully supported within his local community; it was the networked relations mediated by the Internet that led to ongoing peer-based learning and specialization.

Attention to specialized and esoteric knowledge is characteristic of all fandoms, but it is even more accentuated in highly technical fan practices such as machinima, video editing, music making, and fansubbing.

Certain forms of media creation often involve collaboration between different specialties. In describing youth hip-hop creation, Dilan Mahendran (Hip-Hop Music Production) notes how youth develop targeted specialty crafts, such as beat making (Mahendran 2007). Beats are instrumental works created on Reason or other software. For many students at his DJ project, "making beats" was a primary practice. Beat making is a specialty craft that requires an enormous dedication of time if one is to become proficient, and only a couple of the students Mahendran observed mastered this craft at this amateur level. The beat is elemental to a rap song and aspiring rappers want original authentic beats to set themselves off from other rappers. The practice for beginning rappers is to use commercial beats that they sample from CDs of well-known rappers such as Jay-Z or Kanye West, but after they begin to hone their craft they often demand custom beats that they may help produce. Beats are highly valued works that accomplished rappers will seek original versions of.

Mahendran's work, as detailed in box 6.3, highlights the dynamic interplay between specialization and collaboration and the ways in which consumption, fandom, media connoisseurship, and remix are stepping-stones to developing voice and an identity as a member of a creative elite. Hip-hop is a particularly important case, in that it was a genre of music that was ahead of the curve in terms of developing styles of sampling and remix, as well as being grounded within very active amateur production communities where youth develop creative identities and competencies. Within different media and genres of creative production, becoming a creator entails developing either a specialized role in collaborative forms of production, or a signature style that marks an individualized voice. For example, within the fandom surrounding anime, there is a wide range of fan productions, varying from the more individualized mode of fan fiction writing or AMV creation to more collaborative modes such as fansubbing, in which subtitles are added to anime and which require working together as a tight-knit team.

Box 6.3   Making Music Together
Dilan Mahendran

Mistreat was almost nineteen in the summer of 2005 when she joined the Rap Project in the Mission District of San Francisco (see figure 6.4) . Mistreat had moved from El Salvador to the Mission District with her mother, father, and brother when she was ten years old. Years later, she came to participate in a ten-week introduction to hip-hop class being offered that summer in the Mission. Mistreat had never rapped or made music before but she had become a deep listener of hip-hop and rap music. She had not always listened

Figure 6.4
Mistreat rapping in the San Francisco MC Competition. Photo courtesy of www.Uthtv. com, 2006.

to hip-hop, though. In grade school she listened to *salsa romantica* and Latin artists such as Marc Anthony, music that was popular with girls in her home country of El Salvador. In middle school she listened to mainstream hip-hop and R&B such as *NSYNC and 98 Degrees. When she hit high school most other students were into hip-hop, particularly artists such as Messy Marv and San Quinn from the local Bay Area Hyphy[7] music scene. Hip-hop was the music she related to because of its rough edge and its rejection of warm and fuzzy love music. Mistreat wrote often about her life and experience coming to San Francisco and figuring out how she could fit into the teenage scene in junior high and high school.

There was plenty to write about: home life, school life, and friends. So when she came to the Rap Project she had quite a bit of content about everyday matters, which she prodigiously transformed into rap lyrics. Mistreat's first attempt at recording was masterful. It was hard to believe that she had never rapped before nor had previous experience in a recording booth. It was clear that she was a virtuoso and a rare talent. She had a rapid-fire style of rapping that would twist the tongues of most mortals. More important, she had a unique voice that was readily distinguishable from the luminary rappers that she most tried to emulate. She sounded like no one else; she was authentically herself when she rapped. This was rare for most of the rappers who came out of the Rap Project, because as they began rapping they tried to emulate their favorite rappers in tone and style, although some did eventually develop their own voices. Most others who began emulating Jay-Z or Lil Wayne, for example, often continued this mimicry and never got to the next level.

A major pedagogical component of the Rap Project introductory classes was the pairing up of students to work on songs together. Though collaboration was a part of the structure of the class, the pairing up of students was a more organic process and not directly organized by the instructors. Mistreat quickly linked up with Young MIC, another very eager rapper who wrote lyrics prolifically. Young MIC, a nineteen-year-old Puerto Rican, grew up in the African-American San Francisco neighborhood of Lakeside. Young MIC would come each day two hours early with notebooks full of writing. Like Mistreat, he wrote extensively before finding rap as a mode of expression. Young MIC began his personal narrative writing while in the county jail for burglary. One of the other inmates had suggested that he should rap to those stories he wrote. Young MIC said he was kind of surprised when the inmate told him that, because he too had never rapped before, even though he had listened to hip-hop since middle school.

Both Mistreat and Young MIC were inseparable in open-studio time and during classes. They were the most avid of all the students in the areas of rapping and recording. Unlike other students in the class, both Mistreat and Young MIC were determined to become rappers. During the ten-week class,

Young MIC recorded twenty-four of his own songs, several of which ended up on the class's final album, much to the chagrin of some of the other students. Young MIC gave frequent advice to Mistreat, who seemed to see herself as a novice compared to him. They shared ideas on lyric writing and how to improve lyrical flow and cadence. Young MIC often used beats from established rappers such as Jay-Z or Kanye West. Mistreat was willing to work with other students and rap to the beats that they were working on in the studio. Mistreat asked another student named Johnny Quest, a sixteen-year-old African-American who lived in Pittsburgh, California (an hour drive from the Mission), to make a beat for a song that she had been working on. Johnny Quest was an avid sampler and he sampled tracks from the sound track of the movie *Charlie and the Chocolate Factory* (2005) to compose a deep meditative beat for her.

Mistreat's collaboration with Johnny Quest was as much about his unique beats as about including Johnny Quest in the making of the class final album. Because Johnny Quest did not show any interest in rapping, he felt a bit marginalized by the others. Some of the other students who focused on rapping and song making were a bit reluctant to use the beginner beats of other students because the commercially available beats that they sampled were more highly produced. This did not deter Mistreat from using Johnny Quest's beats for a production track on the final class album.

Though the digital technological environment in the studio affords incredible power to individuals to control their own production from beginning to end, it seemed an impossibility to produce a song without the concerted help of others. Each of the students developed special skills, whether in rapping, beat making, or producing final songs. The songs that students loved the most were often songs in which either two or more students rapped on the track or one rapped and another sang the chorus. Though some students enjoyed the camaraderie of coming to the after-school program, most were passionate about making music. Music as the goal of these students' attention was significantly different from the notion of hanging out. Music brought these students together and in some cases, such as the ones described here, close friendships bloomed.

In the case of fansubbing, established groups generally have formal tests and trial periods before admitting a new member, and there is a high degree of specialization within each production team as well as in the community overall. Each fansub group has a "raw provider," who collects the original episode in Japanese; a translator; an editor; a timer, who times the length of time the subtitles should be on screen; a typesetter; an encoder; and usually several quality checkers, who review the final episodes. Although many fansubbers experiment with different roles in a group, they usually have a specialty around which they build their reputations. For example, one encoder described how initially he was attracted to the specialty because of the depth of knowledge that he could pursue within an expert community. "It just got interesting because other encoders were like, 'Here are some tips and tricks.' . . . There were so many tricks in how to handle that stuff that it got pretty interesting." Mastering esoteric knowledge becomes a source of status and reputation. After gaining this status as an expert, a subber will find that his or her services are in great demand in the tight-knit community.

In her study of YouTube video makers and video bloggers (YouTube and Video Bloggers), Patricia Lange found that video production, especially those that spring from video as a form of social activity among offline friends, often relies on the coordination of several individuals who gravitate toward specialized roles within a group. Lange found that in such groups of friends, roles such as director or editor were not particularly fluid. Interviews indicated that friends recruited to be actors did not always express the desire nor did they achieve the mentoring to shift into other aspects of informal video production. Rather, one or at most two people in a small video-making group usually stood out as recognized experts among their local peers in school or in social activities, and it was these individuals, Lange found, who often contributed more intensely to the final product. A production group in many of the genres Lange studied, such as informal comedy sketch and video blogging, emerged from peer groups of friends who get together to make videos. While everyone might contribute by acting and perhaps providing improvised dialogue, not everyone directs and edits. Other members may or may not be encouraged to experiment with taking on new roles. A select few often have the interests and abilities to guide the efforts of a loosely collaborative local group of friends who make movies together as an expression and extension of friendship rather than because all individuals have an equal interest in future video making at the professional level.

Another form of collaboration that online video makers engage in is the "collab" video, in which a maker will collect video from other video creators. In these cases, sometimes well-known or even famous YouTubers may lend footage or become actors or participants in someone else's montage or video compilation. In addition, youth video makers may also attend meet-ups and interview their favorite YouTube personalities. The youth then edit together the videos in ways that show their perspective and interpretation of attending events. In these scenarios, the youth have control over the videos. YouTube celebrities or participants that a kid may respect because of their popularity or technical video-making ability are not in positions of authority or mentorship, but rather they are contributors to the kid's vision. These dynamics show that online spaces such as YouTube and blogs, and real-life gatherings such as meet-ups, provide opportunities for youth to interact with adults as peers, moving out of the age-segregated contexts of the school.

In the case of media production that requires multiple forms of expertise, collaboration is an integral part of the production process. Digital media and networks enable kids to decompose bits of the production process and coordinate their work through a variety of online tools. The current media ecology represents a convergence of a range of social and technical capabilities—the ability to share rich media online, greater processing power in personal computers, accessible video- and image-editing tools, and social media sites—that enable these forms of collaborative creation. We saw the growth of amateur, collaborative digital media production first in music (Russell et al. 2008); now these arrangements are being produced in video making. Through these collaborative arrangements, kids develop close partnerships and friendships and gain opportunities to learn from others with different forms of knowledge and expertise.

Improving the Craft

Creative communities that are organized online provide sources of help, expertise, and collaborative partners as well as a context where creators can get feedback from audiences and fellow creators. We found that trajectories of improvement varied across individual producers and different communities of interest, but in all cases, there were mechanisms in place for creators to learn from one another. Some groups had hierarchical structures, recognized standards, and specific mechanisms for distributing feedback. Others had a more unstructured organization, varying or minimal standards, and more informal mechanisms for providing peer-to-peer advice and assistance.

Certain online sites become a focal point for peer-based learning, sharing, feedback, reviews, and competitions that push young people to improve their craft. Sites such as YouTube, deviantART, and after-school media programs give kids access to peers and experts in the areas of interest to them. It enables access to people who are uniquely placed to evaluate their particular media creation or contribution in ways that people outside a narrow area of specialization could not appreciate. Online sites provide both hard-coded and social mechanisms that enable participants to share their work as well as engage in related commentary and discourse. For example, animemusicvideos.org has numerous mechanisms for feedback and reviews, including discussion forums, simple ratings, competitions, top video lists, and templates for doing full reviews of videos. AbsoluteDestiny[8] (a white twenty-seven-year-old), one of the most well-known editors in the community, describes to Mizuko Ito (Anime Fans) how he initially created AMVs in relative isolation, until he discovered what AMV editors fondly call "the org."

AbsoluteDestiny: I wasn't really being influenced by other communities until I went online for AMVs and found this whole other community already going on. A lot of the work really pushed the boundaries in terms of effort and editing and the kind of level you would go to in order to create effects. It was much more than I had done, and it became a bit of a challenge to see if I could extend my own work to bring it up to that standard.

Mizuko: How did you get famous?

AbsoluteDestiny: It was slow at first. At first I joined the community, asked for feedback, didn't really get any, and discovered that the way to become noticed and to get feedback on your own works was to give feedback to other people. There's a lot of mutual back rubbing going on, and we would do feedback swaps. I would say OK, I'll give you my thoughts on your video if you'd give me your thoughts on mine. By doing that, and by being very active, just having your name out and about, really really helps. . . . So I would leave feedback on quite well known creators and lesser known creators, and just getting into chat conversations on forums with these people, getting to know them. . . . When it finally came about that I made a video that actually did something that people might notice, which was the *Shameless Rock* video, was quite a departure for me. . . . So then because people knew me, but didn't really know my work, they would watch my video and then they would say, "Oh my god, there is this really great video out there that AbsoluteDestiny has done. Go and see it." And I was essentially an overnight success.

In addition to the org, AMV creators also meet up at anime conventions, which participants call "cons," and these meet-ups often define the elite core of the AMV world. For example, the Anime Weekend Atlanta (AWA) convention is widely known to be the central con for the AMV scene, and there will be a dinner meet-up of more than a hundred creators to kick things off. At most cons, the AMV editors will be hanging out in the screening room or the hotel-lobby bar, exchanging opinions about work, or as Darius,[9] a twenty-four-year-old African-American editor, described, "And they'd talk about some other works or—or whatever. Not even their works, but just 'Hey, what's up. How you doing? This is damn good Jack Daniel's.'" Both AMVs and fansubbing are specialized practices relying on deep knowledge of cult media. Creators appreciate feedback from other creators or well-informed members of their public, and they think that there are certain creative standards that have been established by their tight-knit community. The reciprocity between different creators is an important dimension of how learning works in these communities; the core participants occupy the roles of creator, viewer, and critic. For example, fansubbers have ongoing debates about what constitutes quality work, and fansub comparison sites will conduct detailed comparisons of the quality of translation, encoding, editing, and typesetting between competing groups.

These peer evaluation mechanisms are in play in online writing communities as well. In C. J. Pascoe's "Living Digital" study, the case of Clarissa, a seventeen-year-old white female from California, is an example of how this dynamic operates with online creative writing. The role-playing board[10] she participates in is a tight-knit creative community intent on maintaining quality standards. To participate in the board, writers must craft extensive character descriptions and formally apply for admission. Clarissa described how she receives ongoing and substantive feedback from other participants on the site, and she does the same for her peers (Pascoe 2007b). For her story, see box 1.3. In the case of fan fiction, writers and readers have a range of sites that they can go to. As is the case with orangefizzy, a thirteen-year-old Asian-American female from

California, recommendations and social networking play a large part in decisions related to where they read and publish fan fiction (Herr-Stephenson, Harry Potter Fandom).

Becky: Where do you read fanfic?

orangefizzy: at harrypotterfanfiction.com [HPFF] and fictionalley.org [FA].

Becky: and have you published your writing there too?

orangefizzy: not on FA, but on HPFF.

Becky: why did you choose those sites?

orangefizzy: i don't remember why i chose them to begin with looking on FA because it's bigger. I like HPFF, though, because it's small and is not full of people who like to write Snape/Hermione doing extremely x rated things.

Becky: why did you choose to publish on HPFF and not FA?

orangefizzy: because HPFF's forums has more of a "community we all know each other" feeling to it than FA, which is huge. and since i talked to the HPFF people, i preferred to put my work in their archives.

The social aspects of fan fiction communities can be important influences on how readers and writers interact with texts. For example, many communities have norms defining what is and is not appropriate feedback. At times, however, and particularly in larger communities, readers do not always provide what writers perceive as valuable feedback, as ChoMalfoy, a seventeen-year-old female originally from China and now living in Canada, mentioned in her interview (Herr-Stephenson, Harry Potter Fandom):

Becky: you mentioned that you used to write a little bit . . . did you share the stuff you wrote?

ChoMalfoy: Yes. On FanFiction.net and FictionAlley and my LJ [Live Journal].

Becky: did you get a lot of feedback on your pieces?

ChoMalfoy: Yeah, a reasonable amount.

Becky: did it impact your writing at all?

ChoMalfoy: No, the thing with reviews on fanfiction . . . people don't usually do constructive criticism. Mostly, it's encouragement/expressing desire for the author to hurry up with the next chapter.

In other communities, critical feedback is provided by "beta readers," who read fics before they are published and give suggestions on style, plot, and grammar. The relationships between writers and beta readers vary greatly depending on the situation and the people involved, and the expectations for beta readers differ between different sites. Describing her (quite different) relationships with her beta reader and the writer for whom she reads, orangefizzy said: ". . . yeah. i have a beta, and beta for another person. my beta is my best online friend, but i haven't heard from the girl i beta for in MONTHS. i need to poke her soon, see that she hasn't died."

Not all creative groups have a tight-knit community with established standards. YouTube, for example, functions more as an open aggregator of a wide range of video-production genres and communities, and the standards for participation and commentary differ according to the goals of particular video makers and social groups. Although some creative works are targeted for small niche groups, other youth creators we have spoken to wish to take advantage of opportunities to connect with a wide set of dispersed, similarly interested people in order to maximize the potential for receiving feedback, recognition, and critique for their work. Critique and feedback can take many forms, including posted comments on a site that displays works, private message exchanges, offers to collaborate, invitations to join other creators' social groups, and promotion from other members of an interest-oriented group. On YouTube a famous video maker might give a "shout-out" or mention another creator's work he or she admires. Even in the most competitive environments, the collaboration of other participants as promoters is often crucial to determining the critical and popular success of certain works. Viewers and fans who are often producers themselves rate, comment, and promote certain works over others.

In both the more tight-knit niche communities and more open sites such as YouTube, creators distinguished between productive and unproductive feedback. Simple five-star rating schemes, while useful in boosting ranking and visibility, were not valued as mechanisms for actually improving one's craft. Fansubbers generally thought that their audience had little understanding of what constituted a quality fansub and would take seriously only the evaluation of fellow producers. Similarly, AMV creators play down rankings and competition results based on "viewers choice." The perception among creators is that many videos win if they use popular anime as source material, regardless of the merits of the editing. In the YouTube world, many participants are concerned about "haters," or people who leave mean-spirited, discriminatory, or hurtful comments containing images of violence or death. While creators disliked these comments, they did not necessarily think that they should be restricted or excluded from the site. A number of youth creators also mentioned that they deliberately refuse to remove even hurtful comments posted on their pages as a way of showing their support for free speech online (Lange 2007a).

In contrast to these attitudes toward audience feedback, a comment from a respected fellow creator carries a great deal of weight. Darius, the twenty-four-year-old African-American mentioned earlier (Ito, Anime Fans), described some of the challenges he had in getting people to view and comment on his videos, but he was deeply appreciative when one fellow editor did give him feedback on his work.

And so somebody finally watched it at AWA, and was, like, oh, different concept, but it was a pretty cool video. Not necessarily award winning, but it was cool. I can watch it. I was, like, oh, okay. Thank you. I finally got somebody to tell me that, that much. But, like, you know, sometimes trying to get feedback on some of these things is like pulling teeth.

These moments, when young people get validation for their work from a peer, are important stepping-stones to developing an identity as a media creator. While some youth eschew the critiques as less useful because they are telling them what they already know, others highly value finding recognition and acceptance from peers for their work, even when they must endure hurtful commentary or harsh criticism from others. As Frank, a white fifteen-year-old male from Ohio on YouTube, stated, "But then even when you get one good comment, that makes up for fifty mean comments, 'cause it's just the fact of knowing that someone else out there liked your videos and stuff, and it doesn't really matter about everyone else that's criticized you" (Lange, YouTube and Video Bloggers). Edric, a rapper in Dilan Mahendran's study (Hip-Hop Music Production), is a nineteen-year-old Puerto Rican male who was born and grew up in San Francisco. He described the moment when he first stepped into the recording booth and received some recognition from fellow artists.

So I went into the booth. And I was nervous. It took me two times to finally get my words right. And finally I got my words right and did this song. And everyone was like, "Man, that was nice. I liked that." And I was like, for real? I was like, I appreciate that. And ever since then I've just been stuck to writing, developing my style . . .

Almost all creators bounce their ideas off fellow creators and ask specifically for feedback on their work. For example, in his work studying after-school video programs, Dan Perkel (The Social Dynamics of Media Production) observed how Nina (a twelve-year-old girl from a low-income neighborhood in northern California who appeared to be African-American) used LiveType to make a title for a video. After she made a title for her group's show, a few of her other team members came around, happened to see it, and showed their appreciation. One of the boys got very excited upon seeing it and the girl beamed proudly.

This type of ongoing feedback and communication among fellow creators and informed critics is one of the primary mechanisms through which creators improve their craft after their entry into a creative practice. Youth media programs, such as those described by Mahendran and Perkel, can provide the contexts for this kind of peer-based evaluation to happen. Other youth turn to online forums and interest-based communities, with their corresponding infrastructures of meet-ups and screenings. Through these ongoing exchanges, creators develop a sense of shared creative standards, genre conventions, and new forms of literacy. These social practices of evaluation, standard setting, and reputation building, well established in professional art worlds, are now being taken up by a larger swath of amateurs engaged in digital media production and online sharing.

Gaining Audience

Although audiences are not always seen as the best sources of critical feedback, most creators do seek visibility for their works, even if it is with relatively small groups of friends, families, or peers. The desire for sharing, visibility, and reputation is a powerful driver for creative production in the online world. While fellow creators provide the feedback that improves the craft, audiences provide the recognition and validation of the work that is highly motivational.

Although sharing is a motivator for most kinds of media creation we have observed, the boundaries that kids put on the sharing vary by kids and media type. For personal media, though youth may post publicly to sites such as MySpace or YouTube, the work generally is not intended to be circulated beyond friends and family. Many budding media creators also decide to share with only a small group. For example, several fan fiction writers in Mizuko Ito's Anime fans study wrote extensively but shared their writing with only their close friends. In some cases, people produce works for themselves and use their online creative-production spaces as personal sketchbooks in which they can experiment with things. Finally, our study has also identified a number of kids and youth who are reluctant to publicly share their materials. Keke, a sixteen-year-old black female from Los Angeles in danah boyd's study (Teen Sociality in Networked Publics), described conflicting desires to become a music producer and her reluctance and shyness at sharing her work:

danah: So what about writing? What do you write about?

Keke: I write about global warming and the war on Iraq, and I also write songs. I want to be a music producer when I grow up. I do a lot of music. Me and my best friend, London, we do a lot of music. We got a lot of songs that we've written together. So, yeah.

danah: So what do you do when you've written these things? Do you share them with anyone?

Keke: No. They're just . . . 'cause I'm real shy. For my music, I'm real shy. I don't know. I've just been shy. But every time I . . . 'cause I rap, so when I've rapped and stuff, people tell me I'm real good. I'm still shy, but I don't share none of the stuff I write about with other people 'cause some of it is real personal, 'cause I write a lot of stuff about my brother, who died, yeah.

danah: Do you think you'll ever share what you write?

Keke: Nope, never [laughs]. Never will I share it, 'cause everybody I hang out with . . . they don't really pay attention to stuff like I do. Like, I watch the news like it's a channel . . . if I am on the Internet, I'm looking up homeland security, stuff like that.

Young people struggle over their sense of confidence and safety about sharing their work to wider audiences. As creators get more confident and involved in their work, however, they generally will seek out audiences, and the online environment provides a vehicle for publishing and circulation of their work. In Dilan Mahendran's study (Hip-Hop Music Production), the more ambitious musicians would use a MySpace Music template as a way to develop profiles that situate them as musicians rather than a stan-

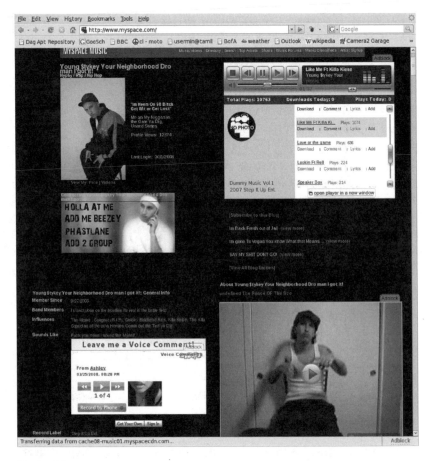

Figure 6.5
An example of a MySpace Music profile. Reprinted with permission from Young
MIC. Screen capture by Dilan Mahendran, 2006.

dard teen personal profile (see figure 6.5). Similarly, video makers who seek
broader audiences gravitate toward YouTube as a site to gain visibility.

More specialized video communities, such as those who do AMVs or
live-action vidding,[11] will often avoid general-purpose video-sharing sites
such as YouTube because they are not targeted to audiences who are well
informed about their genres of media. In fact, on the forums on the org,
any instance of the term "YouTube" is automatically censored. Even within
these specialized groups, however, creators do seek visibility. One AMV
creator in Mizuko Ito's "Anime Fans" study, Xstylus[12] (a white twenty-

eight-year-old), described the moment when his video was shown at Anime Expo (AX), the largest anime convention in the United States:

It was replayed again to an even-more-packed house during Masquerade, AX's most popular event. Never had I ever seen so many people laugh so hard in my entire life. The only people who could ever come close to experiencing such a feeling are Hollywood directors having won an Academy Award for Best Picture. It was the finest, greatest, most moving moment of my entire existence. Nothing will ever top it. Ever.

XStylus received recognition for his work in the context of a formal competition organized by the convention. Most major anime conventions now will include an AMV competition in which the winning works are showcased, in addition to providing venues for fan artists to display and sell their work. The young hip-hop artists Dilan Mahendran spoke to also participated in musical competitions that gave them visibility, particularly if they went home with awards. Even fansubbers who insist that quality and respect among peers are more important than download numbers will admit that they do track the numbers. As one subber in Ito's study described, "Deep down inside, every fansubber wants to have their work watched, and a high amount of viewers causes them some kind of joy whether they express it or not." Fansub groups generally make their "trackers," which record the number of downloads, public on their sites.

Similarly, on YouTube, people have access to "view counts" of particular videos, although these are generally regarded as unreliable (YouTube was sometimes slow and inconsistent in updating them) and easily manipulated (by makers who can create automated refresh programs to reload the video and make it appear as though it is being viewed widely). For youth producers who wish to professionalize or maintain an advanced-amateur status in which they can partner with YouTube, numbers of views and comments are used as a rough metric for granting partnership and promoting their work. Another metric involves the number of "subscribers" that a person on YouTube has. Being a subscriber of someone on YouTube means that you will be alerted (usually via email) when he or she posts new videos. Some YouTubers participate in a kind of "sub-for-sub" reciprocity in which a video maker subscribes to someone else with the expectation or hope that the subscription will be reciprocated. However, many people actively resist this assumption and prefer to subscribe only when content interests them.

Reciprocity agreements and friendships can greatly assist one's visibility. One interviewee in Mizuko Ito's anime study (Anime Fans), SnafuDave (see box 7.1), is a successful web comics creator who hosts a site for his own comics and the comics of several other artists. He described that as he was getting started, the friendships he made with other established artists were instrumental in his gaining audience. Some of these artists ended up using his site to publish their work, which "was a really big pull." Others would mention his site in their own postings, which would also drive audience to his site. He also places advertisements for his site on other established web comics sites. In this case, he generally pays for the advertising, even though there is often a spirit of reciprocity within the web comics world. "I try to pay for all of my advertising, just because I know, say, if they do give me that spot, then they're losing money by not selling it to someone else." In addition to these forms of ad placements, he visits conventions around the country to promote his work and sell related merchandise.

Youth such as SnafuDave who are able to reach wide audiences can parlay their creative work into future careers. Even in the case of youth who stay within recreational and amateur domains of creative production, the ability to connect with audiences is a key part of what drives their participation and learning in creative production. The ability of digital networked media to create new publics and audiences for amateur work is one of the most transformative dimensions of contemporary new media. The ability to define new collectivities and niche publics for culture and knowledge has been the subject of much writing on contemporary digital culture (Anderson 2006; Benkler 2006; Jenkins 2006; Shirky 2008; Varnelis 2008). Examining media production provides a window onto how these dynamics are operating in the everyday lives of youth.

Aspirational Trajectories

In most cases, young people who create digital media are not aspiring to be professionals or to get famous through their creative work. They engage in digital media production as a social activity, a fun extracurricular hobby, or maybe even a serious lifelong one. Most of the dominant forms of fan production—fan fiction, video remix, amateur comics—are not commercially viable. Even older fans who do professional-quality work and who have a substantial following in the fandom generally have no professional or commercial aspirations in the area and have day jobs that are not related to their creative hobbies. For example, doki[13] (a white thirty-two-year-old), one of the leaders of the AMV world, describes himself as "a game designer by day, AMV creator by night." Another anime fan, Scottanime[14] (a white thirty-one-year-old), spends almost all his time off from being a mail carrier organizing anime conventions (Ito, Anime Fans). Even though these activities may not result in economic or vocational outcomes, participants in amateur media creation work hard to improve their craft, and they get tremendous validation from their creative communities and audiences for a job well done. Some researchers refer to a category of creators as "proteurs," or "people who have gained recognition as professionals for their hobbies even if they don't have relevant professional certificates or degrees" (Faulkner and Melican 2007, 53). As discussed in box 6.4, several groups of youth podcasters have achieved recognition for their achievements from fans and from major corporations such as Scholastic (the U.S. publisher of the Harry Potter book series) and Warner Brothers (the studio that produces the films).

Box 6.4   **Spoiler Alert**: Harry Potter Podcasting as Collaborative Production

Becky Herr-Stephenson

Sitting on the floor of a crowded annex of a Los Angeles bookstore, I am just one of nearly two hundred people waiting for an event to start. To my right, a mother and son talk about a theory on time travel. Behind me, a teenage girl scribbles furiously in a well-worn notebook. All around, excited conversation ebbs and flows, at times becoming uncomfortably loud. One can only imagine what the other bookstore patrons are thinking. This place has often served as a quiet space for a cup of tea and a new book; that is certainly not the experience available today. When the event starts, the audience cheers for a group of people making their way to the small stage. It is not a prolific author, nor a band, nor a popular public speaker that we are there to see—it is a group of regular (if geeky) people who have become BNFs (big-name fans) for recording podcasts about Harry Potter. But one would never know that if she were just wandering by the annex on the way to the travel guides section.

Since the publication of *Harry Potter and the Sorcerer's Stone* nearly ten years ago, Harry Potter fans have adopted a variety of technologies for sharing writing, facilitating discussion, creating artwork and computer graphics, and producing audio and video. Podcasting, the production of audio files for download via RSS (really simple syndication), emerged as a popular genre of fan production in August 2005. At the height of the "Summer of Potter," July 2007, more than thirty podcasts were in active production. Podcasting seems a natural fit for this technology-savvy fandom, which has expertly migrated many longtime elements of fandom (including sharing information, media production, and social networking) to online spaces, opening the fandom to geographically diffuse and generationally diverse groups of fans. Podcasting allows for ongoing analysis of canon materials and in-depth, sustained commentary on fans' consumption and production practices, discussions that do not necessarily have a home within other forms of fan production.

Harry Potter podcasts take on a variety of formats, but most contain the same basic elements: news updates, literary analysis and theory building, and commentary on other media within the franchise, such as the films, sound tracks, merchandising, and video games. Some shows focus on a specific interest—such as fan fiction or "Wizard Rock" music, while others focus on a particular character or relationship within the books.

Podcast production can vary from an individual who hosts, records, edits, and publishes the show by herself to a group with a hierarchical organization similar to small video-production collectives or independent bands. In most cases, podcasts are run by a small team of hosts. The hosts prepare a rundown or outline for each episode, usually working with collaboration and communication tools such as instant messaging, Skype conference calls, and Google Docs, which allow simultaneous collaborative editing of texts. Email and phone contact (voice and texting) also frequently play a role in some of the necessary microcoordination around a podcast recording. In addition to the discussion among the hosts, podcasts frequently feature segments recorded by correspondents or specialists in a particular aspect of fandom that are "rolled in" between host discussions. One podcaster, a white nineteen-year-old from Illinois, emphasized the importance of opening the production to contributors aside from the hosts. He said, "The main focus of [our show] is to give other people a chance to be podcasters . . . we want to give them an opportunity to be a podcaster. The first thing we decided was that anyone who wants to be a guest host can be on the show." Since the technological demands for recording a podcast are relatively low, and because there is no need for the hosts to be colocated for recording, it is possible to open up the production process in this way.

Equally important as segment contributors and guest hosts, a third element to Harry Potter podcasts (and podcasting in general) is general audience participation. Shows frequently have voice-mail services where listeners can call in and record questions and comments that are played during the show. Alternatively, some podcasts solicit audience feedback via email, and the hosts read and respond to those comments. The audience participation in podcasting is similar to that of talk-radio programs, but it also reflects the value placed on accessibility, dialogue, and blurring boundaries between producer and consumer that are characteristic of online creative production.

Production does not end with recording. In addition to editing the audio, podcasters must navigate distribution and publicity channels. Unlike mainstream media, in which a separate entity generally would handle the distribution and marketing of a program, podcasters (and most other amateur producers) need to make decisions about the venues in which they will publish and promote their shows. For many podcasts, the first step is creating a show webpage. The show page acts as home base for the show and provides information about the podcast and links to download episodes. Other venues for publication of the show feeds include online music retailers such as iTunes, podcast aggregators such as podcastalley.com, or social network sites such as MySpace. Within the fandom, cross-promotion and linking are a regular practice, as fans tend to exist within small "neighborhoods" of sites that cater to their particular interests and favorite practices. It is not unusual to find out about new episodes of a podcast through one's Friend list on LiveJournal, a MySpace bulletin, or a Friend's Facebook status update before the episode is available on iTunes.

In some cases, promotion extends beyond the fan community. Two podcasts made of geographically dispersed, teenage and mixed-age adults are particularly noteworthy. They are associated with large fan sites that have achieved notoriety within the fandom as well as recognition by corporations such as Scholastic (the U.S. publisher of the series) and Warner Brothers (the studio that produces the films). These two podcasts have produced weekly episodes (with few exceptions) for more than two years, and they continue to put out new episodes even after the final book was released on July 21, 2007. One unique element of these shows is that they regularly record live podcasts at events such as fan conferences, book releases, movie premieres, and occasionally, just because they happen to be traveling together for another event. To support the costs of production (bandwidth, software, on-site production, travel, etc.), both shows feature advertisements in the episodes and on their websites. In a manner very similar to early radio and television, episodes start with advertisements for the shows' sponsors, which vary from website hosting services to major chain booksellers.

Several popular Harry Potter podcasts are winding down production since the release of the last book, releasing sporadic special episodes rather than weekly or monthly episodes. At the same time, some podcasters are beginning to experiment with video podcasts and live streaming technologies. It is a moment of transition for this type of production, just as it is for the fandom as a whole. Harry has grown up and defeated the Dark Lord, and fans, who still have much to say, are looking to new forms of production for expression.

Although aspirations for creative production are quite varied, we have observed a category of children and youth who have plans to become media professionals. These individuals see their creative production as a means to train themselves, improve technical skills, gain visibility and reputation, and develop relevant contacts in appropriate arenas. In some cases, parents lend support to their children's endeavors by helping to provide material and emotional infrastructures that enable them to develop their skills and visibility. In other cases, parents are involved much more directly in children's career paths by participating and coproducing the media productions. For example, as box 6.2 describes, a sixteen-year-old girl, who calls herself a "future filmmaker" on YouTube, and her mother make and post videos to build the daughter's résumé and help her gain the skills that will enable her to become accepted in appropriate media-oriented educational programs.

In most cases, children who express interest in becoming professionals are not necessarily sure which role in media, such as being a director or editor, they wish to take up. Some of them plan to major in artistic or related disciplines in college. A few kids and youth we have spoken to did not necessarily start out with particular plans to pursue media careers, but they found broad success in their communities of interest and changed their majors or started to consider media as a potential career. In his research in after-school video programs (The Social Dynamics of Media Production), for example, Dan Perkel found that several participants planned to pursue media-related careers. However, he stated that it was difficult to tell to what extent participation in the after-school program stimulated this interest or if it was part of a deep prior interest. We have found that hierarchies of recognition and technical specialization often develop among youth in local peer groups and in schools. For example, we observed some experienced youth video makers being asked to contribute to school activities by holding workshops or creating videos to advertise or document school events. Regardless of how many of these kids actually will be able to go on to pursue careers as video makers, we have seen many instances of kids who begin in the amateur space but eventually aspire to a professional track.

In addition to providing new avenues for professionalization, new-media distribution affords different aspirational trajectories. By linking "long tail" (Anderson 2006) niche audiences, online media-sharing sites make amateur-

and youth-created content visible to other creators. Aspiring creators do not need to look exclusively to professional and commercial works for models of how to pursue their craft. Young people can begin by modeling more accessible and amateur forms of creative production. Even if they end there, with practices that never turn toward professionalism, they still can gain status, validation, and reputation within specific creative communities and smaller audiences. The ability to specialize, tailor one's message and voice, and communicate with small publics is facilitated by the growing availability of diverse and niche networked publics. Gaining reputation as a rapper within the exclusive community of Bay Area Hyphy hip-hop, being recognized as a great character writer on a particular role-playing board, or being known as the best comedic AMV editor for a particular anime series are all examples of fame and reputation within specialized communities of interest. These aspirational trajectories do not necessarily resolve into a vision of making it big or becoming famous within the mode of established commercial media production. Yet they still enable young people to gain validation, recognition, and audience for their creative works and to hone their craft within groups of like-minded and expert peers. Gaining recognition in these niche and amateur groups means validation of creative work in the here and now without having to wait for rewards in a far-flung and uncertain future in creative production.

In terms of discourses of fame, some producers straightforwardly claimed they sought fame and widespread recognition for their work. However, others eschewed connections to fame, which is a construct that often is laden with ideological baggage and negative connotations. For example, a group of older male teen producers from California on YouTube (who had won a festival prize for their work) expressed frustration that some of the most famous youth contributors to the site created work that they saw as subpar, uncreative, and not particularly technical (Lange, YouTube and Video Bloggers). Fame is often discussed as a relational construct in which a person who may be considered famous by certain measures denies being as famous as another producer or media maker. For some participants, being famous was not as important as improving their skills and receiving legitimation from a select few peers they deemed capable of understanding their contribution in a meaningful way.

What is significant about contemporary networked publics is that they open up multiple aspirational trajectories for young people. While some may aspire to professionalization and large audiences, others see their creative work as a serious but amateur hobby, pursued for the love of it and not for financial gain. Online distribution may be opening new avenues to fame and professional careers for a small number of creators, but the more radical and broad-based changes are happening at the amateur layer. Unlike professional media production, amateur media can support a proliferating number of creators buoyed by long-tail, small audiences. These niche audiences represent an opportunity for a growing number of youth to engage in media production in the context of public participation.

Conclusion

In this chapter, we describe some of the specificities of how kids engage in creative production and a wide range of practices that might fall under the umbrella of "online content creation." Most of the content creation that youth engage in is a form of personal media creation that is focused on documenting their everyday lives and sharing with friends and family. In some cases, this everyday personal media production serves as a jumping-off point for developing other kinds of creative interests. In other cases, youth express interest in developing highly technical media skills from an early age. Yet both commonplace and exceptional cases in media production share certain commonalities, and the boundary between "casual" and social media production and "serious" media production is difficult to define. Although friendship-driven and hanging out genres of participation are generally associated with more casual forms of media creation, they can transition quickly to messing around and geeking out. Conversely, the relationships that youth foster in interest-driven creative production can become a source of new friendship and collegiality that is an alternative to the kinds of friendships and status regimes that youth must inhabit at school. We can see this in the social energies that young people bring to online discussions with their interest-based friends as well as in conventions and meet-ups where youth are sharing their lives as well as their creative work.

All these cases demonstrate the growing centrality of media creation in the everyday social communication of youth. Whether it is everyday photography or machinima, youth are using media they create as a way of documenting their lives and as a means of self-expression. These cases also demonstrate the centrality of peer-based exchange in motivating creative work and providing a learning context. Peers are fellow creators youth see as knowledgeable audiences who have shared investments in the work, and with whom they have a relation of reciprocity. Peers view and comment on their work and vice versa. This may be the given peer group of local friends or family, or it may be a specialized creative community. Teens consider what their friends will think of their MySpace profiles, and video creators hope fellow makers will appreciate the craft that went in to their work. In both these cases, networked publics enable kids to connect with others in ways that facilitate sharing and peer-based learning. Even when the initial impetus for media production comes from family, school, or after-school programs, a prime motivator for improving the craft lies in the network of peers who serve as audiences, critics, collaborators, and coproducers in the creation of media.

School programs can provide an introduction to creative production practices that kids may not otherwise have exposure to. In most programs, however, the audience for production is limited to the teacher and possibly the class. In addition, most classroom projects are not driven by the interests of the participants themselves. By contrast, the examples we have found in youth recreational and hobby productions indicate a different dynamic. When youth have the opportunity to pursue projects based on their own interests, and to share them within a network of peers with similar investments, the result is highly active forms of learning. In after-school programs where youth have the opportunity to showcase their work to a broader audience of creators and aficionados, they can gain validation for their work in ways similar to what we have observed online. For example, Dilan Mahendran's study (Hip-Hop Music Production) found that youth hip-hop creators in the program he studied distributed their works to larger audiences and participated in a range of public performances and competitions. The case of hip-hop demonstrates the power of amateur and small-scale communities of media production to support aspirational trajectories that rely on reputation in more niche or local contexts. Online networks enable young people to find these niche audiences in ways that were not historically available to youth. Although it is rare for youth to be able to reach a scale of audience that rivals professional media production, many are able to reach beyond the boundaries of home, local activity groups, and families in finding appreciative audiences for their work.

Within all these contexts, whether supported by online groups or local programs, youth are experimenting with new genres of media and new forms of literacy that take advantage of a moment of interpretive flexibility in the contemporary media ecology. This chapter focuses on the social processes of media production. By concentrating on these processes, we have investigated how young people are actively negotiating with one another about standards of quality and craftsmanship. Part of the excitement for young creators is that they can be part of defining new genres and cultural forms, not simply reproducing existing ones. This is an example of some of the specificity of how generational identity, media literacy, and technical change coconstruct one another.

Notes

1. "Prosumer" is a contraction of "producer" and "consumer," or "professional" and "consumer." The term was coined by Alvin Toffler (1980) to describe the blurring of the boundaries between producers and consumers.

2. "Pro-am" refers to "professional amateurs" and was popularized by Charles Leadbeater and Paul Miller (2004). The term refers to the trend toward amateurs creating work to professional standards.

3. "Inertia" is a screen name.

4. Anime music videos (AMVs) are remix fan videos, in which editors will combine footage from anime with other sound tracks. Most commonly, editors will use popular Euro-American music, but some also will edit to movie trailer or TV ad sound tracks or to pieces of dialogue from movies and TV.

5. "Gepetto" is a screen name.

6. "SnafuDave" is a screen name.

7. "Hyphy" is a rap genre that originated in the San Francisco Bay Area and is closely associated with the late rapper Mac Dre and with Fabby Davis Jr. Hyphy music is often categorized as rhythmically up-tempo with a focus on eclectic instrumental beat arrangements, and is tightly coupled with particular dance styles.

8. "AbsoluteDestiny" is a screen name.

9. "Darius" is a real name.

10. Role-playing boards, also know as play-by-post games, are a hybrid between fan fiction and role-playing games. Writers generally take on the role of a character in a fantasy world and post narrative about their character to a web forum to collaboratively create stories or engage in a role-playing game.

11. "Vidding," like AMVs, is a process of remixing footage from TV shows and movies to sound tracks of an editor's choosing. Unlike AMVs, however, the live-action vidding community has been dominated by women.

12. "Xstylus" is a screen name.

13. "Doki" is a screen name.

14. "Scottanime" is a screen name.

7 WORK

Lead Author: Mizuko Ito

In her research on Silicon Valley families, Heather Horst writes about the Smith family, who "view digital media as a tool for their children's personal and professional life." One of the two daughters is a budding musician who writes her own music and performs at local venues. Her older sister is an accomplished dancer and writer who uses the skills learned while attending the Girls Technology Academy to help her father digitally record, edit, burn, and distribute CDs of her sister's performances. For the Smith family, digital media production is a creative hobby that they engage in together as well as an activity that is intimately tied to future career aspirations for the children (Horst 2007). Another Silicon Valley teen, a nineteen-year-old Filipino and Japanese American, has been a leader of a fansubbing group since high school. At the time of our interview, he was taking time off from college to help out in his family technology-related business. "If I didn't stop school and help out, we'd be in serious trouble now," he explained. At the same time, he was still continuing his unpaid work in fansubbing, managing a team of more than a dozen staff who churn out subtitled anime every week for eager fans. His technical expertise serves him across multiple domains of work, some paid, some unpaid (Ito, Anime Fans).

These examples of engagement with new media point to certain domains of practice that are not covered by the other chapters in this volume. The focus of our project has been on learning in relation to youth practices of play, socializing, and creative experimentation. As we have pursued this research, however, we have found that new media also have important implications for how young people engage in activities that they see as serious or productive work, or that have a role in preparing them for jobs in the future. The promotion of new media use among youth is often justified in terms of skills training for "competitiveness" in the twenty-first-century workplace (Drucker 1994; Florida 2003); parents, educators, and kids often describe their relationship to learning and new media in these terms. In addition to this educative, future-oriented role of technology engagement, new media have an important influence on the here-and-now of at least some of the more digitally mobilized youth we have met through our research. One of the important roles that new media play in the lives of youth is in providing access to experiences of volunteerism and work that give them a greater sense of autonomy and efficacy than those avenues of work that previously have been available to U.S. teens.

This chapter describes these different dimensions of new media and work—how new media engagement operates as a site of training and preparatory work as well as how it becomes a vehicle for new forms of volunteerism, nonmarket labor, and new media ventures. The effort is to capture those new media activities characterized by a productive or seriousness of purpose, where play, socializing, and messing around begin to shade into what youth consider "work," "real responsibility," and economic gain. We draw primarily from studies that look at the everyday lives of youth in families (Martínez, High School Computer Club and Animation around the Block; Sims, Rural and Urban Youth; Tripp and Herr-Stephenson, Los Angeles Middle Schools), studies of gaming and fan production (Cody, Final Fantasy XI; Herr-Stephenson, Harry Potter Fandom; Horst and Robinson, Neopets; Ito, Anime Fans; Lange, YouTube and Video Bloggers), and studies of youth media production (Antin, Perkel, and Sims, The Social Dynamics of Media Production; Mahendran, Hip-Hop Music Production). After providing a conceptual framework for our understanding of the relationship between new media, youth, and work, the chapter describes three categories of work-related practice: training, entrepreneurship, and nonmarket work.

Work, Youth, and New Media

Our understandings of what work or labor means in relation to children and youth are diverse and contested within different scholarly communities. Although it is not our intention here to fully review this body of work or to formulate our own definitions, we would like to take a moment to contextualize our descriptions and outline the boundaries of what we address in this chapter. In the United States, youth are largely shut out from the primary labor market, but they still engage in a wide range of activities that could be recognized as work, varying from schoolwork to chores to part-time jobs in the service sector. Researchers have argued that we run the risk of erasing youth contributions to our economy and productive labor if we insist on categorically excluding certain forms of youth activity from our definitions of work (Orellana 2001; Qvortrup 2001). Activities such as "helping" at home or in class often are not counted as work, although they are clearly productive labor (Orellana 2001). Our definitions of work are further complicated by the fact that even play is often defined as "the work of childhood" (Seiter 1993), and "serious" extracurricular activities such as volunteer activities, music lessons, and sports also can be considered "work" by children and parents. Narrow definitions of work would limit the discussion to activity that has clear economic outcomes, while broader definitions could include activity that is more general to any productive or compulsory activity, such as the work of education (Qvortrup 2001). Educational, preparatory work is what Jens Qvortrup (2001) has argued is the most important kind of economically productive activity that children engage in—preparing themselves as future workers. While we might hesitate to call schooling and extracurricular activities "work" in the traditional sense, it is important to acknowledge the ways in which this "prep work" is part of the cultivation of skills and dispositions that will serve youth as they move into jobs and careers. These diverse accounts of what constitutes work are all important reference points in understanding the discourses and practices of work that we encountered in our case studies.

Children and youth represent a special case in discussions of labor and work. As with other industrialized countries, the United States has a well-established set of laws and social norms that limit children's and youth's access to certain categories of work. The shift toward education as defining the primary work of teens was a constitutive element of the definition of adolescence as a unique life stage (Hine 2000). Although teens may have the right to take jobs, they do not always have access to the jobs that they imagine for themselves in their future as adults. Jim McKechnie and Sandy Hobbs have argued that compulsory education did not force adolescents out of employment; rather "it has moved the main forms of employment from full-time to part-time and changed the nature of that employment" (2001, 10). They point out that the majority of youth in industrialized countries works in a part-time capacity, often negotiating tensions with their "primary" occupation as students. Phillip Mizen, Christopher Pole, and Angela Bolton (2001, 19) describe the work available to adolescents today as "unskilled work around the edges of the formal labour market," typically retail, distribution, catering, and fast food. The United States is characterized by a dual track in terms of youth relationships with school and work. While more privileged youth typically engage in "low intensity" work and give priority to an academic pathway, lower-income youth more typically take a pathway that "leads directly from high intensity high school employment to full-time adult employment" (Hansen, Mortimer, and Krüger 2001, 133). These structural conditions of youth and labor are an important backdrop to kids' engagements with new media work.

New media add some unique wrinkles to our understandings of youth and work. For starters, in public debates surrounding education and new media, the issue of job preparation is often central to the discourse. These approaches are framed by the expectation that education should be the primary work of childhood, and new media learning is validated by the expectation that it will translate to job-relevant skills in the future. All the structured educational efforts around new media that we observed are justified, at least in part, by the argument that they are helping to develop job-relevant skills. Programs that have an equity agenda are often funded as efforts to provide disadvantaged children and youth with remedial access to high-tech skills. At the same time, there is a growing recognition that digital media skills are largely cultivated in the home and other more informal and social settings (Seiter 2007). Schools are not the dominant sites of access to these forms of preparatory training with new media and information technology. Privileged homes take new technology for granted, integrating computer use seamlessly into their everyday routines and domestic spaces. They see new media engagement as part of a more general stance of participation in public life, not necessarily those that are focused on job skills. By contrast, low-income families struggle to keep up with the rising bar for participation in an increasingly high-tech ecology of culture and knowledge. These ways in which new media play into practices that participants see as preparatory for jobs and careers is the first descriptive category for this chapter. This is a set of practices we call "training." This includes learning activities that are pursued in both formal and informal educational settings, though our focus is on the latter.

The second set of work practices we have encountered in our case studies are those that are directly tied to economic activity. This would include jobs that rely on digital media and small economic ventures that were started by youth. Because of our focus on informal learning, most of our cases are on the latter—youth-driven forms of economic activity that we call new media "entrepreneurism." Digital and networked media have opened up opportunities for economic activity for young people that are not part of the existing ghettoes of youth labor, but rather involve young people's mobilizing and hustling to market their new media skills in a more entrepreneurial vein. These new forms of accessibility to entrepreneurial opportunity are the second wrinkle that new media add to the landscape of youth and work. While some of these activities are tied to existing genres of youth labor—such as the marketing of youth talent or getting paid for helping in local and community settings, other enterprising youth are disrupting expectations about the categories of economic activity that youth should engage in. In all these cases, though, the substantial technology expertise of some young people challenges the assumption that youth labor is necessarily unskilled or preparatory, demonstrating that they can make contributions that exceed the capacity of many local adults.

Much of the productive labor of childhood is in the domain of what we call "nonmarket work"—volunteerism, helping in the home, noncommercial production, labor in virtual economies, and hobbies. Although not tied to economic gain, these activities involve commitments that participants consider in the vein of "jobs" and "serious responsibilities" to produce work and contribute labor. For kids in lower-income and immigrant households, nonmarket work is often dominated by domestic labor, and girls shoulder a disproportionate amount of these forms of work (Orellana 2001). For more privileged youth, it tends to have a more preparatory dimension. Many of the in-school and organized extracurricular activities that young people engage in are not directly tied to job and career aspirations but are part of what Annette Lareau (2003) has described as "concerted cultivation," as described in chapter 4. These are activities that immerse children and youth in cultures of competition, achievement, and public participation that are key to certain modes of social success. Although this chapter does not deal substantively with school-based work or practices of concerted cultivation, these preparatory activities are a backdrop for and often a trajectory into nonmarket work.

This last set of practices introduces what is perhaps the most intriguing and significant wrinkle that new media bring to young people's experience of work. In digital-culture studies, theorists have been describing the growth of various types of unpaid digital work, including open-source software development (Weber 2003), "nonmarket peer production" (Benkler 2006), "crowdsourcing"[1] (Howe 2006), virtual economies (Castronova 2001; Dibbell 2006), and other forms of noncommercial free culture (Lessig 2004). In many of these kinds of new media work practices, the unpaid labor of youth is a significant factor. Our case studies describe how these practices are being driven forward by the interests and social practices of youth from wired households. The opportunities that youth have to participate in new forms of creative work is discussed in chapter 6.

Here we look more broadly at the range of ways young people work in virtual worlds and with new media, motivated by reputation, learning goals, a sharing ethic, and their own satisfaction rather than economic gain. Although the free time and online activities of youth are certainly not the only factors driving free culture and peer production online, it is one integral component of what theorists have identified as a trend toward exploiting free labor in digital economies (Terranova 2000). Andrew Ross notes how networked media have initiated a process "by which the burden of productive labour is increasingly transferred on to the user or consumer" (Ross 2007, 19). Our ethnographic material describes some of the specificities of these trends by describing the unique alchemy between the marginalized role of youth in the labor market and the development of nonmarket forms of collective work. The story cannot be reduced either to a simple equation of empowerment or exploitation as youth gain nonquantifiable social benefits, though they may not be reaping economic ones.

In many ways, the current practices of youth engaged in new media–related work complicate our existing assumptions about youth, labor, work, and the role of educational institutions to prepare youth for the workplace. First, the cases we describe challenge the assumptions that the appropriate role of youth work is in preparatory educational contexts or in unskilled labor. Youth media production and ventures, when combined with the distribution capacity of the Internet, means that the nonmarket work of childhood is channeled in broader networks that can challenge the authority of existing industry models. New media practices are becoming a vehicle for *some* youth to exercise more agency in defining the terms of their own work practices. The new media skills and talents that these kids are exhibiting make the *productive* labor (as opposed to preparatory work) of childhood more visible (at least in the new media domain), and they challenge the status of educational institutions in defining the training of youth for high-tech work. This in turn is tied to structural changes in certain forms of economic exchange activities, in which the business models of creative industries are being undermined by user-generated content and peer-to-peer (P2P) file sharing. Domains of creative work that were considered almost exclusively the province of commercial efforts are being partially displaced by the work of creative hobbyists who are not necessarily seeking monetary rewards. While we do not see evidence that new media practices are leading to any fundamental reordering of the conditions of economic inequity, we are seeing some indicators that the interfaces between the productive and preparatory work of childhood are being renegotiated through these practices.

Training

Although our work has focused on learning in informal settings, a number of our case studies did examine media-education programs in schools and after-school centers, and we had many opportunities to speak to kids and parents about how they thought computers contributed to their school-work and their future careers. Computers and media-related expertise intersect in complicated and sometimes contradictory ways with how parents, educators, and kids believe young people should be prepared for schooling and jobs. In her analysis of how computer-based pedagogy relates to young people's school performance and future careers, Ellen Seiter argues that educators and technology boosters often fail to take into account the contexts of structural inequity that usually overwhelm the benefits of technology access that educational programs might provide. She points out "the barriers that make the dream of winning something like a 'cool job' in new media a very distant one for working class students" (Seiter 2007, 28). At the same time, she describes how technology-based educational programs are justified by exactly this promise of social mobility. In Seiter's view, the resources that middle-class and elite children have at home, in contexts of concerted cultivation, are what determine cultural and social capital in relation to digital media and the ability to parlay fun engagement with digital media into careers in the "new economy."[1] Educators who are fighting for social-equity agendas have always faced an uphill battle against the entrenched structures of social and cultural distinction that extend well beyond the classroom walls; we have no reason to believe that simply introducing technology to this equation is going to transform these structural conditions. In fact, since some of the most cutting-edge technology practices are learned outside schools, in the private contexts of peer interaction and family life, the equity agenda is made even more challenging from a public-policy perspective.

As we describe in chapter 1, it is difficult to clearly map differences in socioeconomic status to new media fluency. At the same time, we see some patterns in the degree to which computer use is framed in terms of an education-oriented or vocational tool for social mobility, versus one that is an unremarkable and taken-for-granted component of everyday social, recreational, and academic pursuits. "Training" as a genre of computer use tends to be associated with aspirations of upward mobility by less financially privileged families rather than by families who see computers as already deeply embedded in the fabric of the children's everyday lives. In the earlier chapters of this book, we describe some of the informal settings of peer groups, interest groups, and family where much of the basic learning and literacy about new media is supported. In these contexts, parents and youth generally are not mobilizing a discourse of vocational training but rather a discourse of enrichment and creativity.

The day-to-day struggles of educators, parents, and kids to chart trajectories through educational institutions and on to jobs and careers need to be contextualized by these structural conditions and by our cultural imaginings and values around technology, achievement, and work. Even before a consideration of whether kids might get a creative-class job, parents and educators hope that computers will give kids a leg up in their educational performance. This often translates to a parental concern that computers should be used for serious educational purposes and not for socializing or play. We observed this tendency most strongly in less privileged families that saw schooling as their primary hope for upward mobility. One parent in Lisa Tripp and Becky Herr-Stephenson's study (Teaching and Learning with Multimedia) explained how she tries to encourage certain forms of computer use in the home for her thirteen-year-old daughter, Nina, the third of four children. Anita emigrated from Mexico eighteen years ago, and her husband is from El Salvador.

Anita: *[Mi hija] se pone en la computadora y le digo que la computadora es para hacer tarea, no es para estar buscando cosas en la computadora. Y a veces [mis hijas] se me enojan por eso. Y les digo: "No, la computadora yo se las tengo para que hagan tarea." A veces les pregunto: "¿tienen tarea?" O: "estás haciendo tarea." Pero a veces tengo que estar lista a ver qué es lo que están haciendo. Se meten a la Internet y tantas cosas que sale ahí. Y se ponen a mirar sus amigas y eso. . . . Entonces, es lo que no le gusta a ella que yo le diga: "¿sabes qué? La computadora no es para que andes buscando; es para lo de la escuela."*
([My daughter] sits in front of the computer and I tell her that the computer is for doing homework, not for looking around. And sometimes [my daughters] get mad at me because of that. And then I say, "I got this computer so you could do your homework." Sometimes I ask, "Do you have homework?" Or, "Are you actually doing your homework." I have to keep a close eye on them to see what is going on. They get on the Internet, and with so many things there. They look for their girlfriends and all. . . . They don't like me saying, "You know what? The computer is not for you to be looking around. It is for schoolwork.")

Lisa: *¿Qué es lo que más le preocupa a usted acerca de la Internet y sus hijas?*
(What is your main concern with the Internet and your daughters?)

Anita: *Lo que me preocupa . . . ya ve . . . es que salen muchas cosas ahí que se meten con niños, y a veces platican con ellos, y a veces no saben ni qué gente es. Es lo que me preocupa, porque digo "no." Y a ver qué es lo que están mirando ellos y uno tiene que estar siempre listo con ellos. A veces estoy que les quiero quitar la Internet, pero a veces me dice él: "por su tarea está bien. Porque después van a andar que 'me voy a hacer tarea,' 'que no tengo computadora,' 'que no tengo esto.'" Pero es por lo que más peleo ahorita con ellos.*
(My main concern is . . . you see . . . you hear all the time that people try to reach kids and talk to them. Sometimes [kids] don't even know who they are talking to. . . . That is my concern. That is why I say, "No." I need to keep an eye on what they are looking at. I always need to be attentive. Sometimes I feel like canceling the Internet, but my husband says, "It is good to keep it because of their homework. You don't want them saying 'I need to go somewhere else to do my homework,' or 'I don't have a computer,' or 'I don't have this.' But this is mostly what I fight about with them these days.") (Translation by Lisa Tripp)

A father who is raising his two daughters on his own also voices a commitment to educational goals, though he does support his daughters' use of the Internet for personal communication. Juan has been in the United States for almost thirty years, and he is raising his two younger daughters (eleven and twelve years old) on his own while working in a restaurant. They have an older computer at home that they acquired secondhand, but it is not connected to the Internet (though it once was) because of the cost.

Lisa: *¿Usted que cree que la Internet es una buena manera para los niños comunicarse entre ellos, o le preocupa esto?*
(Do you think that the Internet is a good way for your kids to communicate with their friends or are you worried about that?)

Juan: *No, es mejor que se comuniquen de esa forma porque les ayuda más a salir adelante. Y también cómo lo tomen ellos. Si lo van a tomar como un juego, esto y lo otro, no. La cosa es que vayan a la cosa seria, que vayan aprendiendo. Con ánimos de seguir adelante en sus estudios, de salir adelante. Sabes que ahorita sin estudios uno no es nada. No es nada. A trabajar, andar limpiando y haciendo acá, sufriendo más si un día no te necesitan. Salir adelante.*
(No, it is better for them to communicate that way because it is going to help them get ahead. And it also depends how they treat it. If they are going to treat it like a game, then no. But if they take it seriously, they will be learning from it, and it will help them with their studies, and help them get ahead. You know that now without an education you are nothing, nothing. You have to work, clean places, do odd jobs, suffering if one day they don't need you anymore. [It's important to try] to get ahead.)
(Translation by Lisa Tripp)

In their work in Los Angeles, Tripp, Herr-Stephenson, and Martínez interacted with parents, teachers, and kids in both the classroom and at home, affording a rare opportunity to look across multiple contexts of media use for particular kids. In the multimedia course that the kids were engaged in, teachers occasionally spoke of the possibility of careers in media, but their goals were generally more immediate and less ambitious. They saw new media production as a way of keeping kids engaged in the classroom, which could in turn keep them from dropping out. They also thought that one side effect of this engagement was that kids would pick up basic reading and writing skills. One teacher describes his first year with the multimedia curriculum: "I think this year, in terms of behavior and classroom management, was one of my best years because I didn't have to force the kids to be in the classroom." While at a local classroom level, these educators are doing their best to make the most of the opportunities put forward to their students; the risk, as Renee Hobbs (1998) points out, is that media production can be constructed as a curriculum geared for low-achieving students, who "are allowed to 'play' with video-based and computer technologies, while high-ability students get more traditional print-based education." While higher-achieving students are engaged with computers and media production as part of a more general media ecology they inhabit, the classroom becomes a place for a more remedial form of media education for students who do not have this cultural capital.

In contrast to the orientation of classroom teachers, educators in youth media programs had a different view of the potential of media education. Educators in the hip-hop program that Mahendran (Hip-Hop Music Production) observed and the video-production program at the Center where Dan Perkel, Christo Sims, and Judd Antin (The Social Dynamics of Media Production) observed saw their roles more in terms of vocational training than in general or remedial education. Media production is tied explicitly to the hope of employment in creative-class jobs, though educators at the Center struggle daily to instill this ethic of professionalism in the media-production process. At times, the goal of producing work in a vocational vein conflicts with the goal of empowerment and the development of youth voice. Hobbs (1998) describes this as a tension between more expressive and vocational forms of media education. Although youth were encouraged to take charge of their own projects, adults would intervene to focus them and orient them toward the goal of creating a polished work. In contrast to the hip-hop program, where youth were motivated by their existing engagements and knowledge of popular culture, youth in the Center's program had to rely more on the adult educators to set the agenda and provide the cultural capital for their work.

Among youth engaged in youth media programs, we also found some who were deeply pessimistic about what opportunities formal education afforded them, and who saw a more vocational orientation toward digital media as an alternative to a middle-class school-to-work trajectory. One of the participants in Dilan Mahendran's hip-hop music production study, Louis, an eighteen-year-old African-American, describes a moment during his first day of high school, referencing a famous scene in the book and the movie *The Paper Chase*, in which a Harvard Law School dean warns first-year students that most of them will not make it through the program.

Louis: Yeah. When you're a senior, 80 percent of the people you see right now are going to drop out . . . look to the left and look to the right, because they're not going to be here.

Dilan: That's what the teacher said to you?

Louis: Yeah. They set you up for failure. You know what I'm saying? We look to the left and we look to the right, and we laugh about it at that time. We're like . . . ha, ha, ha. I had my best friend Jerell and my best friend Rob. Sure enough . . .

Dilan: You were fourteen?

Louis: We were fourteen, fifteen at the time. Sure enough, Jerell drops out in eleventh grade and Rob drops out somewhere I think in eleventh grade. I dropped out somewhere in the twelfth grade. And it's kind of like they was fucking right. We all dropped out. It was kind of like [inaudible] . . . fuck, they were right. How the fuck did you know? It's a psych trip. First day of school, of course you're going to sit with your friends. Of course you're going to sit with somebody that you identify with. All right, look to your left and look to your right; they ain't going to be here. Then you go to school every day and it's like this—fuck up, fuck up, fuck up. . . . That's how school is.

This same teen is deeply involved in the production of hip-hop in a youth media program. His awareness of certain social structural conditions reflect what Hansen, Mortimer, and Krüger (2001, 133) have described as the differential pathways between school and work that are characteristic of the United States. Rather than focusing on an academic pathway, Louis sees the apprenticeship and mentorship of the media-production program as a compelling alternative. Hansen, Mortiner, and Krüger also note that the United States is distinctive, in comparison to many European countries, in having very few vocational and apprenticeship programs for teens, so they often turn directly toward employment to receive career training. Mahendran notes that the after-school setting is opening the horizon for explicit vocational training in the digital economy, contrasted with high school, which is oriented toward preparation for college. In this way, digital-media training and youth efforts can be compared to traditional vocational training such as auto mechanics or HVAC schooling. The programs are a kind of introduction to vocations associated with creative labor.

All these examples that we encountered in our fieldwork illustrate the ways in which different forms of media and technology engagement are tied to different trajectories of school-to-work for youth. Youth media programs navigate a complicated balance, using media production as a form of remedial classroom work as well as at times framing the programs as vocational training. In both of these stances, it can be a challenge to develop programs that support the development of expressive capacity and voice rather than skills development. In the most promising cases of youth media programs we have observed, these programs can fill a vacuum in apprenticeship and vocational training that is largely absent in the United States. We need to keep in mind, however, that these conditions of engaging with new media differ quite markedly from the opportunities afforded to youth from more highly educated families, who grow up in contexts where high-end technology is within easy reach, and where the adults with whom they regularly interact at home provide expertise and role models for careers in the high-tech workforce. Heather Horst's study of middle-class families in Silicon Valley describes settings where parents are intimately involved in structuring high-tech environments for informal learning in the home; they are not focused on specific vocational outcomes as we see with youth media programs.

Entrepreneurism

Contemporary childhood in the United States is characterized by a primary focus on play and education rather than on economic activity. At the same time, even after child-labor laws were in full effect in the early twentieth century, there has been a role for working children, particularly as they enter their teenage years (Zelizer 1994). In the latter half of the 1900s, it became common for youth to combine part-time work with their schooling, and studies through the 1990s indicate that approximately 70 percent of teens ages sixteen to eighteen have part-time jobs (Hansen, Mortimer, and Krüger 2001). As described earlier, the jobs available to teens are usually part of the unskilled service sector. Historically, paper routes and fast-food jobs are stereotypical forms of teen labor. The high-tech and creative jobs that young people are being prepared for in digital-production programs and through middle-class high-tech cultivation in the home are largely reserved for credentialed adults. This is in line with broader indicators that show that employment in skilled labor is generally unavailable for children and youth (Mizen, Pole, and Bolton 2001). Although our study included many youth with high degrees of technology expertise, we saw only three cases in which they were actually employed in jobs that made use of their technology skills during their teenage years. Technology was more commonly where they spent money; many teens in our study did engage in part-time work, often with the goal of funding their new media habits. The adults in our study who did have new media jobs did not have these jobs in their teenage years; as teens new media was a domain of hobbies and not salaried work.

Our focus for this section follows from this observed reality. We do not delve into the jobs that teens have or the domestic labor that they perform in the home, since this work has at best a tangential relationship to new media practices. It is beyond the scope of this effort to do justice to the complex realities of young people's economic lives. The issues surrounding how young people gain and spend money, particularly on media and communications, is a crucial topic that deserves an even more sustained treatment than we can give in this book. In this section we focus on somewhat more exceptional cases that illustrate the avenues that young people are finding to mobilize new media for economic gain. While the majority of youth in our study did not engage in these innovative new forms of economic activity, the cases that we do have are compelling: they illustrate the emerging potential for activating youth entrepreneurism and real-life learning through online networks of peer-based commerce and media sharing. Unlike training-oriented genres of participation, these entrepreneurial practices involved youth from a variety of socioeconomic backgrounds (though overall these cases were rare). They also involve kids engaged in productive labor in the here and now rather than as a model of preparatory work or training.

Youth with expertise and interests surrounding media and computers often understand that they have skills that can translate to economic gain. At the same time, their avenues for earning money from these abilities and interests are limited. Until they finish with their schooling, they do not have the option of fully entering the competitive marketplace for high-

tech and media jobs. Among youth whose primary occupation is schooling, and who are interested in capitalizing on their new media skills, we
have found three modes of economic activity: publishing and distribution
of creative work, freelancing, and the pursuit of enterprises.

Publishing and Distribution

In our discussion of creative production, in chapter 6, we describe the ways
in which young creators are using online venues as a way of publishing
and disseminating their work. While the vast majority of these efforts are
not oriented toward immediate economic gain, some of the more entrepreneurial young creators are reaping economic benefits from their creative
work. Even if they are not receiving actual revenue, they see online sites
such as MySpace, deviantART, and YouTube as spaces where they can
promote their careers as musicians, artists, or video makers.

A small number of creators we encountered were successfully making
money off their work, either by selling the actual work or by acquiring ad
revenue online. As described in chapter 6, Patricia Lange's study (YouTube
and Video Bloggers) is peppered with cases in which youth were aspiring
to make it big through YouTube and were at times successful in monetizing
their participation or gaining mainstream attention for their work. Perhaps
one of the most visible examples is Caitlin Hill,[2] a nineteen-year-old
Australian woman who is ranked thirty-first among most-subscribed-to
YouTubers of all time. Coming from a modest economic background, she
used her grandmother's digital camera to make videos. Her grandmother
now comes to her when she has computer problems. As her channel page
indicates that she is a YouTube partner, she is presumably receiving a share
of ad revenue from ads placed on her YouTube videos. Another youth in
Lange's study, Max (a white fourteen-year-old), was contacted by ABC
about getting his video shown on television. Although he did not ask to
get paid for this, after the ABC appearance other requests started to come
in. He explained that now "I've gotten pretty good. . . . I'd say 'Oh. I want
to get paid if you're gonna . . . for my video.' And they'd be like, 'Oh. Yeah,
we are expecting to pay you,' and then, we would negotiate about price
and stuff like that." While these cases represent the much-sought-after goal
for youth who aspire to media careers, most will acknowledge that it is
quite difficult to achieve this level of success on one's own as a purely
garage operation.

In Dilan Mahendran's study of young hip-hop musicians (Hip-Hop Music Production), he found a strong entrepreneurial spirit among many of the youth he spoke to. Some of the beat makers sold their creations to rappers who would use them for their own song production. Others produced mix tapes of their own work or that of other artists and sold them on public transportation or in other pedestrian areas. Artists can often feel conflicting loyalties over whether they are pursuing their craft for the love of the work or for economic goals, and this is tied into widely recognized tensions between hip-hop culture and the commercial rap industry (Mahiri et al. 2008). Louis gave voice to this ambivalence:

It shouldn't be about a meal ticket. It's not always about money. I mean, it's two ways to do it. It's either you make music to make music, or you make music to make money. Me? I do both. . . . I know that the music I make, it's not necessarily going to be accepted by all, because not everybody is going to be able to identify and agree with it. But the thing is, is that in order for that to survive, I have to make music that people can identify with, that people are going to listen to.

Although hip-hop may be an example of a form of media in which practitioners have an unusual amount of self-reflexivity regarding the problems of commercialization, many young creators struggle with this boundary between a creative pastime and a more work-oriented commercial stance.

Among the case studies of anime and Harry Potter fans, we have also encountered examples of youth who have successfully capitalized on their creative talents. Although intellectual-property regimes make it difficult for fans to make money off fan-related creative production, there are some niches where economic gain is possible. Becky Herr-Stephenson's study of Harry Potter fans focused in part on podcasters who comment on the franchise. Although most podcasters are hobbyists, a small number have become celebrities in the fandom who go on tours, perform Wizard Rock music, and in some cases, have gained financial rewards. Mizuko Ito, as part of her study on anime fans, spoke to Ian Oji[3], an artist who draws comics as part of a comic writers' collective. Once a year the group self-publishes a comic anthology that it sells at local anime conventions. All the large anime conventions have an "artist's alley" that will feature young aspiring artists selling their artwork, stickers, T-shirts, pins, and bookmarks for a small fee. These same artists generally will also have online sites that promote their work. The peer-based spaces of the convention floor and online sites are closely linked; they are spaces for artists to both promote and sell their work in an informal economy.

These kinds of ventures are examples of ways in which youth can make money from some of their creative talents, even if for relatively small economic gains. Youth recognize that it is highly competitive to make a living off their creative talents, but digital media and distribution provide avenues into online distribution and advertising that enable new possibilities for marketing their talents. As described in chapter 6, most of these ventures stay in the domain of hobbies, but a small number, such as SnafuDave, described in box 7.1, are able to parlay these efforts into successful commercial careers. In many ways, these ventures are examples that are very much in line with the historical position of the work of youth, conducted largely "around the edges of the formal labour market" (Mizen, Pole, and Bolton 2001, 38) and often involving gray zones outside of officially sanctioned forms of work (McKechnie and Hobbs 2001). At the same time, digital distribution is opening a wider range of venues for circulating and monetizing skilled forms of creative work, which have been largely limited to specific professions such as child acting (Zelizer 1994).

Box 7.1   "I'm Just a Nerd. It's Not Like I'm a Rock Star or Anything"
Mizuko Ito

The online world is home to a growing number of successful web comics ventures, including well-established names such as Penny Arcade and xkcd, as well as thousands of others that cater to niche and small audiences. Just as blogs have reinvented the medium of news, web comics are reinventing the comic strip, using digital authoring tools and online publishing to connect to different publics. Although most web comics artists are amateurs who spend more on their hobby than they bring in, there are a handful of artists who bring in significant amounts of revenue through online publishing.

SnafuDave, whom I interviewed as part of my study of anime fans, is one such successful web comics artist. In addition to creating his own web comics, SnafuDave, who is in his early twenties, manages a web comics site, Snafu Comics (snafu-comics.com), which features comics by twelve other artists in addition to his own. The styles and genres of the comics that SnafuDave hosts on his site are diverse, but many reference Japanese popular culture. Snafu-Dave is a regular in the anime convention circuit. We first learned of Snafu-Dave's work in a talk that he gave at an anime convention, where he gave his audience tips on how to launch a successful website.

In our interview, SnafuDave explained how he got started with web comics in his first year of college. He went to school in what he described as a "super, super, super tiny town," and he had been planning to major in math. The summer of his freshman year, he decided to stay for summer school when none of his friends did and was "bored out my mind in this little town." This was when he ran across Penny Arcade, the first web comic that he had read. "I just got obsessed with it. It took me three or four days to go through all of their comics. And I just absolutely loved it." He described how he went on to find other web comics that he liked and then decided to take the plunge himself. He went to the library and checked out *HTML for Dummies*, got a copy of Photoshop from a friend, and got started. After much trial and error, and learning through a variety of online tutorials, he began to hone his craft. "About three years later, I actually started getting semigood at it" (see figure 7.1).

Along the way SnafuDave tried changing majors to suit his new interests, first enrolling in a computer science major and then eventually switching to digital media. He thinks, however, that he learned few of his current skills in the formal educational context. "This whole time, school's more valuable for me to have basically a time frame where I could learn on my own and practice." College also gave him the time to learn how to market his work online and to develop an online network of fellow creators and readers. When he was getting started, he engaged in a wide range of strategies to get his comics noticed. These included asking fans to vote for his comics for top web comics lists, doing link exchanges with other comics sites, doing guest comics for other sites, and posting material to sites such as deviantART and video- and animation-hosting sites. Eventually he began offering to host for other web comics creators, and now, he said, "Literally every day I'll have at least five or six people begging me to put a web comic on my site."

He attributed a large part of his success to the fact that he has good friends in the web comics world and close ties to his fans through his web forums. In addition, he has made full use of the viral properties of the web in driving traffic to his site. This included a "tampon tag" game that he designed in which people could tag each other's forum posts. After seeing the popularity of the game on his own forums, he made a version for MySpace and "it spread like wildfire. . . . Totally, just this viral content the people are spreading around. Yeah. That's kind of how Snafu made it to the top."

Snafu Comics makes a substantial amount of money through online ads, but SnafuDave explained that he uses this revenue to pay for the costs of maintaining and improving the site. Since the site aggregates the work of multiple artists, he does not lay claim to the site revenue for his personal income. Instead, he makes his living as a freelance web designer. The other artists on his site also have day jobs, mostly in graphic design. When I spoke

Figure 7.1
SnafuDave comics. Reprinted from www.snafu-comics.com with permission from David Stanworth. 2006.

to him, SnafuDave had recently launched merchandising ventures such as T-shirts, prints, and buttons to sell at conventions. Now his site hosts a web store where fans can order these items. At the time of our interview, he was not making a living off web comics. "I would really like it to be paying for all of our lifestyles someday. And definitely, right now, I believe it could." I asked if his family and local friends were supportive of these aspirations.

Well, my mom actually thinks I'm a complete waste to society, no matter what. She's all, "Get a real job." Even though, I . . . yeah. Whatever. My dad thinks it's pretty cool. About a third of my friends are really supportive of it. I'd say about two thirds . . . actually, about one third doesn't care at all. And then another third actually despises me for it. Like they hate that I get all this attention online when I'm just a kid from a small town.

I am curious about whether there is a stigma attached to being so involved in comics and anime, and SnafuDave explained that the issue is more personal. "I design websites once or twice a month for clients and then I play online all day. And it drives people crazy. It really does. . . . But I don't think it's *that* envious. I'm sure it is a really cool job, but I'm just a nerd. It's not like I'm a rock star or anything." In a follow-up email, almost two years after the initial interview in 2006, he gave me an update. His merchandising business had started paying off enough that he quit his day job to devote himself full time to web comics. He may not be a rock star, but he is one of a handful of artists who have parlayed their web comics hobby into a professional career.

Freelancing

Another category of paid work that young people can gain access to through new media is different forms of freelance and contract labor. Technically sophisticated youth recognize that they have marketable skills that are in demand from their peers and adults in their vicinity. Most of these kids do not try to profit from this and engage in informal help and sharing with family and friends. This is in the vein of chores and child care, for which youth may receive small financial rewards, but the work also often is framed as household obligation. Altimit, an eighteen-year-old Filipino American told Mac Man, a seventeen-year-old Filiipino American in Katynka Martínez's study (High School Computer Club), that his father often asks him to help out fixing his family's and friend's computers:

Altimit: Yeah, and like my friend's house, usually my family friend, they would say, "Oh, something's broken." So, rather than him coming, he sends me. So, like, "I'm trying to play World of Warcraft." "I don't care.

Go. You're not doing anything anyway." I'm like, "I'm trying to level."
"I don't care. Go."

Mac Man: Do you get paid to do it?

Altimit: No.

Mac Man: Hey, that's sad.

Altimit: I wish I did. Make a lot of money.

The challenge for many youth is to move their labor from a category of
unpaid helping to a category of valued labor, which might be potentially
monetized. Marjorie Faulstich Orellana (2001) describes how adults
often resist describing the children as "workers" and prefer to describe
what the kids are doing as "helping," activities that are good for kids'
social development but not part of the monetized labor economy. This
exchange with Mac Man and Altimit is evidence of how kids may see
this dynamic differently. Altimit understands the economic value of his
technical labor even though his father may not recognize it. We have seen
some cases of a few entrepreneurial kids overcoming these challenges
and making real money off their technology skills. The case of SnafuDave
in box 7.1 is one example of a youth's transitioning into a successful
career as a freelance web designer and later into one that centers on his
own creative work.

One fifteen-year-old white participant in Patricia Lange's YouTube and
video bloggers study described how he has started a small design company.
"I have a couple clients that I do web hosting for. And then, I've done
some programming, but I'm not that good at it. But I've pretty much done
some of every geeky thing that there is out there." He built his client base
from personal connections, beginning with family and then branching out
to friends at school and people he met online.

I have pretty good customer services. Since I have a very small client base, I can
afford to help them make websites and [with] any problems that they have, so a lot
of it is just helping them make websites, fix websites, change things, and basic things
like that.

In a similar vein, as described in box 7.2, sixteen-year-old Zelan built up
a career as a freelance technical expert.

In the gaming world, the most-skilled players can gain sponsorship or
win financial awards through tournaments, and a number of game titles
have a professional gaming scene. The top players can make a living
playing the games on the marketing value they gain as a result. Hundreds,

thousands, and even millions of dollars in prize money are turned out each year for competitors in these tournaments. The most popular tournaments are those run by the Cyberathlete Professional League, the World Cyber Games, the World eSports Games, the Electronic Sports World Cup, the Championship Gaming Series, and Major League Gaming. It was rare to encounter youths in our study who were actually able to make money off their gaming. Even among those who did, none saw gaming as a primary occupation. For example, Altimit described making small amounts of money off his Warcraft play. Scottanime (a white thirty-one-year-old), one of the interviewees in Mizuko Ito's study of anime fans, described how when he was younger he used to be a card-game expert. He would be hired by gaming companies to demonstrate the games at conventions. He did not see this as a sustainable or secure career option, however, and he went on to take a job as a mail carrier. Similarly, MercyKillings,[4] a white thirty-five-year-old in Ito's anime fans study, was also a professional gamer after college, but he maintained a day job working in construction. These stories parallel the kinds of involvements that youth have historically had with sports; gaming is an activity most children and youths participate in regularly, but very rarely does it translate into a career. Although we saw many instances of youth who admired pro gamers, we did not have examples of kids who actually were pursuing pro gaming as a career.

Box 7.2 Technological Prospecting in Rural Landscapes
Christo Sims

About an hour's drive east of Sacramento, the Great Central Valley of California meets the Sierra Nevada range. The valley's end loosely binds one edge of Sacramento's suburbs. As one climbs into the mountains, roads and rivers narrow, towns and neighborhoods become smaller and more far-flung. About 150 years ago these hills were the epicenter of the California Gold Rush. Evidence of this historical prospecting can still be read on the landscape. Ashen ruptures in otherwise pine-green panoramas continue to mark sites where hydraulic mining sluiced ore from the mountainsides. Locals call these barren desertlike patches "diggins." It was in one of these diggins that Zelan was first introduced to video games:

When we lived in Sacramento my parents got me a Game Boy to go out there in the diggins. 'Cause we'd come here on the weekends and go dig for gold. And I never really liked it, so I'd sit in the corner and, you know, play with Batman or whatever I was into.

And so one day they got me a Game Boy for my birthday, and I sat in the corner and played games all day.

As his parents prospected for gold, Zelan began his trajectory of engagement with digital media, one that would lead him far beyond gaming as only a way to kill time. As Zelan recalled it, this incident occurred when he was four or five years old. He was sixteen at the time of our interview, and for the past eleven years his family has lived in a secluded rural town near where they used to go digging for gold. In pursuing his passion for games he has developed pragmatic strategies for making and managing money, he has acquired unique technical skills and knowledge, and, lately, he has configured these resources into his own form of prospecting, one that enacts and imagines modes of work that defy local tendencies and expectations.

This pragmatic sensibility partially stems from, and continues to mix with, his passion for video games and digital media. After immersing himself in the Game Boy, he pursued newer and better consoles. As he did so he also learned how they worked. His parents did not like buying him gaming gear, so he became resourceful. When his neighbors gave him their broken PlayStation 2, he took it apart, fixed it, and upgraded from his PlayStation 1 in the process.

Soon he started devising ways of making money to support his hobby. He learned that the technical knowledge he was developing could be applied as employable labor. When he was in middle school a teacher asked him to help run the audiovisual equipment. He soon transferred this knowledge into a DJ business. In another case he made two hundred dollars fixing a teacher's computer. More recently, the high school has hired him to help maintain "the empire" of more than two hundred computers on the school's network. In addition to selling his labor, Zelan has begun to realize that he can be a valuable broker in markets for used technology goods. In several instances he has acquired broken computer equipment or game consoles, fixed them, and then sold them for a profit (see figure 7.2).

Since these opportunities have built up, he now imagines starting a technology-centric business after he graduates high school:

I wanna start a business about, you know, just like computer repair, gaming, just anything computerwise. So I can get it all started and hopefully start another business and get two businesses going and, or two chains going or whatever, and hopefully just be able to sit back when I'm older. Not to just sit there and do nothing. Have the businesses going around me.

This vision of work differs considerably from the manual labor practiced by his parents and many others in his local community. His town is one of the most remote and blue-collar of those feeding his regional high school. Both of Zelan's parents make money by performing manual labor; his mom cleans houses and his dad is a freelance handyman. Zelan seems to understand that

Figure 7.2
Attic workbench where Zelan and his dad tinker with remote control aircraft and other electronics. Photo by Christo Sims, 2006.

many of his peers will end up in similar careers as his parents and neighbors. In describing his "nerd" identity, Zelan differentiated his work trajectory from the one he imagines for his peers:

But the jocks, they're more into construction. And my group's into the computers, and the computer jobs where you have to do little to nothing to make your money. Everybody else is into hard labor and mechanics. That's what the metalheads are in . . . they're the mechanics. And, you know, the computer nerds, I think they've got the best side of it. 'Cause computers are spreading, if you can see, they're everywhere in this room. You know, everybody's houses are turning into that, and they're just everywhere. And they're gonna be here. Before long houses are gonna be computers.

For Zelan, being a "nerd" is a purposely unconventional path, one deeply entangled with practical economic concerns. In embracing a nerd identity, he imagines an alternative life of work, one that sidesteps the expectation of a career in manual labor. By entwining technology with an entrepreneurial trajectory, Zelan echoes those who brought sluices and shovels, and then hoses and hydraulics, to his region of California nearly 150 years before. With them, technology is implicated in an effort to bypass the gridlock of social mobility, a partner for creatively prospecting the economic landscape.

The examples of entrepreneurism that we present involve young people working to break into established models of publication, distribution, and freelance labor. These practices involve a kind of modeling of adult careers in what might be called creative-class labor (Florida 2003). Young people are developing skills and talents that they can market and contract out to others. The last category of entrepreneurism that we would like to discuss is one that is more closely tied to genres of practice that we associate with the street smarts of a small-business person.

Enterprises

The classic model of a childhood enterprise is the lemonade stand. New media, online distribution, and auction sites such as eBay have expanded the potential for entrepreneurial activity that relies on digital media for the buying and selling of goods.

For example, Gerar, a fifteen-year-old from a mixed Mexican and Salvadorian background, in Katynka Martínez's "Animation Around the Block" study, found a market niche where he could establish his own small-business enterprise. He explained how many of the youth in his neighborhood own an iPod but not a computer. "They pay me to upload some songs for them and depending on how many songs I have to download or upload into their iPod that depends how much I get paid. If I have to download a hundred songs I charge them four bucks or something." He has a corner on the local market, because there is only one other person in his peer group who has a computer. The other person he has heard of who does have a computer, however, "does not have an Internet connection so there's no way he can download music and charge the others." Toni, a twenty-five-year-old who emigrated from the Dominican Republic as a teen (Ito, Anime Fans), described how he was dependent on libraries and schools for his computer access through most of high school. This did not prevent him from becoming a technology expert, however, and he set up a small business selling *Playboy* pictures that he printed from library computers to his classmates. The two cases of sixteen-year-old Zelan and of seventeen-year-old Mac Man, presented in boxes 7.2 and 7.3, provide an illustration of this small-business spirit animating youth digital ventures. These are not privileged kids who are growing up in Silicon Valley households of start-up capitalists, but rather they are working-class kids who embody the street smarts of how to hustle for money. They are able to translate their technical knowledge and expertise into capitalist enterprises that have immediate financial outcomes.

None of these cases represents a major restructuring of the basic financial conditions that youth live under. They replace paid unskilled formalized labor with new financial arrangements in the informal economy, but they are not generating large amounts of new income. The larger impact on kids' lives is perhaps not a financial one but is more about kids being able to develop financial agency that is not fully determined by existing commercial models (such as online ads) or by the more formal school-to-work transitions envisioned by parents and educators. These practices resist the existing normalized pathways for youth labor. They are not part of a future-oriented vocational or preparatory orientation, the model of youth "talent," nor are they framed by the stance of "helping out" that underlies most freelance youth labor. The enterprise genre described in this section does not even appear as a category of youth labor in surveys of youth work (McKechnie and Hobbs 2001). While youth have had small spaces in which to begin their own enterprises, in at least a small number of cases we have found, youth have mobilized online media to expand this genre of participation in new directions.

Box 7.3   Being More Than "Just a Banker": DIY Youth Culture and DIY Capitalism in a High-School Computer Club
Katynka Z. Martínez

The lunch bell rings and a group of high-school boys make their way across campus. They meet at the computer lab, where they view anime on their laptops and play games on computers that they have networked to one another. Although the atmosphere is relaxed, the boys have posted rules for their computer club: "Don't talk loud," "When playing don't scream," and "Five deaths only." Breaking these rules is grounds for having one's computer privileges revoked for a week. The boys take their club quite seriously and hold fund-raisers to buy new equipment. Yet they still have a lot of fun joking around and teasing one another, and sometimes they eat their lunches too.

Mac Man, seventeen, is the president of the computer club, and Altimit, eighteen, is an officer. The boys met in middle school when the two were recent emigrants from the Philippines. Their fathers, who are both computer savvy, introduced the boys to computers. Based on the stories told by the boys, it seems that their fathers introduced computers as toys rather than as educational tools or adult devices. One boy's father used to engage in computer-hacking activities and both men enjoyed using computers when they were younger. However, they are now employed as a banker and as a landlord overseeing apartments in the United States. Altimit and Mac Man romanticize the early days of computer programming and their fathers' participation in that world. The formation of the computer club may be their attempt to tap into some of the renegade spirit that their dads once possessed.

Altimit and Mac Man take on a sense of nostalgia when they remember the first computer games they played. For example, when Mac Man talked about playing an early version of The Sims in which "everything's all pixilated," Altimit sarcastically responded, "Don't remind me of those days." A lot has changed since "those days" and now the boys not only play computer games but they fix computers themselves. Altimit's dad taught him how to use and fix computers and now he expects Altimit to help friends and family members who have computer problems. Altimit recounted the ordeal he goes through when his father asks him to fix someone's computer while he is absorbed in his favorite MMORG:

Yeah, and like my friend's house, right, usually my family friend, they would say, "Oh, something's broken." So, rather than him coming, he sends me. So, like [in child's voice], "I'm trying to play World of Warcraft." [In dad's voice] "I don't care. Go. You're not doing anything anyway!" [In child's voice] "I'm like, I'm trying to level." [In dad's voice] "I don't care. Go!"

When Mac Man heard Altimit tell this story he immediately asked if the boy gets paid for his service. (Altimit does not receive any monetary compensation.) Mac Man, a young entrepreneur, has found a way to develop multiple small businesses—even at school. He heats up water in the computer room during lunchtime and sells ramen to students for a dollar. Also, when he learned that a group of teachers was going to be throwing away their old computers, he asked if he could take them off their hands. Mac Man fixed the computers and put Windows on them. The computer club was started with these computers. Mac Man still comes to school with a small bag that carries the tools he uses to work on computers. Teachers and other adults kept giving him computers that were broken and he had to figure out what to do with them. He fixed them and realized that he could sell them on eBay. He makes a hundred dollars' profit for every computer that he sells.

Mac Man's entrepreneurial spirit is very much influenced by his father's work ethic. When asked what his father thinks of his small business, Mac Man told a story about his father creating the chemical mixture needed to kill cockroaches when he saw that the apartments he managed needed this service. His father also buys beat-up classic Mustangs, refurbishes them with his son, and sells them. Mac Man showed off before-and-after photos of the cars they have worked on and then he said, "My dad and I—we're similar because we're physical people. We like to get our hands dirty, you know? Pull things apart, put them together. See I do the computer things. My dad does the car things. We're very similar."

While Mac Man recognizes that he and his father share a tinkering mentality, Altimit is frustrated by his father's current career. Altimit's father was once a computer hacker and even was friends with one of the people who created the "I Love You" virus. Knowing this history, Altimit finds it hard to understand why his dad does not enjoy using computers anymore. He said that now his dad is "just a banker." Altimit cannot look to his father to learn how to make a living from his interest in computers. However, he has been able to find role models in the world of professional gaming. Altimit is an avid gamer and claims that strangers pay for his World of Warcraft (WoW) subscription just so they can play against him. This interview was conducted during his school's winter break and it was no surprise that he had been spending most of his vacation watching anime and playing WoW. In a discussion of the *South Park* episode that features WoW, Altimit began talking about the fact that the "best gamer in the world" makes his living playing games. Mac Man, the pragmatic one, explained that this gamer is "one in a million."

Altimit: He got a job for it though.

Mac Man: Only a few people get a job.

Altimit: No. Yeah, but he's rich. I mean, come on, just for playing games, he's rich.

Mac Man: There are exceptions.

Altimit: That's just kick ass . . .

Katynka: Is he a pro gamer or . . .

Altimit: He's the best gamer in the world, at shooter games. He can kill anyone and he will not die. And I think some guy picked him up to play for tournaments. He would win all the tournaments, and then he got paid to play games, pretty much. And like make shows, so . . .

Katynka: And does this guy seem like a nerdy guy from *South Park* or . . .?

Altimit: No. He's normal. He's, what he did, it's like, what he's doing is before he played games right, he would wake up, eat, jog, like exercise. Play games for three hours. Play console games for four hours, and then play PC games, eat again, just take a break, three hours again. I do three, four, three.

Mac Man: Is that what you do?

Altimit: Yes.

Mac Man: Why are you not getting paid for it?

Both Altimit and Mac Man are high-end users of new technology. However, they have very different personalities and approach media in different ways.

Altimit said that he is the "software guy" while Mac Man is the "hardware guy." Altimit spends hours playing video games and drawing manga while Mac Man occasionally plays games but does not have aspirations of making a living from this leisure-time activity. He was asked what his ideal job would be and was told that he could make up a profession or job if it did not already exist. He said that his ideal job would "be either in biomedical engineering or in business." He added, "In both of these career choices, I would definitely be using computers."

The boys imagine that computers will continue to be a central part of their lives. Now they are engaging in this technology on their own terms. By starting up a computer club at their high school, they are establishing their own community in a hierarchical environment that can often be hostile for kids who do not conform to mainstream interests and activities. Both boys bring a DIY (do-it-yourself) ethos to the construction of their identities. Altimit participates in a DIY youth culture by drawing his own manga while Mac Man engages in a type of DIY capitalism by selling ramen and refurbished computers. An initial childhood interest in gaming led them to deeper explorations of computer technology. It is unknown whether, as adults, they will be able to find employment opportunities and continue to establish new forms of social organization that hold on to the same inquisitive spirit that drew them to games and computers in the first place.

Nonmarket Work

Although most young people in our study were not engaged in paid work related to digital media, there was a substantial number of kids who were engaged in nonmarket work with new media. Amateur and nonmarket activities historically have been a place for middle-class and elite kids to "practice" work, develop creative talents, and gain experience in self-actualization and responsible work. While formal education can impart knowledge and skills, nonmarket work provides domains where youth can put these to practice in a context of accountability and publicity. Whether that context is a piano recital, helping out at a church, or being part of a soccer team, these activities are domains where young people can develop their identities as productive individuals engaged in serious and consequential work, in contexts where they can build reputations and gain public acknowledgments of their accomplishments. Lareau's argument (2003) is that these activities of concerted cultivation, which are pursued vigorously in privileged families, are a site for the production of class distinction.

Children in working-class and poor families engage in fewer of these kinds of activities, and they are often expected to perform much more domestic work. The domestic work of cooking, cleaning, and child care contributes directly to the household economy but is invisible outside the home. These forms of nonmarket domestic work, while instilling a sense of responsibility and self-efficacy, do not build the broader networks of human relations and skills for navigating various contexts of publicity as you see in activities of concerted cultivation outside the home. While these forms of helping and domestic work can have many benefits to youth who engage with them (Orellana 2001), they are not directly tied to immediate participation in contexts of publicity with new media, with the exception of some of the categories of practice described in the previous section.

The relation between concerted cultivation and vocation is not straight-forward, however. The same families who encourage sports, arts, and music as childhood activities also push their children toward traditional high-status careers with more stable and guaranteed financial rewards. Upper-middle-class youth who are avid fan producers, for example, are still pursuing traditional career paths through elite universities. One accom-plished fan producer seemed puzzled by Mizuko Ito's question as to whether he might consider a career related to anime. "Well, first off, [my parents] would kill me. Secondly, I could probably make more as a biomedi-cal engineer than anything in that neighborhood" (Ito, Anime Fans). By contrast, less privileged families might see creative-class careers as one of their few chances at upward social mobility, what one of Ito's interviewees described as a "pipe dream for a fancy job." In the previous section, we discuss some of the ways in which new media might provide broadened access to new forms of economic networks. We see how youth from a wide range of class backgrounds exploited these networks for economic gain. In the case of nonmarket work, household economic status is a stronger determinant of forms of participation. Here we see youth who choose to engage in unpaid labor in far-flung networks that makes no contribution to their household economy. While they are arguably gaining experience that will help them in their longer-term career aspirations, immersive participation in these activities is predicated on the fact that they do not feel pressures to engage in domestic work or paid work outside the home.

Nonmarket Peer Production

Within the field of digital-culture studies, theorists are debating how to understand the "free" nonmarket labor that supports activities such as open-source software development, citizen science, game modding, fansubbing, and Wikipedia authoring. For example, Yochai Benkler (2006) sees "nonmarket peer production" as part of a fundamental shift from the market mechanisms that characterized cultural production in high capitalism. Other theorists see these processes as exploitation of users and consumers for the commercial gain of media industries (Ross 2007; Terranova 2000). These kinds of practices differ in important ways from traditional forms of volunteerism and community service, yet they may provide some of the same social benefits for youth. When examining youth practice in this domain, we need to negotiate a complicated tension. On one hand, it is important to value these activities as spaces where youth can engage in active forms of social organization and develop a sense of efficacy and leadership. Further, these activities are part of a "free culture" sharing economy that has a unique ethic of civic participation aimed at developing public rather than proprietary goods (Lessig 2004). On the other hand, widespread youth participation in unpaid digital cultural production is part of a resilient structural dynamic in which many constructive activities of youth are not "counted" as a contribution to economic productivity (Qvortrup 2001). The enthusiasm that media-savvy youth are bringing to nonmarket digital production represents a unique twist to these existing dynamics.

As part of Mizuko Ito's case study on anime fans, she has researched the practices of amateur subtitlers, or "fansubbers," who translate and subtitle anime and release it through Internet distribution. Chapter 6 describes some of the ways in which they form tight-knit work teams, with jobs that include translators, timers, editors, typesetters, encoders, quality checkers, and distributors. Although the quality of fansubs differ, most fans think that a high-quality fansub is better than the professional counterpart. Fansub groups often work faster and more effectively than professional localization industries, and their work is viewed by millions of anime fans around the world. Fansubbing, like much of digital-media production, is hard, grinding work—translating dialogue with the highest degree of accuracy, timing how long dialogue appears on the screen down to the split second, fiddling with the minutiae of video encoding to make the highest-quality video files that are small enough to be distributed over the

Internet. They often work on tight deadlines, and the fastest groups will turn around an episode within twenty-four hours of release in Japan. For this, fansubbers receive no monetary rewards, and they say that they pursue this work for the satisfaction of making anime available to fans overseas and for the pleasure they get in working with a close-knit production team.

Similarly, fan conventions are organized entirely by volunteers, who at best might get a free hotel room for months of work in organizing an event for thousands of fans. Some of the most dedicated of convention organizers Ito interviewed described spending almost all of their vacation time and a substantial amount of their own financial resources to act as volunteer organizers. Gamers also pour tremendous amounts of time and energy into organizing online guilds and developing their own content to enhance the gaming experience for others, such as game reviews, walk-throughs, mods, and machinima. Because these activities are constructed as fan or player activities, and there are legal constraints on their monetization, participants are doubly hampered in translating these activities into personal financial gain. The nonmarket ethic of fan-based production is that this work is done "for fellow fans" and not for financial gain. This stance represents a kind of accommodation between fans and commercial media industries, in which the latter tolerates some degree of fan distribution and derivative works, provided they are not framed as commercial work.

Box 7.4   Final Fantasy XI: Trouncing Tiamat
Rachel Cody

According to Wurlpin,[5] a twenty-six-year-old white male in San Diego, "Final Fantasy XI is like a chat room with action in the background." The game is about the people. It is the peer groups—from friends to linkshells[6]—that provide motivation for many to log in to the game and make the game meaningful. The communities and relationships developed within the game extend beyond it into websites, forums, instant messenger programs, and email. The players chat with one another across servers or linkshells in these common spaces, sharing their strategies, advice, and questions. Working collectively allows a level of success in the game that would be impossible to attain individually. One of the most impressive acts of coordination and collaboration during my fieldwork was the slaying of the dragon Tiamat by the linkshell KirinTheDestroyers (KtD).

At the time of our fieldwork, Tiamat was one of the most difficult dragons in Final Fantasy XI. When linkshells were first attempting to kill her, Tiamat would often require more than two alliances[7] (thirty-six players) and four hours of coordinated teamwork to defeat. Strategies to kill Tiamat had to deal with a variety of the dragon's special moves and abilities as well as an increase in difficulty for the last ten percent of her health.

Despite how daunting the fight seemed, KirinTheDestroyers[8] had spent the first half of 2005 taking on progressively harder areas and monsters in the game and wanted a new challenge. As Wurlpin told me, new activities "keep the game interesting" and "keep the challenge on." A successful defeat of Tiamat would demonstrate how far KtD had advanced in the game and the capabilities of its members. When KtD began discussing a linkshell attempt on Tiamat in June 2005, none of the linkshell members had experience fighting anything like Tiamat. KtD members had grown up together in the game as a linkshell, and nearly all the experience the players had with the game had been acquired through linkshell activities. And none of the linkshell activities had been dragons like Tiamat. It took the linkshell two months of effort, frustration, heartache, and brainstorming to be able to conquer the dragon.

The first attempt at Tiamat relied on the advice of a new KtD member, Bokchoi,[9] who joined the linkshell only a few days before the first attempt. Bokchoi, a twenty-two-year-old white male in Florida, came from the only English-speaking linkshell on the server that had successfully killed Tiamat, and brought with him a wealth of knowledge about the fight. Using the linkshell's website forums, Bokchoi provided the strategy that his former linkshell had used in its Tiamat fight. He used a screen shot of the fight, with arrows pointing where people should stand during the fight. Through text and the screen shot, Bokchoi explained where the fight would take place, where people would stand depending on their jobs, where the dragon would be kept throughout the fight, and what each job should do during the fight. Bokchoi warned the linkshell, however, that the strategy would need to be tailored to KtD's strengths and weaknesses:

I would like to say this is by no means the only way to defeat Tiamat and during the course of the fight the strategy can be altered to benefit from the linkshell's strengths and overcome any weaknesses. I would also like to say even going in with a proven strategy it is no easy fight, and in all honesty do not expect to walk away with a win. This fight takes a bit of practice and some reworking of strategies to enhance this basic strat to work for KtD. I think KtD has the numbers and the skill, just needs a bit of practice to get a fight like this down.

After reading Bokchoi's strategy, KtD members used the forums to form groups and discuss their individual moves for the fight. One officer debated between different moves that players could perform in the fight: "Spinning Slash is better for Tiamat. Spiral hell will do more Damage at 300% TP, but it's more efficient to do 3 Spinning Slash in the same amount of time."[10] Other players used the forum thread to organize parties and coordinate their moves with one another. Coordinating with one another before the fight allowed KtD members to discuss ways to maximize their damage and efficiency.

The first Tiamat attempt was not successful. After several hours, the dragon was down to nine percent health before KtD was forced to leave because of sheer exhaustion. Tiamat had become much more difficult in the last ten percent of her health and KtD would need to modify its strategy to be successful. Despite not killing Tiamat, KtD's attempt became part of a larger conversation as many in the server's community watched and discussed the fight. As the linkshell left the fight with heavy hearts, a KtD member logged in and said, "I heard the news, it's all over the server. Lol everyone's talking about us." The mood in the linkshell brightened at this collective support and it started a battle cry of "TIAMAT!"

KtD didn't attempt Tiamat again for another month, using the time to discuss the first attempt, difficulties they had, possible solutions, and new strategies on their forums. More than fifteen players contributed to these brainstorming sessions. Once the main problem was identified—the tanks[11] were dying too fast—players relied on their experiences within the game to suggest solutions. As Fyrie,[12] a seventeen-year-old Asian-American in New York, suggested, "Next time we fight him, we definately need more NINs[13] to tank his last 10% left when he spams mighty strikes.[14] It was doing about 580~ to our PLD[15] and they fell in 2 hits." Some of these suggestions required minor changes, such as using different players for different roles or modifying the spells they would use. Other changes, such as using different subjobs,[16] required some players to spend hours or days leveling a new subjob. For example, Ghostfaced,[17] a nineteen-year-old white male in Oregon, offered to level his white mage subjob so that he could be more versatile in the fight.

Another major contributor to the strategy for the second fight was a new KtD member, Tacoguy.[18] He posted in the brainstorming thread, "Alright well i have a friend on a different server and him and his ls have taken down tiamat many many times and i asked him how do they do it so quick cause it takes them about 1:30 [one hour and thirty] minutes." Tacoguy served as a messenger between KtD and his friends on the other server, asking questions about the fight and posting their strategies onto the KtD forums.

KtD tried Tiamat again in August, armed with their previous experiences with Tiamat, the adaptations to Bokchoi's and Tacoguy's strategies, and their own brainstorming and hard work. The new strategies proved successful for the first half of the fight, but a minor mistake by one player had major consequences and the linkshell lost claim, or ownership, over the dragon, and KtD chose to withdraw rather than start over. Many in the linkshell were frustrated and angry that their hard work had not met with success, but a few remained positive. One member posted on the forums, "One way or another we should all be proud for doing what we have the past two attempts. Grats and a pat on the back to everyone."

In September 2005, KtD attempted Tiamat for the third time using the same strategy as in the second attempt. KtD members were confident that despite the mistake in the second attempt, the strategy would work. Members used the time between the second and third attempts to relax, better their gear, and increase their playing abilities through other activities. Quite experienced with the fight by this point, KtD's third Tiamat fight resembled a choreographed dance. The tanks were rotated out of the fight as they became exhausted; players moved in a cycle of positions as they fought, healed, were attacked, or rested; and the black mages had an elaborately ordered system in which they took turns casting special spells against the dragon. After four hours of this extremely coordinated and intense teamwork, KtD successfully killed Tiamat.

The conquering of Tiamat was a collective success; it was the work of more than fifty players who diligently combed through their experiences, outside videos and screen shots, and the experiences of their friends to create a successful strategy. They brought years of collective experience and ideas to the battles and brainstorming sessions, and their deaths in the dragon pit taught them even more. Screen shots and videos were researched by some members to suggest other successful ideas. Bokchoi became a mentor, and Tacoguy became a resource and messenger of questions for his friends, who had more experience with Tiamat. Throughout their journey, KtD members combined all that they knew or thought, laid it bare, disassembled it, analyzed it from every direction, demolished some parts and polished others, and then reassembled it to be a work of art. It was a strategy that took two months to perfect, but the success was worth the effort.

Another version of nonmarket work is the kind of involvements that youth have with online gaming economies that exhibit many of the same features as real-life economies, but that are quite separate from them. These involvements are most evident in multiplayer online gaming worlds (Castronova 2001; Dibbell 2006), but they also are an important part of sites such as Neopets or games such as Pokémon and Yu-Gi-Oh! that involve the buying and selling of game items. The grind of nonmarket work is familiar to any player in a massively multiplayer online role-playing game (MMORPG). Rachel Cody's case study of a linkshell's defeat of a high-level monster (see box 7.4) documents a culminating moment for players who have poured months of their time into the repetitive labor of "leveling" their characters by battling monsters and engaging in menial craftwork. Laura Robinson's study of Neopets (see section 7.5) illustrates some of the energies that young people bring to these online economies, even though they do not translate to real-life capital.

Fan production and gaming production are not the only examples of practices youth engage in that involve many of the same disciplines of professional media production but that bring none of the financial rewards. Even an activity such as the creation of YouTube videos, which often seems playful and off-the-cuff, involves this kind of grinding labor to create good work. One of the youths Patricia Lange interviewed, Jack, a seventeen-year-old white male (YouTube and Video Bloggers), described a video shoot with a group of fellow homeschooled teens.

The environment was just, you know, torturous. And tempers were flaring 'cause we were all . . . we would be shooting day in and day out for, you know, sometimes for two or three days in a row, and we would just be sitting there and we'd get really mad at one another. And then looking back, we just always laugh at it because it's just so ridiculous that we're all sitting here in this hundred-degree weather with all this stuff around us, and we're just absolutely dying. Reshooting the same scene over and over again, and, you know, and it never just progressed anywhere.

Youth pour their energies into producing videos, writing fan fiction, making music, or recording podcasts, and they most commonly release their work on the Internet for free. At the time of Google's purchase in 2006, YouTube was valued at more than a billion dollars, capitalizing on the economy of freely shared amateur media production, for which creators did not earn a penny from the distribution of their work online. Although business models and terms of service for online sharing sites are changing, and there are more opportunities for amateur creators to gain revenue from online distribution, most amateurs, youth, and fan producers do not see any economic gain from their work.

These practices add a new twist to our existing understanding of volunteerism and civic engagement. Just as with more long-standing forms of youth volunteer work and internships, this nonmarket work is a space for young people to experiment with different work practices before they make commitments to jobs and careers. For example, in Ito's study of fansubbers, some described how poeple "retire" because "it wasn't fun anymore" or it was becoming too much like a "real job." Although the practices resemble market-based labor in many ways, they are still a form of volunteer practice that youth can drop out of with little material consequence. Still, relationships they foster with their peers in these groups provide opportunities for mentorship and for youth to take on identities as leaders and media producers. Further, these activities are often animated by a civic spirit of sharing that takes "free culture" as a rallying point in working toward a cultural commons that is not dominated by commercial interests. At the same time, it is important to keep in mind the broader political economic conditions in which these kinds of engagements occur. Most of these more sophisticated forms of nonmarket online production are the province of relatively privileged youth who are pursuing these activities during college or other times in their lives when they are not under financial and time pressures to engage in domestic or paid work.

While we should look to these youth practices as examples of highly engaged forms of youth mobilization and creativity, we must also recognize how they remain embedded in existing structural conditions of inequity and in a robust set of commercial practices that define the contours of Web 2.0 industry. In many ways, the free-culture movement and industry attention to user-generated content are part of a cultural logic that is growing in salience and that defines a particular historic moment in the evolution of media and communications. We see youth innovation as central to defining these new genres of cultural participation, even as they are very much under flux, through a complicated set of struggles between different media industries and sectors as well as the everyday activity of youth and adults.

Box 7.5 Eddie: Neopets, Neocapital, and Making a Virtual Buck
Laura Robinson

Eddie is a precocious teen from California who is a self-described former Neopets addict. In his words, "I loved the economic stimulation!" Significantly, while some players talk about the social aspects of the site, for Eddie, Neopets was a solitary activity. He explained that while it was okay to play while he was in junior high, by the time he got to high school the younger players would tell the older players that they were too old. So Eddie continued to play, alone, in secret, long after it was "cool" for someone his age to play the game. For Eddie, the excitement of Neopets was rooted in the potential for economic activity; the interest in Neopets was almost solely for its economic ventures. When asked about his relationship with his pet, he said with a laugh, "I think mine all died! I never checked on them." For Eddie, the Neopets connection took place on the site's simulated financial sector through bank accounts and a stock market that absorbed all his attention. He elaborated, "I just wanted to hoard my cash to make more. I wouldn't waste my points feeding my pets. I didn't want to buy them anything—just to play the market."

What is most interesting about this player is that all his activities essentially computed into the following equation: time = money. For him, time did not equal creative output, social relations, or fun. Rather, all activities were aimed toward the single overarching goal of amassing capital—neocapital. Eddie invested time in playing the games, not for the enjoyment of the games but for the economic points to add to his bank account. He relished checking his bank balances on Neopets and reported experiencing great satisfaction in doing so: "I would log in just to see my balances. It was really satisfying." When Eddie engaged in other site activities, it was always with an eye to capital acquisition. He explained that when he built a home or opened a store, he had the same goal in mind: time = labor = points = *money*. "It was simple; if it made me money, I did it." Unlike most of the other players who were interviewed, he stated that all the community-building activities on the site or the informal offline player communities were of little or no interest to him because they served no monetary purpose.

This interview is also interesting in that this player very self-reflectively stated that he "knew" Neopets was teaching an extreme capitalist agenda because all activities—regardless of the players' skills—would likely result in some kind of neopoint financial yield. In Eddie's opinion, the normative environment fostered by Neopets teaches an unrealistic expectation that financial gain will be the "natural" outcome of the varied site activities, which are rooted in making money via stocks, playing for points, and opening stores as financial ventures. Eddie cautioned that the Neopets stock markets taught kids an unrealistic view of the market. In his words, "Yeah, you have to be careful because it creates unrealistic expectations. I mean no stock market has stocks that only go up in value." He further reported that no matter the stocks, all stocks increased in value through time; Neopets players could be sure that if they bought low they would eventually be able to sell high. His own strategy was to always buy low-priced stocks when they first came out because, unlike in the "real" stock market, all neostocks increase in value through time. When asked if the value of stocks on Neopets fluctuated wildly, simulating "real" market activity, he said that in his own experience this was not the case. Rather, Eddie explained that all engagement in capitalist activities on the site produced positive economic yield. "There were highs and lows in the market fluctuations but never any real crashes. No one ever got wiped out."

Eddie further explained his own rationale for investing time and energy in the site. He said that the site whetted his appetite for the kind of stimulus-response created by financial risk. Eddie also believed that his playing was rooted in this extreme interest in the financial aspect of Neopets, an interest that grew through time. While Eddie made these connections regarding his own activities on the site, he did not mention any of the advertising that takes place there. Rather, for Eddie, the relationship to the site was framed as preparation for future financial success. His play on Neopets taught him how to save money, spend wisely, and invest in the future. While his connection with the site became centered on his capital- /revenue- /monetary-seeking activities, he claimed that it was always as a training field for his imagined adult practices. "You know I want to make money someday and playing all the time like that made me feel that it was all real. That everything had real consequences."

Conclusion

An exploration of different forms of work that youth engage in through and with digital media illuminates some important dimensions of youth participation in labor and economic activity. Throughout this discussion we see the resilience of existing forms of class distinction in structuring young people's access to particular job trajectories and their orientations toward labor and work. Further, youth labor has tended to be ghettoized into unskilled labor or informal economies that are generally framed as "helping" rather than activity with clear financial motives. New media participate in the production of these familiar distinctions. While recognizing these conservative tendencies and existing structural divisions, in this chapter we try to highlight the potential of new media engagement in changing some of these conditions by describing somewhat exceptional and innovative cases. If these cases are any indication of broader shifts, we are beginning to see evidence that new media are helping to open new avenues for young people to exercise new forms of agency with regard to labor and work.

Although it is rare for teens to get real jobs that make use of their technical and media expertise, their knowledge of new media can support forms of economic activity and work that were not previously available to them. We discuss this in terms of ways that kids can earn money through distributing their work, freelancing, and entrepreneurism. These forms of grassroots economic mobilization are particularly evident among youth from less privileged backgrounds. By contrast, elite youth, particularly those who spend many years in higher education financed by their parents, often parlay their new media skills into the nonmarket sector. Much like how different forms of volunteerism and internships have functioned historically, networked peer production provides opportunities for kids to experiment with different forms of work and public participation. These activities, varying from creative production to fansubbing to virtual currency trading, are training grounds for participation in the twenty-first-century economy. The difference, however, between these and structured educational and preparatory programs is that youth who participate in these activities engage in work that is immediately consequential; these are not training exercises but activities that provide them immediate gains in the context of a network of peers or a broader audience of viewers and readers. Particularly in the context of the United States, where there are comparatively few high-quality apprenticeship and vocational programs for teens not on an academic track (Hansen, Mortimer, and Krüger 2001), these opportunities fill a social vacuum.

In our discussion, we try to work against the assumption that digital media are opening up opportunities to tech-savvy kids in the same ways. Kids from a wide range of economic and social backgrounds are mobilized around diverse forms of new media work. Though we have seen a general opening up of opportunity for participation in various forms of new media work, the vast majority of these engagements do not translate to paying jobs and successful careers in the creative class. Elite kids have access to the real-world social and cultural capital where they may be able to translate these skills to jobs and paid work, and they have a leg up on kids who do not have this social and cultural capital. Even among privileged kids, we see a tendency for them to see these forms of work as serious hobbies that are separate from their real-life trajectories, which guarantee them a stable future career through standard and well-established forms of education. By contrast, less privileged youth may look toward creative-class careers for new kinds of opportunities, but they may not have the social and cultural capital to translate their talents into careers. In either case, we see a growing space of creative-class work that is not directly tied to the day jobs of the people participating in them. The economies of P2P trading that are flourishing online, and the venues for amateurs to showcase their work, are creating a new media ecology that supports these more informal kinds of work and economic arrangements. Across the class spectrum, we see kids and young adults choosing to participate in creative and technical work because of the pleasure of productive activity that they engage in on their own terms, regardless of whether or not there is economic benefit.

Whether the work is economic activity or nonmarket work, many kids are looking online for sites for exercising autonomy and efficacy and making their labor visible in a public way. Digital-media ventures are more attractive than the unskilled labor usually available for kids. Many motivated kids are not satisfied with a purely preparatory role and look for real-life consequences and responsibilities in the here and now. Many are ready for these responsibilities and launch successful careers online. Youth appreciate the opportunity to be "taken seriously" by their coworkers in forms of work that have clear productive benefits to others and where there is public validation and visibility. For others these activities are a way to experiment in certain forms of work without highly consequential failure. While educators have long noted the importance of learning in situations of real-life work and apprenticeships, there are relatively few examples of these forms of learning in the United States. Studies of Girl Scout cookie sales (Rogoff et al. 2002) give one example that does come from the United States, but many of the most celebrated examples in the literature come from cultural contexts where kids are engaged more directly in economic activity (Lave and Wenger 1991; Nunes, Schliemann, and Carraher 1993). Aside from volunteerism and concerted cultivation, which are framed more as preparatory activities, kids in the United States have few contexts for this kind of learning. The cases we describe, by contrast, are about new media's providing access to high-stakes and real environments where learning has consequences on kids' and others' lives.

The ways in which new media intersect with youth's activities of work are indicative of the complicated role that youth labor has occupied in modern society. Although youth were largely shut out from the formal, high-status labor economy, they have continued to work in a wide variety of forms. New media are making some of these activities more visible and valued, in part because of young people's new media literacy, which can often exceed that of their elders. The examples of youth practice, in turn, are part of a broader restructuring of what counts as work and productive labor, one that sees a greater role for the informal, peer-based economies that have unique affinities with the social positions and cultures of young people. While the relationships between these peer-based economies and existing commercial sectors is still very much under negotiation, we can expect that the activities of youth today will result in resilient changes to the relationships among public engagement, cultural exchange, and economic participation.

Notes

1. "New economy" generally refers to a shift from an industrial and munfacturing-based economy to one centered on services and knowledge production. Information technologies are considered key elements of the infrastructure supporting the new economy.

2. "Crowdsourcing" describes the process in which work that used to be outsourced to a contractor is now performed by an undefined, large group of people in an open environment. Some examples of crowdsourcing are collective citizen-science projects, some of the work of MoveOn.org, or Wikipedia.

3. "Caitlin Hill" is her real-life name.

4. "Ian Oji" is a real pen name.

5. "Mercykillings" is a real screen name.

6. "Wurlpin" is a real character name.

7. "Linkshells" are in-game communities that require invitation, have dedicated chat channels, and often have their own organized activities. They are like the guilds of other MMORPGs.

8. An alliance is a group of three parties.

9. KirinTheDestroyers is the endgame linkshell in the MMORPG Final Fantasy XI with whom I did fieldwork.

10. "Bokchoi" is a real character name.

11. Spinning Slash and Spiral Hell are both moves within the game that can be done using a resource called TP.

12. "Tanks" are players whose role is to "take the hits" of a fight. Certain jobs are more beneficial for this role because of health, abilities, and gear.

13. "Fyrie" is a real character name.

14. "NINs" are ninjas, who have an ability that absorbs damage.

15. A "mighty strike" is a special move of Tiamat that does a lot of damage.

16. "PLDs" are paladins, another tanking job.

17. "Subjobs" are a secondary job that players can have to supplement their primary jobs.

18. "Ghostfaced" is a real character name.

19. "Tacoguy" is a real character name.

Conclusion

The goal of this project and this book is to document the everyday lives of youth as they engage with new media and to put forth a paradigm for understanding learning and participation in contemporary networked publics. Our primary descriptive question is this: How are new media being taken up by youth practices and agendas? We have organized our shared analysis across our different case studies according to categories of practice that correspond to youth experience: media ecologies, friendship, intimacy, families, gaming, creative production, and work. In this way, we have mapped an ecology of different youth practices as well as mapping the broader social and cultural ecologies that contexualize these practices. As we take into account these larger structuring contexts, we remain attentive to the dynamics of youth culture and sociability, seeking to understand new media practices from a youth point of view. We describe the diversity in forms of youth new media practice in terms of genres of participation rather than of categories of youth based on individual characteristics. In this way, we articulate the relationship between broader social and cultural structures and everyday youth activity in ways that take into account the changing and situationally specific nature of youth engagement with particular practices. Although we see our work as essentially exploratory, as among the first steps toward mapping the terrain of youth new media practice, we try to identify some initial landmarks and boundaries that define this area of ethnographic inquiry.

Following from our descriptive focus, we have a central analytic question: How do these practices change the dynamics of youth-adult negotiations over literacy, learning, and authoritative knowledge? We suggest that participation in networked publics is a site of youth-driven peer-based learning that provides important models of learning and participation that are evolving in tandem with changes in technology. We argue that what is distinctive about our current historical moment is the growth of digital media production as a form of everyday expression and the circulation of media and communication in a context of networked publics enabled by the Internet. We see peer-based learning in networked publics in both the mainstream friendship-driven hanging out in sites such as MySpace and Facebook as well as in the more subcultural participation of geeked out interest-driven groups. Although learning in both of these contexts is driven primarily by the peer group, the structure and the focus of the peer group differs substantially, as does the content of the learning and communication. While friendship-driven participation is largely in the mode of hanging out and negotiating issues of status and belonging in local, given peer networks, interest-driven participation happens in more distributed and specialized knowledge networks. We see kids moving between these different genres of participation, often with the mediating practice of experimental messing around with new media. Networked publics provide a space of relative autonomy for youth, a space where they can engage in learning and reputation building in contexts of peer-based reciprocity, largely outside the purview of teachers, parents, and other adults who have authority over them.

These frameworks for understanding the shape of youth participation in networked publics help us understand what may be the most productive levers of change and intervention. Skills and literacies that children and youth pick up organically in their given social worlds are not generally objects of formal educational intervention, though they may require a great deal of social support and energy to acquire. In friendship-driven contexts, young people learn about the opinions and values of their peers through testing of social norms and expectations in everyday negotiations over friendship, popularity, and romantic relationships. These negotiations take place in peer publics that have been largely segregated from adult sociability ever since the establishment of teens as a distinct cultural demographic. On the interest-driven side, gamers and media creators are often motivated by an autodidactic ethic, rejecting or downplaying the value of formal education and reaching out to online networks to customize their own learning practices. Given the centrality of youth-defined agendas in both of these contexts, the challenge is to build roles for productive adult participation that respect youth expertise, autonomy, and initiative.

We believe that one key to productive adult involvement is in taking advantage of this current moment in interpretive flexibility about the nature of public participation. We have an opportunity to define, in partnership with youth, the shape of online participation and expression and new networked, institutional structures of peer-based learning. In this conclusion, we summarize the findings of our research in terms of what we see as potential sites of adult participation and intervention in youth practices. We do this in the spirit of suggesting avenues for future research and programmatic exploration. Our work has not focused on evaluating specific pedagogical approaches or institutional configurations, but we do believe that our work has implications for those seeking to do so. We organize this concluding discussion in relation to current debates over new media literacy, online participation, and the shape of contemporary learning institutions.

Shaping New Media Literacies

In our descriptions of youth expression and online communication, we identify a range of practices that are evidence of youth-defined new media literacies. On the friendship-driven side, we have seen youth developing shared norms for online publicity, including how to represent oneself in online profiles, norms for displaying peer networks online, the ranking of relationships in social network sites, and the development of new genres of written communication such as composed casualness in online messages. The commonplace practices of youth who are not framing themselves as particularly tech or media savvy—creating a MySpace profile, looking around for information online, finding and using a gaming cheat, or knowing how to engage in an appropriately casual IM conversation—are picked up within a networked social ecology widely available to youth today. Chapters 2 an 3, on friendship and intimacy, argue for an appreciation of the social and literacy skills that youth are developing in these ways. A mere decade ago, however, even these kinds of commonplace online competencies were the province of a technology elite of early adopters and certain professional communities.

On the interest-driven side, youth continue to test the limits on forms of new media literacy and expression. Here we see youth developing a wide range of more specialized and sometimes exclusionary forms of new media literacies that are defined in opposition to those developed in more main-stream youth practices. When youth engage in practices of messing around, they are experimenting with established rules and norms for media and technology use and expression. In geeked out interest-driven groups, we have seen youth engage in the specialized elite vocabularies of gaming and esoteric fan knowledge and develop new experimental genres that make use of the authoring and editing capabilities of digital media. These include personal and amateur media that are being circulated online, such as photos, video blogs, web comics, and podcasts, as well as derivative works such as fan fiction, fan art, mods, mashups, remixes, and fansubbing. Chapters 6 and 7, on creative production and work, describe many of these practices. In these geeked out practices, and in the more mainstream prac-tices on the friendship-driven side, we see youth actively negotiating the shape of new media literacies. While standards for literacy are constantly under negotiation in any community of practice, we do believe that the relative newness of digital production and online communication means that we are in a moment of interpretive flexibility, where values, norms, and literacy are particularly malleable.

Although youth online expressions may seem very foreign to those who have not grown up with them, youth *values* in this space are not so far off from those of adults. In our work, contrary to fears that social norms are eroding online, we did not find many youth who were engaging in any more risky behaviors than they did in offline contexts. As we describe in chapter 5, on gaming, those practices most commonly associated with bad behavior, such as play with violent video games, when viewed in a social context are an extension of familiar forms of male bonding. And just like in adult worlds, youth are engaged in ongoing struggles to gain a sense of autonomy and self-efficacy and to develop status and reputation among peers. We think it is important to recognize these commonalities in values that are shared among kids and adults; we see no need to fear a collapse of common culture and values. We do not believe that educators and parents need to bear down on kids with complicated rules and restrictions and heavy-handed norms about how they should engage online. For the most part, the existing mainstream strategies that parents are mobilizing to structure their kids' media ecologies, informed by our ongoing public discourse on these issues, are more than adequate in ensuring that their kids do not stray too far from home.

At the same time, our research does enable us to be a bit more precise about the influence of these technosocial shifts on intergenerational relations. Although the underlying social values may be shared intergenerationally, the actual shape of peer-based communication, and many of its outcomes, are profoundly different from those of an older generation. We found examples of parents who lacked even rudimentary knowledge of social norms for communicating online or any understanding of all but the most accessible forms of video games. Further, the ability for many youth to be in constant private contact with their peers strengthens the force of peer-based learning, and it can weaken adult participation in these peer environments. The simple shift from a home phone to a mobile phone means that parents have lost some of the ambient social contact that they previously had with their children's friends. When you have a combination of a kid who is highly active online and a parent who is disengaged from these new media, we see a risk of an intergenerational wedge. Simple prohibitions, technical barriers, or time limits on use are blunt instruments; youth perceive them as raw and ill-informed exercises of power.

The problem lies not in the volume of access but the quality of participation and learning, and kids and adults need to first be on the same page on the normative questions of learning and literacy. Parents need to begin with an appreciation of the importance of youth's social interactions with their peers, an understanding of their complexities, and a recognition that children are knowledgeable experts on their own peer practices. If parents can trust that their own values are being transmitted through their ongoing communication with their kids, then new media practices can be sites of shared focus rather than sites of anxiety and tension. In the chapter on families, as well as in those on gaming and creative production, we see numerous cases of parents and kids' coming together around new media in ways that exhibit a shared sense of what counts as valuable learning and positive sociability, and where both parents and kids bring interests and expertise to the table. These examples vary from parents who engage playfully in kids' online peer communications, who watch *telenovelas* with their kids in the living room, who work on collaborative media productions with their kids, who will play a social game with a visiting boyfriend, to parents who simply encourage and appreciate kids' self-motivated learning with media and technology, giving them space and time to experiment and tinker. It is important to note that these kinds of engagements do require parents to invest in some basic learning about technology and media, and we believe issues of differential participation and access may be just as important for parents as they are for kids.

We also believe it is important to recognize the diverse genre conventions of youth new media literacy before developing educational programs in this space. Particularly when addressing learning and literacy that grow out of informal, peer-driven practices, we must realize that norms and standards are deeply situated in investments and identities of kids' own cultural and social worlds. Friendship-driven practices of hanging out and interest-driven practices of geeking out mobilize very different genres of new media literacy. While it is possible to abstract some underlying skills, it is important to frame the cultural genre in a way appropriate to the particular context. For example, authoring of online profiles is an important literacy skill on both the friendship- and interest-driven sides, but one mobilizes a genre of popularity and coolness and the other a genre of geek cred. Similarly, the elite-speak of committed gamers involves literacies that are of little, and possibly negative, value for boys looking for a romantic partner in their school peer networks. Following from this, it is problematic to develop a standardized set of benchmarks to measure kids' levels of new media and technical literacy. Unlike academic knowledge, whose relevance is often limited to classroom instruction and assessment, new media literacy is structured by the day-to-day practices of youth participation and status in diverse networked publics. This diversity in youth values means that kids will not fall in line behind a single set of literacy standards that we might come up with, even if those standards are based on the observations of their own practices.

We believe that if our efforts to shape new media literacy are keyed to the meaningful contexts of youth participation, then there is an opportunity for productive adult engagement. Many of the norms that we observed online are very much up for negotiation, and there were often divergent perspectives among youth about what was appropriate, even within a particular genre of practice. For example, as described in chapter 2, the issue of how to display social connections and hierarchies on social network sites is a source of social drama and tension, and the ongoing evolution of technical design in this space makes it a challenge for youth to develop shared social norms. Designers of these systems are central participants in defining these social norms, and their interventions are not always geared toward supporting a shared set of literacy practices and values. More robust public debate on these issues that involves both youth and adults could potentially shape the future of online norms and literacies in this space in substantive ways. On the interest-driven side, we see adult leadership in these groups as central to how standards for expertise and literacy are being defined. For example, the heroes of the gaming world include both teens and adults who define the identity and practice of an elite gamer. The same holds for all the creative production groups that we examined. The leadership in this space, however, is largely cut off from the educators and policy makers who are defining standards for new media literacy in the adult-dominated world. Building more bridges between these different communities of practice could shape awareness on both the in-school and out-of-school side if we could respond in a coordinated and mutually respectful way to the quickly evolving norms and expertises of more geeked out and technically sophisticated experimental new media literacies.

Participation in Networked Publics

At least since the early 1990s, the question of online access and public participation has been on the radar of policy makers in the form of agendas addressing the digital divide (Bikson and Panis 1995; The White House 1993; Wresch 1996). While national context and economic factors have been central to this question, debates over the digital divide also examined factors such as gender and age as structuring differential access to technology-related competencies (Ito et al. 2001; Shade 1998). Throughout the 1990s, policy interventions in the United States focused on providing public access to the Internet through community institutions such as public schools and libraries (Fabos 2004; Henderson and King 1995). Today the picture is much more complex. Basic access to technology, the ability to navigate online information, and the ability to communicate with others online are increasingly central to our everyday participation in public life. At the same time, the range and diversity of networked publics and forms of participation have proliferated dramatically, making the definition of baseline technology access and literacy difficult if not impossible to achieve. Further, commercial online access and Web 2.0 sites have largely overshadowed the public and nonprofit sites and infrastructures of the Internet, even as we have seen a steady growth in user-generated content (Fabos 2004). A digital-divide agenda focused on technology access does not address what Jenkins and his colleagues (2006) have called the "participation gap." The more complex and socially contextualized skills of creating digital media, sharing information and media online, socializing with peers in networked publics, and going online to connect with specialized knowledge communities require both high-end technology access and social and cultural immersion in online worlds (Seiter 2007).

We suggest that the notion of networked publics offers a framework for examining diverse forms of participation with new media in a way that is keyed to the broader social relations that structure this participation. In describing new media engagements, we look at the ecology of social, technical, and cultural conditions necessary for certain forms of participation. When examining the kind of informal, peer-based interactions that are the focus of our work, we find that ongoing, lightweight access to digital-production tools and the Internet is a precondition for participation in most of the networked public spaces that are the focus of attention for U.S. teens. Further, much of this engagement is centered on access to social and commercial entertainment content that is generally frowned upon in formal educational settings. Sporadic, monitored access at schools and libraries may provide sufficient access for basic information seeking, but it is not sufficient for the immersed kind of social engagements with networked publics that we have seen becoming a baseline for participation on both the interest-driven and the friendship-driven sides.

On the friendship-driven side, participation in online communication and gaming is becoming central to youth sociability. As described in chapter 1, youth who are shut out from these networks for technical or economic reasons often develop creative work-arounds, such as going to a friend's house to play games, befriending the computer-lab teacher, or using a digital camera as an MP3 player. The fact that these friendship-driven practices are so widely distributed in youth culture functions as a driver for a kind of bottom-up universal-access agenda. Although there are still kids who are excluded from participation, they get a substantial push of both motivation and peer support because these practices are part of the common currency of youth social communication. For example, as we discuss in chapter 6, although most kids were not well versed in web design and HTML, they generally could find a friend who could help them with setting up their MySpace profile. In many ways, these processes of youth participation in mainstream popular culture are similar to how media such as television, music, and popular games function as a "ticket to play" for kids' communication (Dyson 1997). Economic barriers have continued to be an issue for lower-income kids' participation in commercial cultures (Chin 2001; Seiter 2005). New media accentuate this tendency by requiring more expensive technology and sophisticated forms of technical literacy.

Adult lack of appreciation for youth participation in popular common cultures has created an additional barrier to access for kids who do not have Internet access at home. We are concerned about the lack of a public agenda that recognizes the value of youth participation in social communication and popular culture. When kids lack access to the Internet at home, and public libraries and schools block sites that are central to their social communication, they are doubly handicapped in their efforts to participate in common culture and sociability. These uses of new media for everyday sociability also can be important jumping-off points for messing around and interest-driven learning. Contemporary social media are becoming one of the primary "institutions" of peer culture for U.S. teens, occupying the role that was previously dominated by the informal hanging out spaces of the school, mall, home, or street. Although public institutions do not necessarily need to play a role in instructing or monitoring kids' use of social media, they can be important sites for enabling participation in these activities. Educators and policy makers need to understand that participation in the digital age means more than being able to access "serious" online information and culture; it also means the ability to participate in social and recreational activities online. This requires a cultural shift and a certain openness to experimentation and social exploration that generally is not characteristic of educational institutions.

When we turn to interest-driven practices, we see kids developing more specialized forms of expertise and engaging with esoteric and niche knowledge communities. The chapters on gaming, creative production, and work aim to map some of the characteristics of these interest-driven communities of practice. These are groups that see value in subcultural capital that is not widely distributed in mainstream culture. These are not practices that are amenable to being codified into a baseline set of literacies, standardized bodies of knowledge, or normalized forms of participation. Young people who know how to mess around and pursue self-directed learning with new media have mastered genres of participation that are applicable to different content domains if given the necessary contextual supports. We believe that these genres of participation can generalize across a wide range of cultural and knowledge domains. For example, in both chapters 5 and 6 we note how youth who have been engaged in geeked out practices often participate in multiple technical or creative communities concurrently or serially. As technical and media skills and practices become more mainstream, the kids who are associated with these more specialized groups will compete to differentiate themselves with even more specialized forms of expertise that test the boundaries of technical virtuosity. Because of this, a participation gap in relation to these practices is a structural inevitability, and in fact, drives motivation and aspirations. In this domain we should value diversity rather than standardization to enable more kids to succeed and gain recognition in different communities of interest.

Although we have not systematically analyzed the relation between gender and socioeconomic status and participation in interest-driven groups, our work indicates a predictable participation gap. Particularly in the case of highly technical interest groups and geeked out forms of gaming, the genre itself is often defined as a masculine domain. These differences in access are not simply a matter of technology access but have to do with a more complex structure of cultural identity and social belonging. Girls tend to be stigmatized more if they identify with geeked out practices. While we may recognize that geeked out participation has valuable learning properties, if these activities translate to downward social mobility in friendship-driven networks of status and popularity, many kids are likely to opt out even if they have the technical and social resources at their disposal. The kinds of identities and peer status that accompany certain forms of new media literacy and technical skills (and lack thereof) is an area that deserves more systematic research.

The focus of policy and educational agendas needs to be not on the specific content or skills that kids are engaged in when they pursue interest-driven participation but rather on the genre of participation. We identify a series of peer-based learning dynamics that operate in these contexts, with basic social principles that drive engagement, learning, and the development of expertise. We also describe how youth can transition between different genres of participation by shifting from hanging out forms of media engagement to messing around, to geeking out. Conversely, we have seen youth who use their geeked out interests or marginalized identities to leverage online connections and build friendships with like-minded peers not available to them locally. For example, chapter 6 describes deep friendships built through media production, and chapter 3 describes how for gay youth, online groups can be a lifeline for affiliating with other gay teens. Although not discussed at length in this book, C. J. Pascoe and Natalie Boero's study of pro-anorexia and pro-bulimia groups are also an example of how online spaces can support marginalized identities and practices. This latter case, in particular, argues for the importance of keeping these specialized interest spaces open to participation by experienced and credible leadership that can steer the community in productive directions. These are stories about changing structures of participation that are supported by different social, cultural, and technical ecologies. It is not sufficient to design specific learning environments or pedagogical interventions without considering the overall ecology of social, technical, and cultural support that young people need to navigate these transitions.

For youth who do not have easy access to digital-production tools and the online networks of interest-driven groups, local youth media programs play an important role as a place to connect with like-minded peers. The case studies on local youth media programs that we examine, such as the hip-hop project, the video-production center, the after-school video game–production project, and school computer labs that have opened their doors to kids during breaks and after school, are all examples of adults providing resources and institutional cover for kids to pursue their hobbies and interests in new media. The most successful examples we have seen are programs that bring kids together based on kids' own passionate interests and that have plenty of unstructured time for kids to tinker and explore without being dominated by direct instruction. Unlike classroom teachers, these lab teachers and youth-program leaders are not authoritative figures responsible for assessing kids' competence, but rather they are what Dilan Mahendran has called "co-conspirators," much like the adult participants in online interest-driven groups. In this, our research is in alignment with what Vivian Chávez and Elisabeth Soep (2005) have identified as the "pedagogy of collegiality," which defines adult-youth collaboration in what they see as successful youth media programs. Again, this is an area that we believe deserves further research and attention to pedagogical design. Programs of this kind provide leadership and models for youth to aspire to in addition to the resources for kids to access the means for digital production. These are examples of public institutions not only providing the basic access to technology tools and skills training but also filling a gap in the broader ecology of social, cultural, and technical resources to enable participation in the more informal and social dimensions of networked public life.

Intergenerational Learning Institutions

Adult participation as coconspirators in interest-driven groups provides some hints as to how educators and policy makers can harness these social dynamics for learning agendas that are more keyed to adult social worlds. In many ways, the crucial ingredient in youth engagement and successful adult intervention in these spaces seems to be a stance of mutual respect and reciprocity, where youth expertise, autonomy, and initiative are valued. We describe this in terms of peer-based learning, in which those who youth identify as peers are a crucial determinant of whom they look to for status, affiliation, and competition. In friendship-driven networks, these dynamics are not so different from what their parents grew up with, involving the same growing pains of learning to take responsibility for their actions in a competitive social environment. On the interest-driven side of the equation, the ways in which we have sheltered youth from workplaces and institutionalized them in age-segregated schools means that there are few opportunities for youth to see adults as peers in these ways. As we describe in chapters 6 and 7, when kids have the opportunity to gain access to accomplished elders in areas where they are interested in developing expertise, an accessible and immediate aspirational trajectory that is grounded in an organic social context can be created. In contrast to what they experience under the guidance of parents and teachers, with peer-based learning youth take on more grown-up roles and ownership of their self-presentation, learning, and evaluation of others.

As we point out, adults can have an important role in providing leadership and role models for participants in interest-driven groups, even in contexts of peer-based learning. In friendship-driven practices that center on sociability in given school-based networks, direct adult participation is often unwelcome, but in interest-driven groups there is a much stronger role for more experienced participants to play. Unlike instructors in formal educational settings, however, these adults participate not as educators but as passionate hobbyists and creators, and youth see them as experienced peers, not as people who have authority over them. These adults exert tremendous influence in setting communal norms and what educators might call learning goals, though they do not have direct authority over newcomers. How adult roles are structured in these peer-based interest-driven groups is one element of how the genre of participation is defined, and it could be studied more systematically as a particular pedagogical stance that is grounded in a structure of reciprocity.

This dynamic is fundamentally different from the deferred-gratification model that youth experience in schools, where they are asked to accept that their work in one institutional context (school) will transition at some uncertain time to what they imagine for themselves in the future (work). By contrast, their participation in interest-driven groups and their local friend-based sociability are about status, reputation, and validation in the here and now of their lives. As we describe in chapter 7, less privileged youth can be particularly critical of the aspirational models put forth by schools, because they understand that the odds are stacked against them as far as translating their accomplishments in school into social capital in adulthood. For these youth in particular, the aspirational trajectories offered by more informal economies and flexible forms of creative production in networked publics can be a way out of alienating learning experiences in formal education.

Interest-driven networked publics are often organized by local, niche, and amateur activities that differ in some fundamental ways from the model of professional training and standardized curriculum that is put forth in schools. Just as amateur sports leagues are predicated on a broader base of participation than professional sports, hobby groups and amateur media production lower the barriers to active participation in networked publics. At the same time, kids still can find role models and heroes in these smaller-scale networks, but these role models and heroes are much more accessible than the pros, where the aspirational trajectory is distant and inaccessible. Success and recognition in these niche and local publics can be tremendously validating, and they mark a pathway toward a more civic and participatory public life. Kids from less privileged backgrounds understand that the ideology of equal opportunity through public education does not operate in the same way for them as for more privileged kids. Even those kids who are not going to navigate successfully to adult careers by pleasing their teachers can find alternative pathways toward participation in different kinds of publics that are not defined by structures they see as unfair and oppressive. We see the implications of our work less in the service of reshuffling the deck of who succeeds in professional careers in new media, and more in terms of how educational interventions can support a more engaged stance toward public participation more generally.

Our work across the different domains of practice that we examined queries the changing shape of participation in different kinds of publics, but our focus is on youth-driven publics, not civics as defined by adult agendas. While the latter is something that requires additional research, we believe that some of the most promising directions for encouraging online civic engagement begin from youth-driven bottom-up social energies, an ethic of peer-based reciprocity, and a sense of communal belonging, rather than from a top-down mandate of adult-directed civic activity. We have some examples of this in our research, including the mobilization of kids to immigrant-rights protests through MySpace, connecting with activist groups online, or helping out in school or community institutions as technical and media experts. For the most part, however, local community institutions and activity groups made little use of digital technologies and kids' media interests and did not extend beyond the local given social networks. Few kids we spoke to were interested or involved in traditional politics, even though they might be highly energized by their local politicking among peers on social network sites or in other online groups and games. We did not focus our research on uncovering the more exceptional cases that might function as models in this domain (as we did in the case of creative production), so this is an area that we also believe deserves more research. The gap between the energies that kids bring to their peer-based politics and social engagements, and their participation in more adult-centered civic and political worlds, represents a missed opportunity.

Kids' participation in networked publics suggests some new ways of thinking about the role of public education. Rather than thinking of public education as a burden that schools must shoulder on their own, what would it mean to think of public education as a responsibility of a more distributed network of people and institutions? And rather than assuming that education is primarily about preparing kids for jobs and careers, what would it mean to think of education as a process of guiding kids' participation in public life more generally, a public life that includes social, recreational, and civic engagement? And finally, what would it mean to enlist help in this endeavor from an engaged and diverse set of publics that are broader than what we traditionally think of as educational and civic institutions? In addition to publics that are dominated by adult interests, these publics should include those that are relevant and accessible to kids now, where they can find role models, recognition, friends, and collaborators who are coparticipants in the journey of growing up in a digital age. We end this book with the hope that our research has provoked these questions.

Appendix I: Project Overview

The Digital Youth Project was led by four principal investigators, Peter Lyman, Mizuko Ito, Michael Carter, and Barrie Thorne. During the course of the three-year research grant (2005–2008), seven postdoctoral researchers,[1] six doctoral students,[2] nine M.A. students,[3] one J.D. student,[4] one project assistant,[5] seven undergraduate students,[6] and four research collaborators[7] participated and contributed fieldwork materials for the project. To gain an interdisciplinary understanding of the intersection of youth, new media, and learning, principal investigators sought out individuals with expertise in a wide range of fields including anthropology, communication, political science, psychology, and sociology as well as computer science, engineering, and media studies. Many of the researchers also worked in industry and community organizations and built upon this experience to forge meaningful collaborations across research projects and disciplines.

Just as the examination of young people, new media, and learning called for scholars of diverse disciplinary backgrounds and arenas of expertise, our research agenda also demanded new sites and strategies of investigation. As noted in the introduction, our project was designed to document, from an ethnographic perspective, the learning and innovation that accompany young people's everyday engagements with new media in informal settings. Specifically, our focus on youth-centered practices of play, communication, and creative production located learning in contexts that are meaningful and formative for youth, including friendships and families as well as young people's own aspirations, interests, and passions. In practice, this perspective meant that we maintained a broad commitment to understanding the worlds of our research participants by learning about and engaging in the significant new media practices in young people's lives. Moreover, we recognized that young people's engagements with new media were not necessarily isolated to particular media or locations. For example, social network sites such as MySpace or Facebook are often most meaningful when understood in relation to teenagers and kids at school and at home, with their friends and by themselves. Because these practices move across geographic and media spaces—homes, schools, after-school programs, networked sites, and interest communities—our ethnography incorporated multiple sites and multiple methods (Appadurai 1996; Barron 2006; Gupta and Ferguson 1997; Marcus 1995, 1998).

Alongside participation and observation, the hallmarks of ethnography, we developed questionnaires, surveys, semistructured interviews, diary studies, observation and content analyses of media sites, profiles, videos, and other materials to gain insights into the qualitative dimensions of youth's engagement in digital media and technologies. Where appropriate and relevant, we also interviewed teachers, program organizers, parents, and individuals working in specific media industries. A series of pilot projects conducted by M.A. students at the University of California, Berkeley, were completed in 2005. The bulk of the fieldwork for this project was conducted in 2006 and 2007 by postdoctoral researchers and Ph.D. students. Collectively, we conducted 659 semistructured interviews, 28 diary studies, and focus group interviews with 67 individuals. Interviews were conducted informally with at least 78 individuals and we participated in more than 50 research-related events, such as conventions, summer camps, award ceremonies, or other local events. Complementing our interview-based strategy, we carried out more than 5,194 observation hours, which were chronicled in regular field notes, and we have collected 10,468 profiles, transcripts from 15 online discussion group forums, and more than 389 videos as well as numerous materials from classroom and after-school contexts. The majority of the participants in our research were recruited through snowball sampling in person, via emails, and through institutions, as well as through the placement of recruitment scripts on websites and local community newsletters.

In addition to interviews, we administered paper and online questionnaires to develop a comparative portrait of our participants. The general questionnaire was completed by 363 respondents.[8] Based on the survey material of a significant subset of our research participants, we know that the population we have examined is distinctive in some important ways. Our survey population ranged in age from 7 to 25, with a median age of 16; 86 percent of these respondents fell between 12 and 19 years of age. Our respondents were evenly split in terms of gender identification. In terms of the ethnic identities designated by our participants, we skew from national averages in having a larger proportion of Asian participants and a smaller proportion of whites.[9] These proportions were influenced by the location of many of our research sites, such as online interest groups and in the large metropolitan centers of California.[10] The focus of our work has been to develop a series of in-depth case studies of youth practice, not in developing a nationally representative sample. Many of our studies focused on online interest groups and youth media programs that represented media-savvy youth at the forefront in innovation of new media literacy and practice. We also sought to counterbalance this focus by developing case studies that were centered on mainstream youth and their friendship-driven practices as well as on lower-income communities with members who do not all have the same access to technical resources. The survey material on its own does not permit us to make generalizations for the overall population we have looked at, but it does enable an understanding of some of the key variations in the different populations that we have explored, and how they are situated in relation to other broader quantitative indicators. (See section 1.1 for more on how our study relates to quantitative studies on youth, media, and technology.)

Notes

1. The seven postdoctoral researchers include Sonja Baumer (University of California, Berkeley), Matteo Bittanti (University of California, Berkeley), Heather A. Horst (University of Southern California/University of California, Berkeley), Patricia G. Lange (University of Southern California), Katynka Z. Martínez (University of Southern California), C. J. Pascoe (University of California, Berkeley), and Laura Robinson (University of Southern California).

2. The six doctoral students include danah boyd (University of California, Berkeley), Becky Herr-Stephenson (University of Southern California), Mahad Ibrahim (University of California, Berkeley), Dilan Mahendran (University of California, Berkeley), Dan Perkel (University of California, Berkeley), and Christo Sims (University of California, Berkeley).

3. The nine master's students include Judd Antin (University of California, Berkeley), Alison Billings (University of California, Berkeley), Megan Finn (University of California, Berkeley), Arthur Law (University of California, Berkeley), Annie Manion (University of Southern California), Sarai Mitnick (University of California, Berkeley), Paul Poling (University of California, Berkeley), David Schlossberg (University of California, Berkeley), and Sarita Yardi (University of California, Berkeley).

4. Judy Suwatanapongched is a J.D. student at the University of Southern California.

5. Rachel Cody was a project assistant at the University of Southern California.

6. The seven undergraduates are Max Besbris (University of California, Berkeley), Brendan Callum (University of Southern California), Allison Dusine (University of California, Berkeley), Sam Jackson (Yale University), Lou-Anthony Limon (University of California, Berkeley), Renee Saito (University of Southern California), and Tammy Zhu (University of Southern California).

7. The collaborators include Natalie Boero, an assistant professor of sociology at San Jose State University; Scott Carter, a Ph.D. candidate at the University of California, Berkeley, who now works at FXPal; Lisa Tripp, assistant professor of school media and youth services, College of Information, Florida State University; and Jennifer Urban, clinical assistant professor of law at the University of Southern California.

8. These respondents were those we conducted interviews with in our homes- and neighborhood-focused studies as well as in a number of other projects, including Patricia G. Lange's study "Thanks for Watching: A Study of Video-Sharing Practices on YouTube," Mizuko Ito's "Transnational Anime Fandoms and Amateur Cultural Production," Becky Herr-Stephenson's "Mischief Managed," and danah boyd's "Teen Sociality in Networked Publics." While some parents and other adults participated in the survey, all statistics reported here are based on survey participants who are 25 years old or younger, of which there were 363 respondents.

9. We presented respondents with 14 ethnicity categories (one being "other") and asked them to choose all that apply to them—49.3 percent of our population identified as white and 10.5 percent of our participants self-identified as African-American or black; 9.6 percent self-identified as Other Spanish-American/Latino and another 5.2 percent self-identified as Mexican/Mexican American/Chicano; 6.3 percent of our participants declared themselves Chinese/Chinese American; and just over 8 percent of our respondents identified themselves as Asian, a category that in the United States incorporates East Indian/Pakistani, Filipino/Filipino American, Japanese/Japanese American, Korean/Korean American, Vietnamese/Vietnamese American, and Other Asian. Another 5.2 percent identified as Other. Because respondents were able to choose more than one category, the percentages did not add up to 100 percent. Our participants diverged from the averages calculated by the 2000

U.S. census, particularly in terms of the density of the Asian population (which is quite a bit above the national proportion of 3.6 percent) and white populations (which is far below the national proportion of approximately 75 percent).

10. In California, whites make up approximately 60 percent of the population and Asians constitute around 11 percent of the state's population, according to the 2000 U.S. Census. Latinos, who are not clearly defined in the 2000 U.S. Census, are also prominent in California.

Appendix II: Project Descriptions

In this appendix, we provide an overview of the research sites included in the Digital Youth Project. We have organized the sites into four general categories: homes and neighborhoods, institutional spaces, networked sites, and interest groups. While the categories are primarily organizational, they do help to emphasize the range of sites of inquiry that we draw upon for the analysis here—twenty distinctive research projects in total[1]—as well as the epistemology that shaped the ways we approached our effort to understand youth's engagement with new media from an ethnographic perspective. As is evident in our descriptions, many projects moved among different categories of research sites. For example, Lisa Tripp and Becky Herr-Stephenson's study of Los Angeles middle schools and Katynka Martínez's study of Pico Union families followed students at school and within their homes and neighborhoods. The points of intersection and divergence between the kids in the different studies were of great interest, such as when a researcher in the neighborhood cluster of studies discovered an anime fan, or conversely, when interest-based new media hobbies were notably absent among kids in a particular study. In this book, we describe practices that we observed in multiple case studies that emerged through collaborative analysis, and the specificities of the research sites and projects have largely been erased. In this appendix, we introduce the individual projects to provide the reader with some of the context that readers may feel is missing in previous chapters. Each study comprises an ethnographic analysis of new media in the lives of a particular population; taken as a whole, they offer a broader ecological perspective of how new media practices are distributed among diverse youth in diverse contexts.

Homes and Neighborhoods

We focused on homes and families in urban, suburban, and rural contexts to understand how new media and technologies shaped the contours of kids' home lives and, in turn, how different family structures and economic and social positions may structure young people's media ecologies (Bourdieu 1984; Holloway and Valentine 2003; Livingstone 2002; Silverstone and Hirsch 1992; see also chapter 4 in this book). Working in the context of multicultural California (among other sites), we have taken seriously the need to understand the influence of ethnic, racial, gender, and class distinctions on many young people's media and technology practices (Chin 2001; Escobar 1994; Pascoe 2007a; Seiter 2005; Thorne 2008). Indeed, one of the advantages of this large-scale ethnographic project is the diversity of sites that we have been able to access.

In their study of middle-school students and their families in Los Angeles, Lisa Tripp, Becky Herr-Stephenson, and Katynka Martínez conducted participant observation in the classrooms of teachers involved in a professional-development program for media arts and technology as well as participant observation in after-school programs (Martínez, Animation Around the Block; Martínez, High School Computer Club; Martínez, Pico Union Community Center). In addition to the work in institutionalized settings, this study also incorporated interviews with kids, their siblings, and their parents. The interviews were conducted in English and Spanish and took place, when possible, at students' homes, which allowed the researchers to better understand the rich contexts of neighborhood and family life, such as Martínez's study "Pico Union Families". In a similar vein, but with a very different population, Heather Horst's study "Silicon Valley Families" examined the appropriation of new media and technology in Silicon Valley, California. Recruiting her research participants from parents' email lists at schools in the region, she focused her studies on the role of new media in kids' communication, learning, knowledge, and play in families with children between the ages of eight and eighteen to understand the gendered and generational dynamics of the incorporation of new media at home.

In their study "Living Digital," C. J. Pascoe and Christo Sims conducted a multisited ethnographic project in order to analyze how teenagers communicate, negotiate social networks, and craft a unique teen culture using new media. In C. J. Pascoe's case, she introduced herself to students in a local digital-arts program in an ethnically diverse suburban area of the East Bay, near San Francisco, where she later interviewed many of the high school–aged teenagers outside of school. Christo Sims (Rural and Urban Youth) carried out research in homes in an area near the Sierra Nevada range of rural California with a population of primarily white working- and middle-class families. In addition, he conducted work in Brooklyn, New York, an area that boasts a significant Caribbean, African-American, and Latino population; he gained access to the community with the help of a local after-school program. By looking at teens across a variety of geographic locations (rural, urban, and suburban) and socioeconomic statuses, Pascoe and Sims aimed to understand how new media have been folded into teens' friendship and romance practices.

Megan Finn, David Schlossberg, Judd Antin, and Paul Poling's study "Freshquest" also focused on the role of media and technologies in the lives of teenagers through an examination of technology-mediated communication habits of freshman students at the University of California, Berkeley. Using a survey administered to 3,161 first-year students between 2005 and 2006, their primary goal was to understand how students adopt and use information and communication technologies and how they talk about growing up with technology, both in relation to their socioeconomic status and social networks. Finn and her colleagues also administered 140 surveys and conducted focus-group interviews with first-year students at a community college in a suburb of the San Francisco Bay Area in 2006. As noted throughout this book, most of the material described is derived from the focus groups conducted with undergraduates at the University of California, Berkeley.

Along with interviews, surveys, and questionnaires, many of the projects in our homes-and-neighborhoods studies experimented with different ways of engaging young people, using the media in kids' everyday lives to narrate and explain their varying engagements and commitments to new media. Dan Perkel and Sarita Yardi's project "Digital Photo-Elicitation with Kids" used digital-photography diary studies to show the technology practices of kids entering middle school. Moving from an after-school program in the San Francisco Bay Area to the context of family life, Perkel and Yardi looked at the kinds of technologies participants used in their homes and in their summer activities, who they used them with, and what these activities meant to kids. With the assistance of Scott Carter, a doctoral student at the University of California, Berkeley, we also developed a diary study that used digital cameras and cell phone cameras (camphones) (Carter 2007). Building upon recent use of diary studies to document everyday media use (cf. Dourish and Bell 2007; Horst and Miller 2005; Ito, Okabe, and Anderson forthcoming; Okabe and Ito 2006; Van House et al. 2005), participants used mobile phones and digital cameras to chronicle their use of new media. Combined with other interviews, observations, and participation in many arenas of young people's neighborhood and home lives, this methodology enabled researchers to develop a deeper understanding of the media ecologies that young people create and inhabit.

Learning Institutions: Media-Literacy Programs and After-School Programs

Over the past two decades, researchers interested in "informal learning" have increasingly turned their attention to institutions such as libraries, after-school programs, and museums as sites that structure learning experiences that differ from those in school (see Barron 2006; Bekerman, Burbules, and Silberman-Keller 2006). As institutions temporally and spatially situated between the dominant institutions in kids' lives—school and family—after-school programs and spaces offered potential for observing instances of informal learning, particularly given the increasing importance of after-school and enrichment programs in American public education.

In light of the possibilities of these spaces, a number of our projects focused on after-school programs in an effort to understand how they fit into the lives of young people. For example, Judd Antin, Dan Perkel, and Christo Sims investigated media-production classes at a San Francisco technology center. Assuming roles as volunteer program helpers for their project "The Social Dynamics of Media Production," Antin, Perkel, and Sims looked at how the students from low-income neighborhoods negotiate and appropriate the structured and unstructured aspects of the program to learn new technical skills, socialize with new groups of friends, and take advantage of the unique access to both technical and social resources that often are lacking in their homes and schools. In this case, researchers participated regularly in the program. In some instances, researchers conducted interviews with the participants in their homes or outside the program in an effort to understand how the program—and, more broadly, new media—shaped their lives.

Although we primarily focused on learning spaces outside formal school contexts, we also carried out two research projects in structured learning contexts. Moving beyond binary questions of access, such as digital divides (Compaine 2001; Servon 2002), Lisa Tripp and Becky Herr-Stephenson's study "Los Angeles Middle Schools" examined the complex relationships between the multimedia-production projects that were undertaken in middle-school classrooms and the students' out-of-school experiences with multimedia. Contextualizing these in-class observations with interviews in homes and schools throughout urban Los Angeles, Tripp and Herr-Stephenson aimed to understand the gaps and overlaps of media use within the contexts of homes and schools. Similarly, Laura Robinson's study "Wikipedia and Information Evaluation" examined the role material resources played in everyday information-seeking contexts among economically disadvantaged youth at a high school in an agricultural region of central California. Project researchers primarily focused on the school sites in an effort to think about how digital and online media may facilitate productive learning environments. In addition, our work in schools and after-school programs was motivated by a desire to get to know young people across the multiple contexts of their lives. In all of our institutional projects, researchers carried out observations in the programs and provided formal and informal feedback to the organizations that provided them with access and support.

Networked Sites

Rather than restricting our focus to bounded spaces or locales (Appadurai 1996; Basch, Schiller, and Szanton-Blanc 1994; Gupta and Ferguson 1997), as researchers we wanted to acknowledge the "world of infinite interconnections and overlapping contexts" (Amit-Talai 2000, 6) that young people inhabit through new media. Often working in tandem with other forms of media and communication, new media provide communication venues that individuals incorporate into their lives to form, maintain, and strengthen social ties and relationships (Boase 2007; di Gennaro and Dutton 2007; Hampton 2007; Hampton and Wellman 2003; Miller and Slater 2000; Panagakos and Horst 2006; Wellman et al. 2003; Wilding

2006). Recent scholarship of online communities illustrates that significant relationships and community can be formed, even in the absence of physical copresence (Baym 2000; Constable 2003; Hine 2000; Kendall 2002; Rheingold 2000; Smith and Kollock 1999; Varnelis 2008; Wilson and Peterson 2002). Indeed, the Digital Future Project reveals that the percentage of individuals who report membership in an online community has more than doubled in the past three years (USC Center for the Digital Future 2008), indicating the growing importance of new media in facilitating social groupings and community in the United States. For this reason, a significant part of our research focused on a number of the most prominent online websites with the aim of understanding the inner workings of online groups and emerging practices surrounding community formation.

Exploring a series of sites that dominated young people's media ecologies between 2005 and 2007, we concentrated our efforts on understanding practices as they spanned online and offline settings, without privileging one context as more or less authentic, or more or less virtual (Kendall 2002). We were not interested in establishing a boundary between online participation as distinct from offline; rather, we saw specific online sites as an entry point into a varied set of hybrid practices that flowed through these sites. For example, in the discussion of social network sites that became popular in 2005 (such as Bebo, Facebook, and MySpace), we argue that the online contexts are largely a mirror and extension of sociability in teens' local school-based relations. In her study "Teen Sociality in Networked Publics," danah boyd examined the ways in which teens use sites such as MySpace and Facebook to negotiate identity, socialize with friends, and make sense of the world around them. Her project addresses teens' friendship-driven practices and contextualizes their use of networked publics in their lives more broadly. Dan Perkel's study "MySpace Profile Production" investigated how young people create MySpace pages. Whereas boyd examined the sociality of MySpace, Perkel concentrated on the socio-technical practices and infrastructure of profile making, including getting started with the help of friends, finding visual and audio material online, and copying and pasting snippets of code. The project revealed how a MySpace profile is produced through the socially and technically distributed activity of many people and is intimately tied to the specific, local communities that the profile owner inhabits.

Two of our researchers examined the phenomenon of YouTube, the video-sharing site that became popular in 2006. Patricia G. Lange analyzed how children and youth interactively negotiate aspects of the self by creating, sharing, and watching videos on the site. In her study "YouTube and Video Bloggers," Lange examines how and what participants learn by making videos and providing feedback. She argues that through social interaction and self-comparison to other video makers, YouTubers learn how to represent themselves and their work in order to become accepted members of groups who share similar media-based affinities. In addition to conducting interviews and analyzing videos, Lange became a video blogger and received feedback on her videos posted (and featured) on YouTube and on her own research website. Sonja Baumer focused on identity practices of American youth on YouTube in her study "Self-Production through YouTube." Baumer's study emphasizes self-production as an agentive act that expresses the fluidity of identity achieved through forms of semiotic action and through practices such as self-presentation, differentiation and integration, self-evaluation, and cultural commentary.

Just as social network sites and YouTube emerged as central to a wide range of young people's participation in online sites during the course of our research, gaming sites also piqued the interests of kids and teens. Heather Horst and Laura Robinson's study "Neopets" explored cultural products and knowledge creation surrounding a popular children's website. Looking at practices varying from authoring relatively simple web pages, participating in online auctions, writing stories, and creating galleries to showcase collections of specialized items, the study used questionnaires and interviews to examine how participants develop notions of reputation, expertise, and other forms of identification. Rachel Cody examined a very different kind of online game in her study of the massively multiplayer online role-playing game Final Fantasy XI. By becoming a member of a linkshell, the communities through which players organize their game playing, Cody's research examined how the social activity extended beyond the game into websites, message boards, and instant-messenger programs. This contact strengthened the relationships formed within the game and encouraged a level of collaboration that is impossible within the game, allowing players to create strategies through videos, screen shots, and community experiences. Throughout all of the online-based research, a commitment to participation and engagement through these sites remained central to developing an understanding of these sites and practices.

Interest-Based Groups

Although social scientists have studied youth subcultures for some time, the relationship between media and youth culture emerged most cogently in the British cultural studies movement in the 1970s and 1980s.[2] Ranging from music, fashion, hairstyles, language, lifestyle, and other forms of popular culture, research emphasized youth cultural forms and agency (Hall and Jefferson 1975). Looking at differences in practices across age, class, ethnicity, race, gender, and other measures of difference and power (Hebdige 1979; Jenkins 1983; McRobbie 1980; Willis 1977), cultural studies scholars examined youth, popular culture, media, and the creation of alternative publics, with particular attention to the ways in which the meaning, or texts, resisted and subverted normative practices and structures in society (Amit-Talai and Wulff 1995; Bucholz 2002; Maira and Soep 2004; Snow 1987). For example, rebellion and the development of an alternative lifestyle was pervasive in the do-it-yourself (DIY) ethos of punk culture, one of the first groups to market and circulate its own music outside mainstream society and, in turn, to challenge traditional sites of production, consumption, and copyright (Hebdige 1979). This DIY ethic continues in the remix culture of the early hip-hop and DJ movements (Gilroy 1987; Hebdige 1987; Sharma 1999). This attention to the relationship between media and popular culture and the changing relationships among production, consumption, and participation continues in much of the work on youth and the ethnography of media (e.g., Askew and Wilk 2002; Ginsburg, Abu-Lughod, and Larkin 2002).

Recognizing the tremendous transformations in the empirical and theoretical work on youth subcultures, new media, and popular culture through the past decades, researchers across our project focused on the modes of expression, circulation, and mobilization of youth subcultural forms in and through new media. For example, Dilan Mahendran's project "Hip-Hop Music Production," explored the practices of amateur music-making against the background of hip-hop culture in the San Francisco Bay Area's after-school settings. Mahendran's research illuminated the centrality of music-listening and -making by both enthusiasts and youth in general as world-disclosing practices that challenge the assumption that youth are simply passive consumers. The commodification of digital media technologies focused on the low-cost private or personal-computing model has enabled DIY music makers to create, produce, and distribute both highly collaborative and individual works of art. Following the DIY theme inherent in many subcultural artistic communities, Mizuko Ito's "Anime Fans" examined a highly distributed network of overseas fans of Japanese animation. She focused on how the fandom organized and communicated online and how it engaged in creative production through the transformative reuse of commercial media. Becky Herr-Stephenson's study "Harry Potter Fandom" investigated multimedia production undertaken by young Harry Potter fans and the role technology plays in facilitating production and distribution of fan works. Herr-Stephenson's research situates young fans' media production at the intersection of interest-driven and friendship-driven participation, calling attention to the unique characteristics of this large, vibrant, and prolific fandom.

Where much of the early work on subcultures and media focused on creative and artistic modes of expression, we are only just beginning to understand the scope and scale of other subcultural practices. C. J. Pascoe and Natalie Boero's study "Pro–Eating Disorder Discussion Groups" examined the construction of online eating-disorder communities by analyzing pro-anorexia ("ana") and pro-bulimia ("mia") discussion groups. Based on participants' characterizations of anorexia as a lifestyle choice rather than a disease, the project attempts to move beyond dominant clinical narratives of eating disorders, instead highlighting participants' ambivalence regarding gender, body size, and offline relationality. Pascoe and Boero reveal how the ana and mia lifestyles are produced and reproduced in these online spaces. Moreover, their study demonstrates the ways new media bring to the fore other practices that previously existed but remained underground or outside the purview of mainstream society.

Like the anorexic and bulimic communities that have found new modes of expression in online venues, gaming cultures and communities have become more public in the new media ecology. Focusing on a local gathering place for gamers in the San Francisco Bay Area, Arthur Law's study "Team Play" explored the social context in which teenagers are made use of video games at a cyber café. The study highlighted two styles of game play at the café: solo teenagers playing a real-time strategy game by themselves and groups of teenagers playing first-person shooters together. Despite their differences, each style is highly social and demonstrates that online video games can be seen as a venue for maintaining friendships across vast distances or providing additional social activities on top of traditional ones such as basketball or football. Looking at the emergence of networked gaming, Matteo Bittanti's study "Game Play" examines the complex relationship between teenagers and video games. Bittanti focused on the ways in which gamers create and experiment with different identities; learn through informal processes; form peer groups; develop a variety of cognitive, social, and emotional skills; and produce significant textual artifacts (e.g., information, comments, reviews, music videos, and game videos) through digital play. Electronic gaming has become a focus for young people's social interaction, interest-driven learning, and creative production.

Notes

1. Three pilot projects that we do not discuss at length in this report book were formative in structuring our research methodologies and attention to informal learning. The first, Dan Perkel and Sarita Yardi's project "Searching for Count Whistleboy: Explorations in Collaborative Storytelling through Design Research" used a design research approach to explore the possibilities of collaborative storytelling among fifth graders. Through design activities, games, group discussion, and interviews, Perkel and Yardi examined the topics of collaboration, appropriation, and social dynamics around the kids' creative productions. The second project, Sarita Yardi and Sarai Mitnick's study "Media Literacy Education: Understanding Technology and Online Media in the Lives of Middle-School Girls," investigated the role of technology and online media in the lives of girls in an after-school technology program for middle-school girls in Oakland, California. The third project, Alison Billings's "Wondering, Wandering, and Wireless: An Ethnography of the Explainers and Their Brief Affair with a Mobile Technology," examined the ways in which technology could be incorporated more effectively for technology literacy. Billings explored how "Explainers," or young people who are front-line educators to the visitors at a science and technology museum in the San Francisco Bay Area, used a new mobile device in an effort to improve the quality of their work by providing them access to on-the-fly resources.

2. Mintz (2004) argues that youth subcultures did not emerge until in the 1950s.

Appendix III: Project Index

Study's Short Title	Study's Full Title	Study's Authors
Animation Around the Block	Animation Around the Block: After-School Game Design	Katynka Z. Martínez
Anime Fans	Transnational Anime Fandoms and Amateur Cultural Production	Mizuko Ito
Collaborative Storytelling	Searching for Count Whistleboy: Explorations in Collaborative Storytelling through Design Research	Dan Perkel and Sarita Yardi
Digital Photo-Elicitation with Kids	Discovering the Social Context of Kids' Technology Use	Dan Perkel and Sarita Yardi
Final Fantasy XI	Life in the Linkshell: The Everyday Activity of a Final Fantasy Community	Rachel Cody
Freshquest	Freshquest	Megan Finn, David Schlossberg, Judd Antin, and Paul Poling
Game Play	Game Play	Matteo Bittanti
Harry Potter Fandom	Mischief Managed: Multimedia Production in the Harry Potter Fandom	Becky Herr-Stephenson
High School Computer Club	The Student-Led Startup: One High School's Computer Club	Katynka Z. Martínez
Hip-Hop Music Production	Hip-Hop Music and Meaning in the Digital Age	Dilan Mahendran
Living Digital	Living Digital: Teens' Social Worlds and New Media	C. J. Pascoe and Christo Sims
Los Angeles Middle Schools	Teaching and Learning with Multimedia	Lisa Tripp and Becky Herr-Stephenson

Bibliography

Abbott, Chris. 1998. "Making Connections: Young People and the Internet," pp. 84–105 in *Digital Diversions: Youth Culture in the Age of Multimedia*, edited by J. Sefton-Green. London: University College London Press.

Adams, Natalie Guice, and Pamela J. Bettis. 2003. *Cheerleader! An American Icon.* New York: Palgrave MacMillan.

Alanen, Leena, and Berry Mayall. 2001. *Conceptualizing Child-Adult Relationships.* New York: Routledge/Farmer.

Allan, Graham. 1998. "Friendship, Sociology and Social Structure." *Journal of Social and Personal Relationships* 15(5):685–702.

Alters, Diane F. 2004. "The Family in U.S. History and Culture," pp. 51–68 in *Media, Home, and Family*, edited by S. M. Hoover, L. S. Clark, and D. Alters. New York: Routledge.

Alters, Diane F., and Lynn Schofield Clark. 2004a. "Conclusion: The 'Intentional and Sophisticated' Relationship," pp. 171–80 in *Media, Home, and Family*, edited by S. M. Hoover, L. S. Clark, and D. Alters. New York: Routledge.

Alters, Diane F., and Lynn Schofield Clark. 2004b. "Introduction," pp. 3–18 in *Media, Home, and Family*, edited by S. M. Hoover, L. S. Clark, and D. Alters. New York: Routledge.

Amit-Talai, Vered. 2000. *Constructing the Field: Ethnographic Fieldwork in the Contemporary World.* London, UK, and New York: Routledge.

Amit-Talai, Vered, and Helena Wulff, eds. 1995. *Youth Cultures: A Cross-Cultural Perspective.* London, UK, and New York: Routledge.

Anderson, Chris. 2006. *The Long Tail: Why the Future of Business Is Selling Less of More.* New York: Hyperion.

Anderson, Craig A., Douglas A. Gentile, and Katherine E. Buckley. 2007. *Violent Video Game Effects on Children and Adolescents: Theory, Research, and Public Policy.* Oxford, UK: Oxford University Press.

Appadurai, Arjun. 1996. *Modernity at Large: Cultural Dimensions of Globalization.* Minneapolis, MN: University of Minnesota Press.

Appadurai, Arjun, and Carol A. Breckenridge. 1988. "Why Public Culture." *Public Culture* 1(1):5–9.

Appadurai, Arjun, and Carol A. Breckenridge. 1995. "Public Modernity in India," pp. 1–20 in *Consuming Modernity: Public Culture in a South Asian World*, edited by C. A. Breckenridge. Minneapolis, MN: University of Minnesota Press.

Aries, Phillipe. 1962. *Centuries of Childhood.* New York: Vintage.

Ashcraft, Catherine. 2006. "Ready or Not? Teen Sexuality and the Troubling Discourse of Readiness." *Anthropology and Education Quarterly* 37(4):328–46.

Askew, Kelly, and Richard R. Wilk. 2002. *The Anthropology of Media Reader.* Malden, MA: Blackwell.

Austin, Joe, and Michael Nevin Willard, eds. 1998. *Generations of Youth: Youth Cultures and History in Twentieth-Century America.* New York: New York University Press.

Bakardjieva, Maria. 2005. *Internet Society: The Internet in Everyday Life.* London: Sage Publications.

Banet-Weiser, Sarah. 2007. *Kids Rule! Nickelodeon and Consumer Citizenship.* Durham, NC: Duke University Press.

Baron, Naomi S. 2008. *Always On: Language in an Online and Mobile World.* Oxford, UK, and New York: Oxford University Press.

Barron, Brigid. 2006. "Interest and Self-Sustained Learning as Catalysts of Development: A Learning Ecology Perspective." *Human Development* 49(4):193–224.

Basch, Linda, Nina Glick Schiller, and Christina Szanton-Blanc. 1994. *Nations Unbound: Transnational Projects, Postcolonial Predicaments, and Deterritorialized Nation-States.* London, UK: Routledge.

Baym, Nancy. 2000. *Tune In, Log On: Soaps, Fandom, and Online Community.* Thousand Oaks, CA: Sage Publications.

Bazalgette, Cary. 1997. "An Agenda for the Second Phase of Media Literacy Development," pp. 69–78 in *Media Literacy in the Information Age*, edited by R. Kubey. New Brunswick, NJ: Transaction.

Beck, John C., and Mitchell Wade. 2004. *Got Game: How the Gamer Generation Is Reshaping Business Forever.* Cambridge, MA: Harvard Business School Press.

Becker, Howard S. 1982. *Art Worlds.* Berkeley, CA: University of California Press.

Bekerman, Zvi, Nicholas C. Burbules, and Diana Silberman-Keller, eds. 2006. *Learning in Places: The Informal Education Reader.* New York: Peter Lange.

Ben-Amos, Ilana Krausman. 1995. "Adolescence as a Cultural Invention: Philippe Aries and the Sociology of Youth." *History of the Human Sciences* 8(2):66–89.

Benitez, Jose Luis. 2006. "Transnational Dimensions of the Digital Divide among Salvadoran Immigrants in the Washington DC Metropolitan Area." *Global Networks* 6(2):181–99.

Benkler, Yochai. 2006. *The Wealth of Networks: How Social Production Transforms Markets and Freedom*. New Haven, CT: Yale University Press.

Berndt, Thomas J. 1996. "Exploring the Effects of Friendship Quality on Social Development," pp. 346–65 in *The Company They Keep: Friendship in Childhood and Adolescence*, edited by W. M. Bukowski, A. F. Newcomb, and W. W. Hartup. Cambridge, UK, and New York: Cambridge University Press.

Best, Amy. 2000. *Prom Night: Youth, Schools and Popular Culture*. New York: Routledge.

Bettie, Julie. 2003. *Women without Class: Girls, Race, and Identity*. Berkeley, CA, and Los Angeles, CA: University of California Press.

Bijker, Weibe E., Thomas P. Hughes, and Trevor Pinch. 1987. *The Social Construction of Technological Systems: New Directions in Sociology and History of Technology*. Cambridge, MA: The MIT Press.

Bikson, Tora K., and Constantijn W. A. Panis. 1995. "Computers and Connectivity: Current Trends," pp. 13–40 in *Universal Access to E-Mail: Feasibility and Societal Implications*, edited by R. H. Anderson, T. K. Bikson, S. A. Law, and B. M. Mitchell. Santa Monica, CA: RAND.

Bloustein, G. 2003. *Girl Making: A Cross-Cultural Ethnography on the Processes of Growing Up Female*. New York and Oxford, UK: Berghahn Books.

Boase, Jeffrey. 2007. "Personal Networks and the Personal Communication System: Using Multiple Media to Connect." American Sociological Association Annual Meeting, New York.

Boase, Jeffrey. 2008. "Personal Networks and the Personal Communication System." *Information, Communication & Society* 11(4):490–508.

Bogle, Kathleen A. 2008. *Hooking Up: Sex, Dating and Relationships on Campus*. New York: New York University Press.

Bourdieu, Pierre. 1972. *Outline of a Theory of Practice*, edited by J. Goody. Translated by R. Nice. New York: Cambridge University Press.

Bourdieu, Pierre. 1984. *Distinction: A Social Critique of the Judgment of Taste*. Cambridge, MA: Harvard University Press.

Bovill, Moira, and Sonia Livingstone. 2001. "Bedroom Culture and the Privatization of Media Use," pp. 179–200 in *Children and Their Changing Media Environment: A*

European Comparative Study, edited by S. Livingstone and M. Bovill. Mahwah, NJ: Lawrence Erlbaum Associates.

boyd, danah. 2006. "Friends, Friendsters, and Top 8: Writing Community into Being on Social Network Sites." *First Monday* 11(12). Retrieved June 26, 2008 (http://www.firstmonday.org/issues/issue11_12/boyd).

boyd, danah. 2007. "Why Youth (Heart) Social Network Sites: The Role of Networked Publics in Teenage Social Life," pp. 119–42 in *Youth, Identity, and Digital Media*, edited by D. Buckingham. John D. and Catherine T. MacArthur Foundation Series on Digital Media and Learning. Cambridge, MA: The MIT Press.

boyd, danah m. 2008. "Taken Out of Context: American Teen Sociality in Networked Publics." PhD Dissertation, School of Information, University of Califonia, Berkeley.

boyd, danah m., and Nicole B. Ellison. 2007. "Social Network Sites: Definition, History, and Scholarship." *Journal of Computer-Mediated Communication* 13(1). Retrieved June 26, 2008 (http://jcmc.indiana.edu/vol13/issue1/boyd.ellison.html).

Brandt, Deborah, and Katie Clinton. 2002. "Limits of the Local: Expanding Perspectives on Literacy as a Social Practice." *Journal of Literacy Research* 34(3):337–56.

Brown, B. Bradford. 1999. "'You're Going Out with Who?' Peer Group Influences on Adolescent Romantic Relationships," pp. 291–329 in *The Development of Romantic Relationships in Adolescence*, edited by W. Furman, B. B. Brown, and C. Feiring. Cambridge, UK: Cambridge University Press.

Brown, John Seely, Allan Collins, and Paul Duguid. 1989. "Situated Cognition and the Culture of Learning." *Educational Researcher*, 18(1):32–42.

Bryant, Jennings, and Dolf Zillman. 2002. *Media Effects: Advances in Theory and Research*. Mahwah, NJ: Lawrence Erlbaum.

Bucholz, Mary. 2002. Youth and Cultural Practice. *Annual Review of Anthropology*, 31(1):525–52.

Buckingham, David. 1993. *Children Talking Television: The Making of Television Literacy*. London: Taylor & Francis.

Buckingham, David. 2000. *After the Death of Childhood: Growing Up in the Age of Electronic Media*. Cambridge, UK: Polity.

Buckingham, David. 2003. *Media Education: Literacy, Learning, and Contemporary Culture*. Cambridge, UK: Polity.

Buckingham, David. 2006. "Is There a Digital Generation?" pp. 1–18 in *Digital Generations: Children, Young People, and New Media*, edited by D. Buckingham. Mahway, NJ: Lawrence Erlbaum.

Buckingham, David. 2007. *Beyond Technology: Children's Learning in the Age of Digital Culture*. Malden, MA: Polity.

Buckingham, David, ed. 2008. *Youth, Identity, and Digital Media*. John D. and Catherine T. MacArthur Foundation Series on Digital Media and Learning. Cambridge, MA: The MIT Press.

Buckingham, David, Shaku Banaji, Andrew Burn, Diane Carr, Sue Cranmer, and Rebekah Willett. 2005. *The Media Literacy of Children and Young People*. London, UK: Ofcom.

Buckingham, David, Pete Fraser, and Julian Sefton-Green. 2000. "Making the Grade: Evaluating Student Production in Media Studies," pp. 129–53 in *Evaluating Creativity: Making and Learning by Young People*, edited by J. Sefton-Green and R. Sinker. London, UK, and New York: Routledge.

Buckingham, David, and Julian Sefton-Green. 2004. "Structure, Agency, and Pedagogy in Children's Media Culture," pp. 12–33 in *Pikachu's Global Adventure: The Rise and Fall of Pokémon*, edited by J. Tobin. Durham, NC: Duke University Press.

Carter, Scott. 2007. "Supporting Early-Stage Ubicomp Experimentation." Ph.D. dissertation, Computer Science, University of California, Berkeley.

Cassell, Justine, and Meg Cramer. 2007. "High Tech or High Risk? Moral Panics about Girls Online," pp. 53–75 in *Digital Youth, Innovation, and the Unexpected*, edited by T. MacPherson. The John D. and Catherine T. MacArthur Foundation Series on Digital Media and Learning. Cambridge, MA: The MIT Press.

Cassell, Justine, and Henry Jenkins. 1998. *From Barbie to Mortal Kombat: Gender and Computer Games*. Cambridge, MA: The MIT Press.

Castells, Manuel. 1996. *The Rise of the Network Society*. Vol. 1, *The Information Age: Economy, Society and Culture*. Cambridge, MA, and Oxford, UK: Blackwell.

Castronova, Edward. 2001. *Virtual Worlds: A First-Hand Account of Market and Society on the Cyberian Frontier*. Munich, Germany: CESifo.

CDC. 2007. "YRBSS: Youth Risk Behavior Surveillance System." *Healthy Youth! Data and Statistics*. Vol. 2008. Retrieved June 22, 2008 (http://www.cdc.gov/healthyyouth/yrbs/index.htm).

Chávez, Vivian, and Elisabeth Soep. 2005. "Youth Radio and the Pedagogy of Collegiality." *Harvard Educational Review* 75(4):409–34.

Chin, Elizabeth. 2001. *Purchasing Power: Black Kids and American Consumer Culture*. Minneapolis, MN: University of Minnesota Press.

Chudacoff, Howard P. 1989. *How Old Are You? Age Consciousness in American Culture*. Princeton, NJ: Princeton University Press.

Clark, Lynn Schofield. 2004. "Being Distinctive in a Mediated Environment: The Ahmeds and the Paytons," pp. 79–102 in *Media, Home, and Family*, edited by S. M. Hoover, L. S. Clark, and D. Alters. New York: Routledge.

Clark, Lynn Schofield, and Diane F. Alters. 2004. "Developing a Theory of Media, Home, and Family," pp. 35–50 in *Media, Home, and Family*, edited by S. M. Hoover, L. S. Clark, and D. Alters. New York: Routledge.

Clarke, Alison J. 2004. "Maternity and Materiality: Becoming a Mother in Consumer Culture," pp. 55–71 in *Consuming Motherhood*, edited by J. S. Taylor, L. L. Layne, and D. F. Wozniak. New Brunswick, NJ: Rutgers University Press.

Clarke, Alison J. 2007. "Coming of Age in Suburbia: Gifting the Consumer Child," pp. 253–68 in *Designing Modern Childhoods*, edited by M. Gutman and N. de Coninck-Smith. New Brunswick, NJ: Rutgers University Press.

Cohen, Jere M. 1977. "Sources of Peer Group Homogeneity." *Sociology of Education* 50(4):227–41.

Cohen, Stanley. 1972. *Folk Devils and Moral Panics*. London, UK: MacGibbon and Kee.

Cole, Michael. 1997. *Cultural Psychology: A Once and Future Discipline*. Cambridge, MA: Harvard University Press.

Collins, James. 1995. "Literacy and Literacies." *Annual Review of Anthropology* 24(1):75–93.

Collins, W. Andrew, and L. Alan Sroufe. 1999. "Capacity for Intimate Relationships: A Developmental Construction," pp. 125–47 in *The Development of Romantic Relationships in Adolescence*, edited by W. Furman, B. B. Brown, and C. Feiring. Cambridge, UK: Cambridge University Press.

Compaine, Benjamin. 2001. *The Digital Divide: Facing a Crisis or Creating a Myth?* Cambridge, MA: The MIT Press.

Connolly, Jennifer, and Adele Goldberg. 1999. "Romantic Relationships in Adolescence: The Role of Friends and Peers in Their Emergence and Development," pp. 266–90 in *The Development of Romantic Relationships in Adolescence*, edited by W. Furman, B. B. Brown, and C. Feiring. Cambridge, UK: Cambridge University Press.

Consalvo, Mia. 2007. *Cheating: Gaining Advantage in Videogames*. Cambridge, MA: The MIT Press.

Constable, Nicole. 2003. *Romance on a Global Stage: Pen Pals, Virtual Ethnography, and "Mail-Order" Marriages*. Berkeley, CA: University of California Press.

Coontz, Stephanie. 1992. *The Way We Never Were: American Families and the Nostalgia Trap*. New York: Basic Books.

Corsaro, William A. 1985. *Friendship and Peer Culture in the Early Years.* Norwood, NJ: Ablex.

Corsaro, William A. 1997. *The Sociology of Childhood.* Thousand Oaks, CA: Pine Forge Press.

Cotterell, John. 1996. *Social Networks and Social Influences in Adolescence.* London, UK, and New York: Routledge.

Cross, Gary. 1997. *Kids' Stuff: Toys and the Changing World of American Childhood.* Cambridge, MA: Harvard University Press.

Darrah, Charles N., James M. Freeman, and Jan A. English-Lueck. 2007. *Busier Than Ever! Why American Families Can't Slow Down.* Palo Alto, CA: Stanford University Press.

di Gennaro, Corinna, and William Dutton. 2007. "Reconfiguring Friendships: Social Relationships and the Internet." *Information, Communication & Society* 10(5): 591–618.

Diamond, Lisa M., Ritch Savin-Williams, and Eric M. Dube. 1999. "Sex, Dating, Passionate Friendships, and Romance: Intimate Peer Relations among Lesbian, Gay, and Bisexual Adolescents," pp. 175–210 in *The Development of Romantic Relationships in Adolescence,* edited by W. Furman, B. B. Brown, and C. Feiring. Cambridge, UK: Cambridge University Press.

Dibbell, Julian. 2006. *Play Money: Or How I Quit My Job and Made Millions Trading Virtual Loot.* New York: Basic Books.

Donath, Judith, and danah boyd. 2004. "Public Displays of Connection." *BT Technology Journal* 22(4):71–82.

Dourish, Paul, and Genevieve Bell. 2007. "The Infrastructure of Experience and the Experience of Infrastructure: Meaning and Structure in Everyday Encounters with Space." *Environment and Planning B: Planning and Design* 34(3):414–30.

Drucker, Peter F. 1994. "The Age of Social Transformation." *The Atlantic Monthly* 274(5):53–80.

Dyson, Anne Haas. 1997. *Writing Superheroes: Contemporary Childhood, Popular Culture, and Classroom Literacy.* New York: Teachers College Press.

Eagleton, Maya B., and Elizabeth Dobler. 2007. *Reading the Web: Strategies for Internet Inquiry.* New York: Guilford Press.

Eckert, Penelope. 1989. *Jocks and Burnouts: Social Categories and Identity in the High School.* New York: Teachers College.

Eckert, Penelope. 1996. "Vowels and Nail Polish: The Emergence of Linguistic Style in the Preadolescent Heterosexual Marketplace," pp. 183–90 in *Gender and Belief*

Systems, edited by N. Warner, J. Ahlers, L. Bilmes, M. Olver, S. Wertheim, and M. Chen. Berkeley, CA: Berkeley Women and Language Group.

Eco, Umberto. 1979. *The Role of the Reader: Explorations in the Semiotics of Texts.* Bloomington, IN: Indiana University Press.

Edley, Nigel, and Margaret Wetherell. 1997. "Jockeying for Position: The Construction of Masculine Identities." *Discourse & Society* 8(2):203–21.

Edwards, Paul. 1995. "From 'Impact' to Social Process: Computers in Society and Culture," pp. 257–285 in *Handbook of Science and Technology Studies*, edited by S. Jasanoff, G. E. Markle, J. C. Petersen, and T. Pinch. Thousand Oaks, CA: Sage Publications.

Entertainment Software Association. 2007. *Facts and Research. Top Ten Facts 2007.* Retrieved June 28, 2007 (http://www.theesa.com/facts/index.asp).

Epstein, Jonathon S., ed. 1998. *Youth Culture: Identity in a Postmodern World.* Malden, MA: Blackwell Press.

Escobar, Arturo. 1994. "Welcome to Cyberia: Notes on the Anthropology of Cyberculture." *Current Anthropology* 35(3):211–31.

Fabos, Bettina. 2004. *Wrong Turn on the Information Superhighway: Education and the Commercialization of the Internet.* New York: Teachers College Press.

Faulkner, Susan, and Jay Melican. 2007. "Getting Noticed, Showing-Off, Being Overheard: Amateurs, Authors and Artists Inventing and Reinventing Themselves in Online Communities," pp. 46–59 in *EPIC Proceedings*. The National Association for the Practice of Anthropology conference, October 3–6, 2007, Keystone, CO.

Fine, Gary Alan. 2004. "Adolescence as Cultural Toolkit: High School Debate and the Repertoires of Childhood and Adulthood." *The Sociological Quarterly* 24(1):1–20.

Florida, Richard. 2003. *The Rise of the Creative Class: And How It's Transforming Work, Leisure, Community, and Everyday Life.* New York: Basic Books.

Frank, Thomas. 1997. *The Conquest of Cool: Business Culture, Counterculture, and the Rise of Hip Consumerism.* Chicago, IL: University of Chicago Press.

Gee, James Paul. 1990. *Social Linguistics and Literacies: Ideology in Discourses.* London, UK: Falmer Press.

Gee, James Paul. 2003. *What Video Games Have to Teach Us about Learning and Literacy.* New York: Palgrave Macmillan.

Gee, James Paul. 2008. *Getting Over the Slump: Innovation Strategies to Promote Children's Learning.* New York: The Joan Ganz Cooney Center at Sesame Workshop.

Giddens, Anthony. 1986. *The Constitution of Society: Outline of the Theory of Structuration.* Berkeley, CA: University of California Press.

Gilbert, James Burkhart. 1986. *A Cycle of Outrage: America's Reaction to the Juvenile Delinquent in the 1950s.* New York: Oxford University Press.

Gilroy, Paul. 1987. *"There Ain't No Black in the Union Jack": The Cultural Politics of Race and Nation.* Chicago, IL: University of Chicago Press.

Ginsburg, Faye D., Lila Abu-Lughod, and Brian Larkin, eds. 2002. *Media Worlds: Anthropology on New Terrain.* Berkeley and Los Angeles, CA: University of California Press.

Goggin, Gerard. 2006. *Cell Phone Culture: Mobile Technology in Everyday Life.* London, UK: Routledge.

Goldman, Shelley. 2005. "A New Angle on Families: Connecting the Mathematics of Life with School Mathematics," pp. 55–76 in *Learning in Places: The Informal Education Reader,* edited by Z. Bekerman, N. C. Burbules, and D. Silberman-Keller. New York: Peter Lang.

Goodman, Steven. 2003. *Teaching Youth Media: A Critical Guide to Literacy, Video Production and Social Change.* New York: Teacher's College Press.

Graves, Michael, Connie Juel, and Bonnie Graves. 2001. *Teaching Reading in the 21st Century.* 2nd ed. Boston, MA: Allyn & Bacon.

Gray, Mary L. 2009. *Out in the Country: Youth, Media, and the Queering of Rural America.*

Greeno, James. 1997. "On Claims That Answer the Wrong Question." *Educational Researcher* 26(1):5–17.

Griffith, Maggie, and Susannah Fox. 2007. "Hobbyists Online." *Pew Internet & American Life Project.* Washington, DC: Pew/Internet.

Grinter, Rebecca E., Leysia Palen, and Margery Eldridge. 2006. "Chatting with Teenagers: Considering the Place of Chat Technologies in Teen Life." *ACM Transactions on Computer-Human Interaction* 13(4):423–47.

Gunter, Barrie, and Jill McAleer. 1997. *Children and Television.* 2nd ed. New York: Routledge.

Gupta, Akhil, and James Ferguson. 1997. *Culture Power Place: Explorations in Critical Anthropology.* Durham, NC, and London, UK: Duke University Press.

Gutman, Marta, and Ning de Coninck-Smith, eds. 2007. *Designing Modern Childhoods: History, Space and the Material Culture of Children.* New Brunswick, NJ: Rutgers University Press.

Haddon, Leslie. 2004. *Information and Communication Technologies in Everyday Life.* Oxford, UK, and New York: Berg.

Hall, Stuart, and Tony Jefferson, eds. 1975. *Resistance through Rituals: Youth Subcultures in Post-War Britain.* Milton Keynes, UK: Open University Press.

Hampton, Keith. 2007. "Neighborhoods in the Network Society: The e-Neighbors Study." *Information, Communication & Society* 10(5):714–48.

Hampton, Keith, and Barry Wellman. 2003. "Neighboring in Netville: How the Internet Supports Community and Social Capital in a Wired Suburb." *City and Community* 2(4):277–311.

Hansen, David, Jeylan Mortimer, and Helga Krüger. 2001. "Adolescent Part-Time Employment in the United States and Germany: Diverse Outcomes, Contexts, and Pathways," pp. 121–38 in *Hidden Hands: International Perspectives on Children's Work and Labour*, edited by P. Mizen, C. Pole, and A. Bolton. New York: Routledge Farmer.

Hargittai, Eszter. 2004. Do You "Google"? Understanding Search Engine Popularity Beyond the Hype. *First Monday.* 9(3) Retrieved June 28, 2008 (http://firstmonday. org/issues/issue9_3/hargittai/index.html).

Hargittai, Eszter. 2007.The Social, Political, Economic, and Cultural Dimensions of Search Engines: An Introduction. *Journal of Computer-Mediated Communication* 12(3), article 1. Retrieved June 28, 2008 (http://jcmc.indiana.edu/vol12/issue3/hargittai. html).

Hargittai, Eszter, and Amanda Hinnant. 2006. "Toward a Social Framework for Information Seeking," pp. 55–70 in *New Directions in Human Information Behavior*, edited by A. Spink and C. Cole. New York: Springer.

Hartup, William. 1999. "Forward," pp. xi–xv in *The Development of Romantic Relationships in Adolescence*, edited by W. Furman, B. B. Brown, and C. Feiring. Cambridge, UK: Cambridge University Press.

Hebdige, Dick. 1979. *Subculture: The Meaning of Style*. London, UK: Routledge.

Hebdige, Dick. 1987. *Cut 'n Mix: Culture, Identity and Caribbean Music*. London, UK: Routledge.

Henderson, Carol C., and Frederick D. King. 1995. "The Role of Public Libraries in Providing Public Access to the Internet," pp. 154–74 in *Public Access to the Internet*, edited by B. Kahin and J. Keller. Cambridge, MA: The MIT Press.

Hertz, J. C. 2002. "Harnessing the Hive: How Online Games Drive Networked Innovation." *Release 1.0.* 20(9):1–24.

Hijazi-Omari, Hiyam, and Rivka Ribak. 2008. "Playing with Fire: On the Domestication of the Mobile Phone among Palestinian Girls in Israel." *Information, Communication & Society* 11(2):149–66.

Hillier, Lynne, and Lyn Harrison. 2007. "Building Realities Less Limited Than Their Own: Young People Practising Same-Sex Attraction on the Internet." *Sexualities* 10(1):82–100.

Hillier, Lynne, Lyn Harrison, and Kate Bowditch. 1999. "'Neverending Love' and 'Blowing Your Load': The Meanings of Sex to Rural Youth." *Sexualities* 2(1):69–88.

Hinduja, Sameer, and Justin Patchin. 2008. "Cyberbullying: An Exploratory Analysis of Factors Related to Offending and Victimization." *Deviant Behavior* 29(2):129–56.

Hine, Christine. 2000. *Virtual Ethnography.* Thousand Oaks, CA: Sage Publications.

Hine, Thomas. 1999. *The Rise and Fall of the American Teenager.* New York: Perennial.

Hird, Myra J., and Sue Jackson. 2001. "Where 'Angels' and 'Wusses' Fear to Tread: Sexual Coercion in Adolescent Dating Relationships." *Journal of Sociology* 37 (1):27–43.

Hobbs, Renee. 1998. "The Seven Great Debates in the Media Literacy Movement." *Journal of Communication* 48(1):16–32. Retrieved June 26, 2008 (http://reneehobbs. org/renee's%20web%20site/Publications/final%20seven%20great%20debates.htm).

Hochschild, Arlie Russell. 2003. *The Second Shift.* New York: Penguin.

Holloway, Sarah L., and Gill Valentine. 2003. *Cyberkids: Children in the Information Age.* London, UK: RoutledgeFalmer.

Hood, Lee, Lynn Schofield Clark, Joseph G. Champ, and Diane Alters. 2004. "The Case Studies: An Introduction," pp. 69–78 in *Media, Home, and Family*, edited by S. M. Hoover, L. S. Clark, and D. Alters. New York: Routledge.

Hoover, Stewart M., Lynn Schofield Clark, and Diane Alters (with Joseph G. Champ and Lee Hood). 2004. *Media, Home, and Family.* New York: Routledge.

Horst, Heather A. 2006. "The Blessings and Burdens of Communication: The Cell Phone in Jamaican Transnational Social Fields." *Global Networks: A Journal of Transnational Affairs* 6(2):143–59.

Horst, Heather A. 2007. "Mediating the Generation Gap: Creativity and Communitas in the Digital Family." Society for the Social Studies of Science Conference, October 11, 2007, Montreal, Canada.

Horst, Heather A., and Daniel Miller. 2005. "From Kinship to Link-Up: Cell Phones and Social Networking in Jamaica." *Current Anthropology* 46(5):755–78.

Horst, Heather A., and Daniel Miller. 2006. *The Cell Phone: An Anthropology of Communication.* Oxford, UK, and New York: Berg.

Howe, Jeff. 2006. "The Rise of Crowdsourcing." *Wired* 14(6). Retrieved May 14, 2008 (http://www.wired.com/wired/archive/14.06/crowds.html).

Howes, Carollee. 1996. "The Earliest Friendships," pp. 66–86 in *The Company They Keep: Friendship in Childhood and Adolescence*, edited by W. M. Bukowski, A. F.

Newcomb, and W. W. Hartup. Cambridge, UK, and New York, NY: Cambridge University Press.

Hull, Glynda. 2003. "At Last Youth Culture and Digital Media: New Literacies for New Times." *Research in the Teaching of English* 38(2):229–33.

Hull, Glynda, and Katherine Schultz. 2002a. "Negotiating Boundaries between School and Non-School Literacies," pp. 1–10 in *Schools Out! Bridging Out-of-School Literacies with Classroom Practice*, edited by G. Hull and K. Schultz. New York: Teachers College Press.

Hull, Glynda, and Katherine Schultz, eds. 2002b. *School's Out!: Bridging Out-of-School Literacies with Classroom Practice*. New York: Teachers College Press.

Hutchins, Edwin. 1995. *Cognition in the Wild*. Cambridge, MA: The MIT Press.

Ito, Mizuko. 2003. "Engineering Play: Children's Software and the Productions of Everyday Life." Ph.D. dissertation, Anthropology, Stanford University, Palo Alto, CA.

Ito, Mizuko. 2006. "Japanese Media Mixes and Amateur Cultural Exchange," pp. 49–66 in *Digital Generations*, edited by D. Buckingham and R. Willett. Mahweh, NJ: Lawrence Erlbaum.

Ito, Mizuko. 2007. "Education V. Entertainment: A Cultural History of Children's Software," pp. 89–116 in *The Ecology of Games: Connecting Youth, Games, and Learning*, edited by K. Salen. The John D. and Catherine T. MacArthur Foundation Series on Digital Media and Learning. Cambridge, MA: The MIT Press.

Ito, Mizuko. 2008a. "Introduction," pp. 1–14 in *Networked Publics*, edited by K. Varnelis. Cambridge, MA: The MIT Press.

Ito, Mizuko. 2008b. "Mobilizing the Imagination in Everyday Play: The Case of Japanese Media Mixes," pp. 397–412 in *The International Handbook of Children, Media, and Culture*, edited by K. Drotner and S. Livingstone. Thousand Oaks, CA: Sage Publications.

Ito, Mizuko. 2009. *Engineering Play: A Cultural History of Children's Software*. Cambridge, MA: The MIT Press.

Ito, Mizuko, Vicki O'Day, Annette Adler, Charlotte Linde, and Elizabeth Mynatt. 2001. "Making a Place for Seniors on the Net: SeniorNet, Senior Identity, and the Digital Divide." *Computers and Society* 31(3):15–21.

Ito, Mizuko, and Daisuke Okabe. 2005a. "Intimate Connections: Contextualizing Japanese Youth and Mobile Messaging," pp. 127–43 in *Inside the Text: Social, Cultural and Design Perspectives on SMS*, edited by R. Harper, L. Palen, and A. Taylor. New York: Springer.

Ito, Mizuko, and Daisuke Okabe. 2005b. "Technosocial Situations: Emergent Structurings of Mobile Email Use," pp. 257–73 in *Personal, Portable, Pedestrian: Mobile*

Phones in Japanese Life, edited by M. Ito, D. Okabe, and M. Matsuda. Cambridge, MA: The MIT Press.

Ito, Mizuko, Daisuke Okabe, and Ken Anderson. 2009. "Portable Objects in Three Global Cities: The Personalization of Urban Places," pp. 67–88 in *The Mobile Communication Research Annual Volume 1: The Reconstruction of Space and Time through Mobile Communication Practices*, edited by R. Ling and S. Campbell. Edison, NJ: Transaction Publishers.

Ito, Mizuko, Daisuke Okabe, and Misa Matsuda, eds. 2005. *Personal, Portable, Pedestrian: Mobile Phones in Japanese Life*. Cambridge, MA: The MIT Press.

Jackson, Stevi. 1998. "Heterosexuality and Feminist Theory," pp. 21–38 in *Theorising Heterosexuality*, edited by D. Richardson. Buckingham, UK: Open University Press.

James, Allison, Chris Jenks, and Alan Prout. 1998. *Theorising Childhood*. Cambridge, UK: Polity Press.

James, Allison, and Alan Prout, eds. 1997. *Constructing and Reconstructing Childhood: Contemporary Issues in the Sociological Study of Childhood*. Philadelphia, PA: RoutledgeFarmer.

Jansz, Jeroen, and Lonneke Martens. 2005. "Gaming at a LAN Event: The Social Context of Playing." *New Media Society* 7(3):33–55.

Jenkins, Henry. 1992. *Textual Poachers: Television Fans and Participatory Culture*. New York and London, UK: Routledge.

Jenkins, Henry. 1998. "Introduction: Childhood Innocence and Other Modern Myths," pp. 1–37 in *The Children's Culture Reader*, edited by H. Jenkins. New York: New York University Press.

Jenkins, Henry, ed. 2000. *The Children's Culture Reader*. New York: New York University Press.

Jenkins, Henry. 2006. *Convergence Culture: Where Old and New Media Collide*. New York: New York University Press.

Jenkins, Henry, Ravi Purushotma, Katherine Clinton, Margaret Weigel, and Alice J. Robinson. 2006. *Confronting the Challenges of Participatory Culture: Media Education for the 21st Century*. Retrieved July 16, 2007 (http://www.projectnml.org/files/working/NMLWhitePaper.pdf).

Jenkins, Richard. 1983. *Lads, Citizens, and Ordinary Kids: Working-Class Youth Life-Styles in Belfast*. London, UK: Routledge & Kegan Paul.

Kafai, Yasmin B., Carrie Heeter, Jill Denner, and Jennifer Y. Sun. 2008. *Beyond Barbie and Mortal Kombat: New Perspectives on Gender and Gaming*. Cambridge, MA: The MIT Press.

Karaganis, Joe. 2007. "Presentation," pp. 8–16 in *Structures of Participation in Digital Culture*, edited by J. Karaganis. New York: Social Science Research Council.

Katz, James. 2006. *Magic in the Air: Mobile Communication and the Transformation of Social Life*. Edison, NJ: Transaction Publishers.

Kearney, Mary Celeste. 2006. *Girls Make Media*. London, UK: Routledge.

Kendall, Lori. 2002. *Hanging Out in the Virtual Pub*. Berkeley and Los Angeles, CA: University of California Press.

Kinder, Marsha. 1991. *Playing with Power in Movies, Television, and Video Games*. Berkeley, CA: University of California Press.

Kinder, Marsha. 1999. "Kids' Media Culture: An Introduction," pp. 1–12 in *Kids' Media Culture*, edited by M. Kinder. Durham, NC: Duke University Press.

Kinder, Marsha, ed. 1999. *Kids' Media Culture*. Durham, NC: Duke University Press.

Kinney, David A. 1993. "From Nerds to Normals: The Recovery of Identity among Adolescents from Middle School to High School." *Sociology of Education* 66(1):21–40.

Kline, Stephen. 1993. *Out of the Garden: Toys and Children's Culture in the Age of TV Marketing*. New York: Verso.

Kline, Stephen, Nick Dyer-Witheford, and Gerig de Peuter. 2003. *Digital Play: The Interaction of Technology, Culture, and Marketing*. Montreal, Canada: McGill-Queen's University Press.

Korobov, Neill, and Avril Thorne. 2006. "Intimacy and Distancing: Young Men's Conversations about Romantic Relationships." *Journal of Adolescent Research* 21(1):27–55.

Koskinen, Ilpo, Esko Kurvinen, and Turo-Kimmo Lohtonen. 2002. *Mobile Image*. Helsinki, Finland: Edita Prima.

Kutner, Lawrence, and Cheryl K. Olson. 2008. *Grand Theft Childhood: The Surprising Truth about Violent Video Games*. New York: Simon and Schuster.

Lally, Elaine. 2002. *At Home with Computers*. Oxford, UK, and New York: Berg.

Lange, Patricia G. 2003. "Virtual Trouble: Negotiating Access in Online Communities." Ph.D. dissertation, Anthropology, University of Michigan, Ann Arbor, MI.

Lange, Patricia G. 2007a. "Commenting on Comments: Investigating Responses to Antagonism on YouTube." Society for Applied Anthropology Annual Conference, March 31, 2007, Tampa, FL. Retrieved May 13, 2008 (http://sfaapodcasts.files .wordpress.com/2007/04/update-apr-17-lange-sfaa-paper-2007.pdf).

Lange, Patricia G. 2007b. "How 'Tubers Teach Themselves: Narratives of Self-Teaching as Technical Identity Performance on YouTube." Society for the Social Studies of Science Conference, October 11, 2007, Montreal, Canada.

Lareau, Annette. 2003. *Unequal Childhoods: Class, Race and Family Life*. Berkeley, CA: University of California Press.

Lave, Jean. 1988. *Cognition in Practice*. New York: Cambridge University Press.

Lave, Jean, and Etienne Wenger. 1991. *Situated Learning: Legitimate Peripheral Participation*. Cambridge, UK, and New York: Cambridge University Press.

Leadbeater, Charles, and Paul Miller. 2004. *The Pro-Am Revolution: How Enthusiasts Are Changing Our Economy and Society*. London, UK: Demos.

Leander, Kevin M., and Kelly K. McKim. 2003. "Tracing the Everyday 'Sitings' of Adolescents on the Internet: A Strategic Adaptation of Ethnography across Online and Offline Spaces." *Education, Communication & Information* 3(2):211–40.

Lenhart, Amanda, Sousan Arafeh, Aaron Smith, and Alexandra Rankin Macgill. 2008. "Writing, Technology and Teens." *Pew Internet & American Life Project*. Washington, DC: Pew/Internet. Retrieved June 6, 2008 (http://www.pewinternet .org/pdfs/PIP_Writing_Report_FINAL3.pdf).

Lenhart, Amanda, and Mary Madden. 2007. "Social Networking Websites and Teens: An Overview." *Pew Internet & American Life Project*. Washington, DC: Pew/ Internet.

Lenhart, Amanda, Mary Madden, and Paul Hitlin. 2005. "Teens and Technology: Youth Are Leading the Transition to a Fully Wired and Mobile Nation." *Pew Internet & American Life Project*. Retrieved June 6, 2008 (http://www.pewinternet.org/pdfs/ PIP_Teens_Tech_July2005web.pdf).

Lenhart, Amanda, Mary Madden, Alexandra Rankin Macgill, and Aaron Smith. 2007. "Teens and Social Media: The Use of Social Media Gains a Greater Foothold in Teen Life as They Embrace the Conversational Nature of Interactive Online Media." *Pew Internet & American Life Project*. Washington, DC: Pew/Internet. Retrieved June 6, 2008 (http://www.pewinternet.org/pdfs/PIP_Teens_Social_Media_Final.pdf).

Lenhart, Amanda, Lee Rainie, and Oliver Lewis. 2001. "Teenage Life Online: The Rise of the Instant-Message Generation and the Internet's Impact on Friendships and Family Relationships." *Pew Internet & American Life Project*. Retrieved June 6, 2008 (http://www.pewinternet.org/pdfs/PIP_Teens_Report.pdf).

Lesko, Nancy. 2001. *Act Your Age! A Cultural Construction of Adolescence*. New York: Routledge Farmer.

Lessig, Lawrence. 2004. *Free Culture: How Big Media Uses Technology and the Law to Lock Down Culture and Control Creativity*. New York: Penguin Press.

Levine, Judith. 2002. *Harmful to Minors: The Perils of Protecting Children from Sex.* Minneapolis, MN: University of Minnesota Press.

Lewis, George H. 1990. "Community through Exclusion and Illusion: The Creation of Social Worlds in an American Shopping Mall." *The Journal of Popular Culture* 24(2):121–36.

Lievrouw, Leah A., and Sonia Livingstone, eds. 2002. *Handbook of New Media: Social Shaping and Social Consequences of ICTs.* London, UK: Sage.

Ling, Rich. 2004. *The Mobile Connection: The Cell Phone's Impact on Society.* San Francisco, CA: Morgan Kaufman.

Ling, Rich. 2008. *New Tech, New Ties: How Mobile Communication Is Reshaping Social Cohesion.* Cambridge, MA: The MIT Press.

Ling, Rich, and Tom Julsrud. 2005. "Grounded Genres in Multimedia Messaging," pp. 329–38 in *A Sense of Place: The Global and the Local in Mobile Communication*, edited by K. Nyiri. Vienna, Austria: Passagen Verlag.

Ling, Rich, and Brigitte Yttri. 2006. "Control, Emancipation, and Status: The Mobile Telephone in Teens' Parental and Peer Relationships," pp. 219–35 in *Computers, Phones, and the Internet: Domesticating Information Technology*, edited by R. Kraut, M. Brynin, and S. Kiesler. Cambridge, MA: Oxford University Press.

Livingstone, Sonia. 2002. *Young People and New Media.* London, UK, and Thousand Oaks, CA: Sage Publications.

Livingstone, Sonia. 2003. "Mediated Childhoods: A Comparative Approach to the Lifeworld of Young People in a Changing Media Environment," pp. 207–26 in *The Wired Homestead: An MIT Sourcebook on the Internet and the Family*, edited by J. Turow and A. L. Kavanaugh. Cambridge, MA: The MIT Press.

Livingstone, Sonia. 2008. "Taking Risky Opportunities in Youthful Content Creation: Teenagers' Use of Social Networking Sites for Intimacy, Privacy, and Self-Expression." *New Media & Society* 10(3):393–411.

Livingstone, Sonia, and Moira Bovill. 2001. *Children and their Changing Media Environment: A European Comparative Study.* Mahwah, NJ: Lawrence Erlbaum Associates.

Low, Setha. 2003. *Behind the Gates: Life, Security, and the Pursuit of Happiness in Fortress America.* London, UK: Routledge.

Low, Setha. 2008. "Incorporation and Gated Communities in the Greater Metro-Los Angeles Region as a Model of Privatization of Residential Communities." *Home Cultures: The Journal of Architecture, Design and Domestic Space* 5(1):85–108.

Lowood, Henry. 2007. "Found Technology: Players as Innovators in the Making of Machinima," pp. 165–96 in *Digital Youth, Innovation, and the Unexpected*, edited by T. McPherson. Cambridge, MA: The MIT Press.

Lunenfeld, Peter. 2000. *The Digital Dialectic: New Essays on New Media*. Cambridge, MA: The MIT Press.

Lusted, David. 1991. *The Media Studies Book*. London, UK, and New York: Routledge.

Macgill, Alexandra R. 2007. "Parent and Teenager Internet Use." *Pew Internet & American Life Project*. Retrieved June 6, 2008 (http://www.pewinternet.org/pdfs/PIP_Teen_Parents_data_memo_Oct2007.pdf).

Maczewski, Mechthild. 2002. "Exploring Identities through the Internet: Youth Experiences Online." *Child & Youth Care Forum* 31(2):111–29.

Mahendran, Dilan. 2007. "Music Listening/Making as Non-Representational Learning." Digital Youth Presentation, May 10, 2007, Los Angeles, CA.

Mahiri, Jabari, ed. 2004. *What They Don't Learn in School: Literacy in the Lives of Urban Youth*. New York: Peter Lang.

Mahiri, Jabari, Malik Ali, Allison Lindsay Scott, Bolota Asmerom, and Rick Ayers, R. 2008. "Both Sides of the Mic: Community Literacies in the Age of Hip-Hop," pp. 279–87 in *Research on Teaching Literacy through the Communicative, Visual, and Performing Arts*, Vol. 2, edited by J. Flood, S. B. Heath, and D. Lapp. International Reading Association. Mahwah, NJ: Lawrence Erlbaum.

Maira, Sunaina and Elisabeth Soep, eds. 2004. *Youthscapes: The Popular, the National, the Global*. Philadelphia, PA: University of Pennsylvania Press.

Mankekar, Purnima. 1999. *Screening Culture, Viewing Politics: An Ethnography of Television, Womanhood, and Nation in Postcolonial India*. Durham, NC: Duke University Press.

Marcus, George E. 1995. "Ethnography in/of the World System: The Emergence of Multi-Sited Ethnography." *Annual Review of Anthropology* 24(1):95–117.

Marcus, George E. 1998. *Ethnography through Thick and Thin*. Princeton, NJ: Princeton University Press.

Martin, Karin. 1996. *Puberty, Sexuality and the Self: Girls and Boys at Adolescence*. New York: Routledge.

Marwick, Alice. 2008. "To Catch a Predator? The MySpace Moral Panic." *First Monday* 13(6). Retrieved June 26, 2008 (http://www.uic.edu/htbin/cgiwrap/bin/ojs/index.php/fm/article/view/2152/1966).

Matsuda, Misa. 2005. "Mobile Communication and Selective Sociality," pp. 123–42 in *Personal, Portable, Pedestrian: Mobile Phones in Japanese Life*, edited by M. Ito, D. Okabe, and M. Matsuda. Cambridge, MA: The MIT Press.

Matsuda, Misa. 2007. "Mobile Media and the Transformation of Family." Keynote speech presented at Mobile Media 2007, July 4, University of Sydney, Australia.

Mazzarella, Sharon, ed. 2005. *Girl Wide Web: Girls, the Internet, and the Negotiation of Identity.* New York: Peter Lang.

McKechnie, Jim, and Sandy Hobbs. 2001. "Work and Education: Are They Compatible for Children and Adolescents?" pp. 9–23 in *Hidden Hands: International Perspectives on Children's Work and Labour,* edited by P. Mizen, C. Pole, and A. Bolton. New York: Routledge Farmer.

McLuhan, Marshall. 1994. *Understanding Media: The Extensions of Man.* Cambridge, MA: MIT Press.

McPherson, Miller, Lynn Smith-Lovin, and James M. Cook. 2001. "Birds of a Feather: Homophily in Social Networks." *Annual Review of Sociology* 27(1):415–44.

McRobbie, Angela. 1980. "Settling Accounts with Subcultures: A Feminist Critique." *Screen Education* 34(1):37–49.

McRobbie, Angela, and Jenny Garber. [1978] 2000. "Girls and Subcultures," pp. 12–25 in *Feminism and Youth Subcultures,* edited by A. McRobbie. 2nd ed. London, UK: Routledge.

Medrano, Luisa. 1994. "AIDS and Latino Adolescents," pp. 100–14 in *Sexual Cultures and the Construction of Adolescent Identities: Health, Society and Policy,* edited by J. M. Irvine. Philadelphia, PA: Temple University Press.

Mesch, Gustavo S., and Ilan Talmud. 2007. "Similarity and the Quality of Online and Offline Social Relations among Adolescents in Israel." *Journal of Research in Adolescence* 17(2):455–66.

Meyrowitz, Joshua. 1986. *No Sense of Place: The Impact of Electronic Media on Social Behavior.* New York: Oxford University Press.

Miller, Brent C., and Brad Benson. 1999. "Romantic and Sexual Relationship Development during Adolescence," pp. 99–121 in *The Development of Romantic Relationships in Adolescence,* edited by W. Furman, B. B. Brown, and C. Feiring. Cambridge, UK: Cambridge University Press.

Miller, Daniel. 2001. *Home Possessions: Material Culture behind Closed Doors.* Oxford, UK, and New York: Berg.

Miller, Daniel, and Don Slater. 2000. *The Internet: An Ethnographic Approach.* Oxford, UK: Berg Publishers.

Milner, Murray, Jr. 2004. *Freaks, Geeks, and Cool Kids: American Teenagers, Schools, and the Culture of Consumption.* New York: Routledge.

Mintz, Steven. 2004. *Huck's Raft: A History of American Childhood.* Cambridge, MA: Harvard University Press.

Mintz, Steven, and Susan Kellogg. 1988. *Domestic Revolutions: A Social History of American Family Life*. New York: The Free Press.

Miyaki, Yukiko. 2005. "*Keitai* Use among Japanese Elementary and Junior High School Students," pp. 277–99 in *Personal, Portable, Pedestrian: Mobile Phones in Japanese Life*, edited by M. Ito, D. Okabe, and M. Matsuda. Cambridge, MA: The MIT Press.

Mizen, Phillip, Christopher Pole, and Angela Bolton. 2001. "Why Be a School Age Worker?" pp. 37–54 in *Hidden Hands: International Perspectives on Children's Work and Labour*, edited by P. Mizen, C. Pole, and A. Bolton. New York: Routledge Farmer.

Modell, John. 1989. *Into One's Own: From Youth to Adulthood in the United States, 1920–1975*. Berkeley, CA: University of California Press.

Montemayor, Raymond, and Roger Van Komen. 1980. "Age Segregation of Adolescents In and Out of School." *Journal of Youth and Adolescence* 9(5):371–81.

Moran, Jeffrey. 2000. *Teaching Sex: The Shaping of Adolescence in the 20th Century*. Cambridge, MA: Harvard University Press.

Morley, David. 2000. *Home Territories: Media, Mobility and Identity*. London, UK: Routledge.

Morrell, Ernest, and Jeff Duncan-Andrade. 2004. "What They Do Learn in School: Hip Hop as a Bridge to Canonical Poetry," pp. 247–68 in *What They Don't Learn in School: Literacy in the Lives of Urban Youth*, edited by J. Mahiri. New York: Peter Lange.

Netting, Robert, Richard Wilk, and Eric Arnould. 1984. *Households: Comparative and Historical Studies of the Domestic Group*. Berkeley, CA, and London, UK: University of California Press.

Newcomb, Andrew F., and Catherine L. Bagwell. 1996. "The Developmental Significance of Children's Friendship Relations," pp. 289–321 in *The Company They Keep: Friendship in Childhood and Adolescence*, edited by W. M. Bukowski, A. F. Newcomb, and W. W. Hartup. Cambridge, UK, and New York: Cambridge University Press.

New London Group, The. 1996. "A Pedagogy of Multiliteracies." *Harvard Educational Review* 66(1):60–92.

Nocon, Honorine, and Michael Cole. 2005. "School's Invasion of "After School": Colonization, Rationalization, or Expansion of Access?" pp. 99–122 in *Learning in Places: The Informal Education Reader*, edited by Z. Bekerman, N. C. Burbules, and D. Silberman-Keller. New York: Peter Lang.

Nunes, Terezinha, Analucia Dias Schliemann, and David William Carraher. 1993. *Street Mathematics and School Mathematics*. Cambridge, UK: Cambridge University Press.

Okabe, Daisuke, and Mizuko Ito. 2006. "Everyday Contexts of Camera Phone Use: Steps Toward Techno-Social Ethnographic Frameworks," pp. 79–102 in *Mobile Communications in Everyday Life: Ethnographic Views, Observations, and Reflections*, edited by J. Höflich and M. Hartmann. Berlin, Germany: Frank & Timme.

Oksman, Virpi, and Jussi Turtainen. 2004. "Mobile Communication as a Social Stage." *New Media & Society* 6(3):319–39.

O'Reilly, Tim. 2005. "What Is Web 2.0: Design Patterns and Business Models for the Next Generation of Software." Retrieved June 23, 2008 (http://www.oreillynet.com/pub/a/oreilly/tim/news/2005/09/30/what-is-web-20.html).

Orellana, Marjorie Faulstich. 2001. "The Work Kids Do: Mexican and Central Immigrant Children's Contributions to Households and Schools in California." *Harvard Educational Review* 71(3):366–89.

Ortiz, Steven M. 1994. "Shopping for Sociability in the Mall," pp. 193–99 in *The Community of the Streets*, Supplement 1, *Research in Community Sociology*, edited by S. E. Cahill and L. H. Lofland. Greenwich, CT: JAI Press.

Osgerby, Bill. 2004. *Youth Media*. London, UK, and New York: Routledge.

Pahl, Ray. 2000. *On Friendship*. Cambridge, UK: Polity.

Panagakos, Anastasia N., and Heather A. Horst. 2006. "Return to Cyberia: Technology and the Social Worlds of Transnational Migrants." *Global Networks* 6(2): 109–24.

Pascoe, C. J. 2007a. *"Dude, You're a Fag": Masculinity and Sexuality in High School*. Berkeley and Los Angeles, CA: University of California Press.

Pascoe, C. J. 2007b. "You Have Another World to Create": Teens and Online Hangouts. *Digital Youth Research*, January 12, 2007. Retrieved May 13, 2008 (http://digitalyouth.ischool.berkeley.edu/node/65).

Perkel, Dan. 2008. "Copy and Paste Literacy? Literacy Practices in the Production of a MySpace Profile," pp. 203–24 in *Informal Learning and Digital Media: Constructions, Contexts, Consequences*, edited by K. Drotner, H. S. Jensen, and K. C. Schroeder. Newcastle, UK: Cambridge Scholars Press.

Perry, Pamela. 2002. *Shades of White: White Kids and Racial Identities in High School*. Durham, NC: Duke University Press.

Portelli, Alessandro. 1991. *The Death of Luigi Trastulli and Other Stories: Form and Meaning in Oral History*. Albany, NY: State University of New York Press.

Postman, Neil. 1993. *Technopoly: The Surrender of Culture to Technology*. New York: Vintage.

Prensky, Mark. 2006. *Don't Bother Me, Mom—I'm Learning!* St. Paul, MN: Paragon House.

Qvortrup, Jens. 2001. "School-Work, Paid Work, and the Changing Obligations of Childhood," pp. 91–107 in *Hidden Hands: International Perspectives on Children's Work and Labour*, edited by P. Mizen, C. Pole, and A. Bolton. New York: RoutledgeFarmer.

Radway, Janice. 1984. *Reading the Romance: Women, Patriarchy, and Popular Literature*. Chapel Hill, NC, and London, UK: The University of North Carolina Press.

Rainie, Lee. 2008. "Video Sharing Websites." *Pew Internet & American Life Project*. Washington, DC: Pew/Internet.

Renninger, K. Ann, and Wesley Shumar, eds. 2002. *Building Virtual Communities: Learning and Change in Cyberspace*. New York: Cambridge University Press.

Rheingold, Howard. 2000. *The Virtual Community: Homesteading on the Electronic Frontier*. Cambridge, MA: The MIT Press.

Rideout, Victoria, Donald F. Roberts, and Ulla G. Foehr. 2005. *Generation M: Media in the Lives of 8–18 Year Olds*. Menlo Park, CA: Kaiser Family Foundation. Retrieved June 12, 2008 (http://www.kff.org/entmedia/upload/Executive-Summary-Generation-M-Media-in-the-Lives-of-8–18-Year-olds.pdf).

Roberts, Donald F., and Ulla G. Foehr. 2008. Trends in Media Use. *The Future of Children* 18(1):11–38.

Robinson, Laura. 2007. "Information the Wiki Way: Cognitive Processes of Information Evaluation in Collaborative Online Venues." International Communication Association Conference, May 24–28, San Francisco, CA.

Rogers, Everett M. [1962] 2003. *Diffusion of Innovations*. 5th ed. New York: Free Press.

Rogoff, Barbara. 2003. *The Cultural Nature of Human Development*. New York: Oxford University Press.

Rogoff, Barbara, Karen Topping, Jacquelyn Baker-Sennett, and Pilar Lacasa. 2002. "Mutual Contributions of Individuals, Partners, and Institutions: Planning to Remember in Girl Scout Cookie Sales." *Social Development* 11(2):266–89.

Ross, Andrew. 2007. "Nice Work If You Can Get It: The Mercurial Career of Creative Industry Policy," pp. 17–41 in *My Creativity Reader*, edited by G. Lovink and N. Rossiter. Amsterdam, the Netherlands: Institute of Network Culture.

Russell, Adrienne, Mizuko Ito, Todd Richmond, and Marc Tuters. 2008. "Culture: Media Convergence and Networked Participation," pp. 43–76 in *Networked Publics*, edited by K. Varnelis. Cambridge, MA: The MIT Press.

Salen, Katie. 2007. "Toward an Ecology of Gaming," pp. 1–20 in *The Ecology of Games: Connecting Youth, Games, and Learning*, edited by K. Salen. The John D.

and Catherine T. MacArthur Foundation Series on Digital Media and Learning. Cambridge, MA: The MIT Press.

Schalet, Amy. 2000. "Raging Hormones, Regulated Love: Adolescent Sexuality and the Constitution of the Modern Individual in the United States and the Netherlands." *Body & Society* 6(1):75–105.

Sefton-Green, Julian. 2000. "Introduction: Evaluating Creativity," pp. 1–15 in *Evaluating Creativity: Making and Learning by Young People*, edited by J. Sefton-Green and R. Sinker. London, UK, and New York: Routledge.

Sefton-Green, Julian. 2004. *Literature Review in Informal Learning with Technology Outside School*. London, UK: Futurelab.

Sefton-Green, Julian, and David Buckingham. 1996. "Digital Visions: Children's 'Creative' Uses of Multimedia Technologies." *Convergence* 2(2):47–79.

Seiter, Ellen. 1993. *Sold Separately: Parents and Children in Consumer Culture*. New Brunswick, NJ: Rutgers University Press.

Seiter, Ellen. 1999a. "Power Rangers at Preschool: Negotiating Media in Child Care Settings," pp. 239–62 in *Kids' Media Culture*, edited by M. Kinder. Durham, NC: Duke University Press.

Seiter, Ellen. 1999b. *Television and New Media Audiences*. Oxford, UK: Oxford University Press.

Seiter, Ellen. 2005. *The Internet Playground: Children's Access, Entertainment and Mis-Education*. New York: Peter Lang.

Seiter, Ellen. 2007. "Practicing at Home: Computers, Pianos, and Cultural Capital," pp. 27–52 in *Digital Youth, Innovation, and the Unexpected*, edited by T. McPherson. The John D. and Catherine T. MacArthur Foundation Series on Digital Media and Learning. Cambridge, MA: The MIT Press.

Servon, Lisa. 2002. *Bridging the Digital Divide: Technology, Community and Public Policy*. Melbourne, Australia: Blackwell Publishing.

Shade, Leslie Regan. 1998. "A Gendered Perspective on Access to the Information Infrastructure." *The Information Society* 14(1):33–44.

Shaffer, David. 2006. *How Computer Games Help Children Learn*. New York: Palgrave MacMillan.

Shakib, Sohalia. 2003. "Female Basketball Participation: Negotiating the Conflation of Peer Status and Gender Status from Childhood through Puberty." *American Behavioral Scientist* 46(10):1405–22.

Sharma, Nitasha. 1999. "Down by Law: Responses and Effects of Sampling Restrictions on Rap." *PoLAR: Political and Legal Anthropology Review*, 22(1):1–13.

Sheff, David. 1993. *Game Over: How Nintendo Zapped an American Industry, Captured Your Dollars, and Enslaved Your Children*. New York: Random House.

Shirky, Clay. 2008. *Here Comes Everybody: The Power of Organizing without Organizations*. New York: Penguin Press.

Silverstone, Roger, and Eric Hirsch. 1992. *Consuming Technologies: Media and Information in Domestic Spaces*. London, UK, and New York: Routledge.

Silverstone, Roger, Eric Hirsch, and David Morley. 1992. "Information and Communication Technologies and the Moral Economy of the Household," pp. 15–31 in *Consuming Technologies: Media and Information in Domestic Spaces*, edited by R. Silverstone and E. Hirsch. London, UK, and New York: Routledge.

Sims, Christo. 2007. "Composed Conversations: Teenage Practices of Flirting with New Media." Society for the Social Studies of Science Conference, October 11, 2007, Montreal, Canada.

Singer, Dorothy G., and Jerome L. Singer, eds. 2001. *Handbook of Children and the Media*. Thousand Oaks, CA: Sage Publications.

Singleton, John, ed. 1998. *Learning in Likely Places: Varieties of Apprenticeship in Japan*. New York: Cambridge University Press.

Skelton, Tracey, and Gill Valentine. 1998. *Cool Places: Geographies of Youth Cultures*. London, UK, and New York: Routledge.

Smith, Marc, and Peter Kollock, eds. 1999. *Communities in Cyberspace*. London, UK: Routledge.

Snow, Robert. 1987. "Youth, Rock 'n' Roll and Electronic Media." *Youth & Society* 18(4):326–43.

Spigel, Lynn. 1992. *Make Room for TV: Television and the Family Ideal in Postwar America*. Chicago, IL: University of Chicago Press.

Spigel, Lynn. 2001. *Welcome to the Dreamhouse: Popular Media and Postwar Suburbs (Console-ing Passions)*. Durham, NC: Duke University Press.

Squire, Kurt. 2006. "From Content to Context: Videogames as Designed Experience." *Educational Researcher* 35(8):19–29.

Steele, Jeanne R., and Jane D. Brown. 1995. "Adolescent Room Culture: Studying Media in the Context of Everyday Life." *Journal of Youth and Adolescence* 24(5):551–76.

Stevens, Reed, Tom Satwicz, and Laurie McCarthy. 2007. "In-Game, In-Room, In-World: Reconnecting Video Game Play to the Rest of Kids' Lives," pp. 41–66 in *The Ecology of Games: Connecting Youth, Games, and Learning*, edited by K. Salen. The John D. and Catherine T. MacArthur Foundation Series on Digital Media and Learning. Cambridge, MA: The MIT Press.

Strasburger, Victor, and Barbara Wilson. 2002. *Children, Adolescents, and the Media.* Thousand Oaks, CA: Sage Publications.

Street, Brian V., ed. 1993. *Cross-Cultural Approaches to Literacy.* Cambridge, UK: Cambridge University Press.

Street, Brian V. 1995. *Social Literacies.* London, UK: Longman.

Strunin, Lee. 1994. "Culture, Context, and HIV Infection: Research on Risk Taking among Adolescents," pp. 71–87 in *Sexual Cultures and the Construction of Adolescent Identities: Health, Society and Policy,* edited by J. M. Irvine. Philadelphia, PA: Temple University Press.

Subrahmanyam, Kaveri, and Patricia Greenfield. 2008. "Online Communication and Adolescent Relationships." *The Future of Children* 18(1):119–46.

Tapscott, Don. 1998. *Growing Up Digital: The Rise of the Net Generation.* New York: McGraw-Hill.

Terranova, Tiziana. 2000. "Free Labor: Producing Culture for the Digital Economy." *Social Text* 18(2):33–58.

Thomas, Douglas. 2002. *Hacker Culture.* Minneapolis, MN: University of Minnesota Press.

Thomas, Douglas, and John Seely Brown. 2007. "Why Virtual Worlds Can Matter." Working Paper. University of Southern California, Institute for Network Culture. Retrieved May 12, 2008 (http://www.johnseelybrown.com/needvirtualworlds.pdf).

Thompson, Michael, Catherine O'Neill Grace, and Lawrence J. Cohen. 2001. *Best Friends, Worst Enemies: Understanding the Social Lives of Children.* New York: Ballantine Books.

Thorne, Barrie. 1993. *Gender Play: Girls and Boys in School.* New Brunswick, NJ: Rutgers University Press.

Thorne, Barrie. 2008. "'The Chinese Girls' and 'The Pokémon Kids': Children Negotiating Differences in Urban California," pp. 73–97 in *Global Comings of Age: Youth and the Crisis of Representation,* edited by J. Cole and D. Durham. Santa Fe, NM: School for American Research.

Toffler, Alvin. 1980. The Third Wave. New York: Bantam.

Trudell, Bonnie Nelson. 1993. *Doing Sex Education: Gender Politics and Schooling.* New York: Routledge.

Tyner, Kathleen. 1998. *Literacy in a Digital World: Teaching and Learning in an Age of Information.* Mahwah, NJ: Lawrence Erlbaum.

USC Center for the Digital Future. 2004. *Ten Years, Ten Trends: The Digital Future Report Surveying the Digital Future, Year Four.* Los Angeles, CA: USC Annenberg School Center for the Digital Future. Retrieved June 28, 2008 (http://www.digitalcenter.org/downloads/DigitalFutureReport-Year4-2004.pdf)

USC Center for the Digital Future. 2008. *Surveying the Digital Future, Year Seven.* USC Annenberg School Center for the Digital Future. Retrieved June 29, 2008 (http://www.digitalcenter.org/pdf/2008-Digital-Future-Report-Final-Release.pdf).

Valentine, Gill. 2004. *Public Space and the Culture of Childhood* Hants, UK: Ashgate.

Van House, Nancy, Marc Davis, Morgan Ames, Megan Finn, and Vijay Viswanathan. 2005. "The Uses of Personal Networked Digital Imaging: An Empirical Study of Cameraphone Photos and Sharing," pp. 1853–6 in *Extended Abstracts: Proceedings of the ACM Conference on Human Factors in Computing Systems, CHI 2005, Portland, Oregon, USA, April 2–7, 2005*, edited by G. C. van der Veer and C. Gale. New York: Association for Computing Machinery.

Varenne, Hervé, and Ray McDermott. 1998. *Successful Failure: The School America Builds.* Boulder, CO: Westview Press.

Varnelis, Kazys, ed. 2008. *Networked Publics.* Cambridge, MA: The MIT Press.

Viacom. 2007. "MTV Networks' Nickelodeon Kids and Family Group Commits $100 Million to Its Online Casual Games Business," July 18, 2007. Retrieved May 12, 2008 (http://www.viacom.com/news/pages/newstext.aspx?rid=1027518).

Walther, Joseph B. 1996. "Computer-Mediated Communication: Impersonal, Interpersonal, and Hyperpersonal Interaction." Communication Research 23(3): 3–43.

Weber, Steven. 2003. *The Success of Open Source.* Cambridge, MA: Harvard University Press.

Weiss, Joel, Jason Nolan, Jeremy Hunsinger, and Peter Trifonas. 2006. *International Handbook of Virtual Learning Environments.* New York: Springer.

Wellman, Barry, and Bernie Hogan. 2004. "The Immanent Internet," pp. 54–80 in *Netting Citizens: Exploring Citizenship in a Digital Age*, edited by J. R. McKay. Edinburgh, UK: St. Andrews Press.

Wellman, Barry, Anabel Quan-Haase, Jeffrey Boase, Wenhong Chen, Keith Hampton, Isabel Isla de Diaz, and Kakuko Miyata. 2003. "The Social Affordances of the Internet for Networked Individualism." *Journal of Computer-Mediated Communication* 8(3). Retrieved June 28, 2008 (http://jcmc.indiana.edu/vol8/issue3/wellman.html).

Wenger, Etienne. 1998. *Communities of Practice: Learning, Meaning, and Identity.* Cambridge, UK, and New York: Cambridge University Press.

White House, The. 1993. "The National Information Infrastructure: Agenda for Action." Washington, DC: The White House.

Wilding, Raelene. 2006. "'Virtual' Intimacies? Families Communicating across Transnational Contexts." *Global Networks*, 6(2):125–42.

Willis, Paul. 1977. *Learning to Labor: How Working Class Kids Get Working Class Jobs*. New York: Columbia University Press.

Willis, Paul. 1990. *Common Culture: Symbolic Work at Play in the Everyday Cultures of the Young*. New York: Open University Press.

Wilson, Samuel M., and Leighton C. Peterson. 2002. "The Anthropology of Online Communities." *Annual Review of Anthropology* 31(1):449–67.

Wolak, Janis, David Finkelhor, Kimberly J. Mitchell, and Michele L. Ybarra. 2008. "Online 'Predators' and Their Victims: Myths, Realities, and Implications for Prevention and Treatment." *American Psychologist* 63(2):111–28.

Wolak, Janis, Kimberly Mitchell, and David Finkelhor. 2006. *Online Victimization of Youth: Five Years Later*. Alexandria, VA: National Center for Missing and Exploited Children.

Wolak, Janis, Kimberly Mitchell, and David Finkelhor. 2007. "Does Online Harassment Constitute Bullying? An Exploration of Online Harassment by Known Peers and Online-Only Contacts." *Journal of Adolescent Health* 41(6):S51–S58.

Woodard, Christopher/Gamasutra. 2006. "GDC: Creating a Global MMO: Balancing Cultures and Platforms in *Final Fantasy XI*." Retrieved June 28, 2008 (http://www.gamasutra.com/features/20060324/woodard_01.shtml).

Wresch, William. 1996. *Disconnected: Information Haves and Have-Nots in the Information Age*. New Brunswick, NJ: Rutgers University Press.

Wyness, Michael. 2006. *Childhood and Society: An Introduction to the Sociology of Childhood*. Basingstoke, UK: Palgrave.

Ybarra, Michele L., Marie Diener-West, and Philip J. Leaf. 2007. "Examining the Overlap in Internet Harassment and School Bullying: Implications for School Intervention." *Journal of Adolescent Health* 41(6):S42–S50.

Young, Diana. 2005. "The Colours of Things." Pp. 173–85 in *Handbook of Material Culture*, edited by P. Spyer, C. Tilley, S. Kuechler, and W. Keane. Thousand Oaks, CA: Sage Publications.

Zelizer, Viviana A. 1994. *Pricing the Priceless Child: The Changing Social Value of Children*. Princeton, NJ: Princeton University Press.

Index